Crop Variety Improvement and its Effect on Productivity

The Impact of International Agricultural Research

Crop Variety Improvement and its Effect on Productivity

The Impact of International Agricultural Research

Edited by

R.E. Evenson
Economic Growth Center
Yale University
New Haven
Conncoticut, USA

and

D. Gollin
Department of Economics
Williams College
Williamstown
Massachusetts, USA

CABI *Publishing*

CABI *Publishing* is a division of CAB *International*

CABI Publishing
CAB International
Wallingford
Oxon OX10 8DE
UK

Tel: +44 (0)1491 832111
Fax: +44 (0)1491 833508
E-mail: cabi@cabi.org
Web site: www.cabi-publishing.org

CABI Publishing
44 Brattle Street
4th Floor
Cambridge, MA 02138
USA

Tel: +1 617 395 4056
Fax: +1 617 354 6875
E-mail: cabi-nao@cabi.org

A catalogue record for this book is available from the British Library, London, UK.

Library of Congress Cataloging-in-Publication Data

Crop variety improvement and its effect on productivity : the impact of international agricultural research / edited by R.E. Evenson and D. Gollin.
p. cm.
Includes bibliographical references (p.).
ISBN 0-85199-549-7 (alk. paper)
1. Field crops--Breeding. 2. Field crops--Genetics. 3. Crop improvement. 4. Agriculture--Research--International cooperation. 5. Agricultural productivity. I. Evenson, Robert E. (Robert Eugene), 1934-
II. Gollin, Douglas
SB185.7 .C76 2002
631.5′2--dc21

2002011122

ISBN 0 85199 549 7

Typeset in Melior by Columns Design Ltd, Reading
Printed and bound in the UK by Biddles Ltd, Guildford and King's Lynn

Contents

Contributors

S.O. Ajala, *International Institute for Tropical Agriculture (IITA), c/o L.W. Lambourn & Co., Carolyn House, 26 Dingwall Road, Croydon CR9 3EE, UK.*

F.A. de Almeida, *Av. Getulio Vargas, 1130-Centro, CEP 38400-434, Uberlandia, MG, Brazil.*

A.F.D. Avila, *Secretaria de Administracao, Embrapa, Parque Esracao Biologica – PqEB, Av. W3 Norte (Final), CEP 70770-901, Brasilia, DF, Brazil.*

A. Aw-Hassan, *International Center for Agricultural Research in the Dry Areas (ICARDA), PO Box 5466, Aleppo, Syria.*

M.C.S. Bantilan, *International Crops Research Institute for the Semi-Arid Tropics (ICRISAT), Patancheru 502 324, Andhra Pradesh, India.*

Y.P. Bi, *Shandong Academy of Agricultural Sciences, No. 28 Sangyuan Road, Jinan, Shandong Province, China.*

V. Cabanilla, *International Rice Research Institute, MCPO Box 3127, 1271 Makati City, Philippines.*

E. Cabrera, *International Rice Research Institute, MCPO Box 3127, 1271 Makati City, Philippines.*

S. Ceccarelli, *International Center for Agricultural Research in the Dry Areas (ICARDA), PO Box 5466, Aleppo, Syria.*

T.J. Dalton, *Department of Resource Economics and Policy, 5782 Winslow Hall, University of Maine, Orono, ME 04469-5782, USA.*

S. De Silva, *Department of Economics, Bard College, Annandale-on-Hudson, NY 12504, USA.*

U.K. Deb, *International Crops Research Institute for the Semi-Arid Tropics (ICRISAT), Patancheru 502 324, Andhra Pradesh, India.*

A.G.O. Dixon, *International Institute for Tropical Agriculture (IITA), c/o L.W. Lambourn & Co., Carolyn House, 26 Dingwall Road, Croydon CR9 3EE, UK.*

H.J. Dubin, *CIMMYT Economics Program, Apdo. Postal 6-641, 06600 Mexico, D.F., Mexico.*

W. Erskine, *International Center for Agricultural Research in the Dry Areas (ICARDA), PO Box 5466, Aleppo, Syria.*

R.E. Evenson, *Economic Growth Center, Yale University, 27 Hillhouse Ave, PO Box 208269, New Haven, CT 06520-8269, USA.*

P.C. Gaur, *Central Potato Research Institute, Indian Council of Agricultural Research, Shimla – 171001, Himachal Pradesh, India.*

R. Gerpacio, *Economics Program, IRRI, DAPO Box 7777, Metro Manila, the Philippines.*

D. Gollin, *Department of Economics, Williams College, Fernald House, Williamstown, MA 01267, USA.*

E. Grande, *International Potato Center (CIP), Av. La Molina 1895, Apartado 1558, Lima 12, Peru.*

S. Grando, *International Center for Agricultural Research in the Dry Areas (ICARDA), PO Box 5466, Aleppo, Syria.*

R.G. Guei, *WARDA, 01 BP 2551, 01 Bouake, Côte D'Ivoire.*

P.W. Heisey, *USDA-Economic Research Service, Room S4178, 1800 M Street, NW, Washington, DC 20036-5831, USA.*

M. Hossain, *International Rice Research Institute, MCPO Box 3127, 1271 Makati City, Philippines.*

J. Huang, *Director and Professor, Center for Chinese Agricultural Policy, Chinese Academy of Sciences, Building 917, Datun Road, Anwai, Beijing 100101, China.*

R. Hu, *Center for Chinese Agricultural Policy, Institute of Geographical Sciences and Natural Resources Research, Chinese Academy of Sciences, Beijing, China.*

S. Jin, *Department of Agricultural and Resource Economics, University of California, One Shields Avenue, Davis, CA 95616, USA.*

N. Johnson, *International Center for Tropical Agriculture (CIAT), AA 6713, Cali, Colombia.*

G.S. Khush, *International Rice Research Institute, MCPO Box 3127, 1271 Makati City, Philippines.*

J.G. Kling, *Crop and Soil Science Department, Oregon State University, Crop Science Building 107, Corvallis, OR 97331-3002, USA.*

M.A. Lantican, *CIMMYT Economics Program, Apdo. Postal 6-641, 06600 Mexico, D.F., Mexico.*

J.H. Li, *Gansu Agricultural University, 1 Yingmencum, Lanzhou 730070, Gansu Province, China.*

K.O. Makinde, *International Institute for Tropical Agriculture (IITA), c/o L.W. Lambourn & Co., Carolyn House, 26 Dingwall Road, Croydon CR9 3EE, UK.*

V.M. Manyong, *International Institute for Tropical Agriculture (IITA), c/o L.W. Lambourn & Co., Carolyn House, 26 Dingwall Road, Croydon CR9 3EE, UK.*

J.W. McKinsey, *212 Main Street, North Easton, MA 02356, USA.*

G. McLaren, *International Rice Research Institute, MCPO Box 3127, 1271 Makati City, Philippines.*

M. Mekuria, *Economics Program, CIMMYT, Int. PO Box MP 163, Mount Pleasant, Harare, Zimbabwe.*

A. Menkir, *International Institute for Tropical Agriculture (IITA), c/o L.W. Lambourn & Co., Carolyn House, 26 Dingwall Road, Croydon CR9 3EE, UK.*

M. Morris, *CIMMYT Economics Program, Apdo. Postal 6-641, 06600 Mexico, D.F., Mexico.*

S.N. Nigam, *International Crops Research Institute for the Semi-Arid Tropics (ICRISAT), Patancheru 502 324, Andhra Pradesh, India.*

D. Pachico, *International Center for Tropical Agriculture (CIAT), AA 6713, Cali, Colombia.*

M. Rosegrant, *Environment and Production, Technology Division, International Food Policy Research Institute, 2033 K Street, N.W., Washington, DC 20006, USA.*

S. Rozelle, *Department of Agricultural and Resource Economics, University of California, One Shields Avenue, Davis, CA 95616, USA.*

A. Sarker, *International Center for Agricultural Research in the Dry Areas (ICARDA), PO Box 5466, Aleppo, Syria.*

K. Shideed, *MM National Coordinator, IPA Agricultural Research Center, PO Box 39094, Baghdad, Iraq.*

R. Tutwiler, *International Center for Agricultural Research in the Dry Areas (ICARDA), PO Box 5466, Aleppo, Syria.*

T.S. Walker, *International Potato Center (CIP), Av. La Molina 1895, Apartado 1558, Lima 12, Peru. New address: 207 Agriculture Hall, Department of Agricultural Economics, Michigan State University, East Lansing, MI 48824-1039, USA.*

C.S. Wortmann, *203 South Oak Street, Hartington, NE 68739-6004, USA.*

Foreword

H. Gregersen
(*Chair, CGIAR Standing Panel on Impact Assessment*)

The development of improved, fertilizer-responsive high-yielding varieties of wheat and rice during the early 1960s and their widespread adoption by farmers, first in Asia and then in Latin America, marked the beginning of what is known as the 'Green Revolution'. Much has been written about this technological breakthrough and its impacts – both positive and negative – in the years since its effects were first felt in farmers' fields across Mexico, India, the Philippines and beyond. Since then, improving crop germplasm and the development of new varieties through well focused crop breeding programmes have been extended to many other food and feed crops in developing countries. Today, there are few crops of major economic importance that have not benefited from the application of scientific crop breeding. Each crop has its own story to tell, as evident in this book.

There are many critics of the Green Revolution – those who insist that the impacts have not been that large or that, on balance, the impacts have not been positive (due to adverse environmental effects). Anecdotal evidence and specific case study examples are often cited in support of large positive effects as well as negative ones. The core of the debate centres on the nature and size of the impacts from improvements in the crop germplasm. The total gains achieved have relied on the joint efforts of the national agricultural research systems (NARS) and the international agricultural research centres (IARCs) of the Consultative Group for International Agricultural Research (CGIAR). Commercial crops also have relied heavily on active private sector breeding programmes.

While individual CGIAR centres in the past have undertaken isolated case studies of these impacts, there has, to-date, been no comprehensive crop-wise regional analysis of the aggregate impacts of the IARC and NARS germplasm improvement efforts over the years back to the early days of the Green Revolution.

In 1998, the CGIAR's independent Standing Panel on Impact Assessment (SPIA), which was then called the Impact Assessment and Evaluation Group (IAEG), initiated a major study of the impact of CGIAR's germplasm improvement activities since the beginning of the Green Revolution. SPIA was fortunate to get Professor Robert Evenson of Yale University to coordinate the study in collaboration with the eight crop-based IARCs of the CGIAR, namely CIAT, CIMMYT, CIP, ICARDA, ICRISTAT, IITA, IRRI and WARDA.[1] Professor Evenson was joined by Professor Douglas Gollin of Williams College in producing the overall study and the synthesis of the individual crop assessments. The study remains a collaborative effort of the CGIAR centres and Professors Evenson and Gollin. It builds on the impact assessment work undertaken by the CGIAR centres and their NARS partners to monitor and document the released varieties and the corresponding adoption rates and production gains for individual crop commodities. In addition, country case studies were undertaken for China, India and Brazil that provide further insights into the impacts of the CGIAR and NARS crop germplasm improvement (CGI) activities in some of the major countries where Green Revolution technologies have been applied. The overall study took more than 5 years to complete. Progress reports were presented at almost every CGIAR Annual and Mid Term meeting since 1999.

SPIA and the CGIAR owe a tremendous vote of thanks and gratitude to Professors Evenson and Gollin and to the individual authors from the CGIAR centres and the NARS for their hard and time-consuming work in completing this study.

Quite aside from the authors, many people and organizations gave of their time and resources to make the study possible. SPIA would in particular like to thank Alexander von der Osten and Ravi Tadvalkar and the CGIAR Secretariat for their significant support of the study from its inception. Major credit goes to FAO, the World Bank and UNDP, the Co-sponsors of the CGIAR, who so generously supported the study over its lifetime.

SPIA also thanks Dr Tina David, who served as SPIA focal point for the IARCs and study coordinators on behalf of SPIA; and it wishes to give special recognition to Dr W. James Peacock, the chair of the IAEG when the study was initiated and to Dr Tim Healy, who acted as manager of the project in its early stages. SPIA also gives a strong vote of

[1] See list of acronyms on p. xxiii for the full names of these centres.

thanks to Dr Peter Matlon, then of the UNDP and now with the Rockefeller Foundation, who gave support, guidance and outstanding advice that helped to shape the study and move it along to completion. For several years, Dr Guido Gryseels, formerly of the CGIAR/TAC Secretariat and Executive Secretary to SPIA, managed the details of the program; and SPIA gives special thanks to him. The design proposal was prepared by Professor Greg Traxler of Auburn University in 1997. SPIA gives thanks to him for preparing a proposal that stressed common methodology and rigorous analysis. In fact, throughout the 5 years in which the study was active, SPIA and the study directors have insisted on maintaining the standards called for in the initial proposal and agreement. Annual meetings were held with the teams to monitor progress, and to exchange views, information and insights developed along the way to study completion.

Finally, SPIA also wishes to express its appreciation to the three anonymous external referees who provided significant and helpful suggestions on earlier drafts, particularly with respect to the testing of different specifications of the models for the country studies and synthesis chapters, the inclusion/exclusion of specific exogenous variables in a number of equations, the use of appropriate statistical tests where relevant and, finally, in highlighting a number of analytical difficulties, limitations and qualifications of the study. While the authors could not respond directly to all the concerns raised by the reviewers, SPIA is confident that the authors have responded to the key issues in a positive and satisfactory manner and have in the study acknowledged others that could not be dealt with adequately, given the limitations in data, time and resources.

While certainly not exhaustive, this work provides the most comprehensive documentation of crop genetic improvement impacts to-date. The study covers both the production and diffusion of improved crop varieties for 11 important CGIAR mandate food and feed crops in developing countries over the period from 1960 through to the 1990s.

Through this study, Professors Evenson and Gollin and the team of centre and NARS colleagues have been able to bring together a wealth of data and information to address some long-standing questions regarding the impacts of the CGIAR System and its partner researchers around the globe. Taken as a whole the chapters provide a major milestone in the analysis and documentation of the impacts of crop genetic improvement work over the last 40 years.

To meet the study's major objectives, Evenson and his colleagues had to trace through five interlinked steps:

- First, they needed to establish the nature and magnitude of outputs of the various CGI programmes, including those of the NARS and private sector, and the associated costs of those investments.

- Second, they needed to estimate the varietal make-up of the released varieties in order to establish the direct and indirect CGIAR content thereof, providing an estimate of the CGIAR contribution to all released varieties.
- Third, they needed to estimate adoption rates and production gains between the new varieties and those replaced. This gave them a measure of the production gains related to the CGIAR contribution.
- Fourth, by introducing these production gains into various economic market models, they were able to estimate the economic gains (impacts) on consumers and producers through changes in prices, production, trade, consumption and nutrition.
- Fifth, they needed to assess IARC effects on NARS and private sector investments in CGI programmes in order to establish the appropriate counterfactual situation, i.e. what would have happened without any CGIAR input.

While the results and conclusions of the study are detailed in the chapters that follow, it is worthwhile to summarize the important ones here. Keeping in mind that there are marked differences in results between crops and between regions, the basic conclusions of the overall study can be summarized as follows:

- NARS and the IARCs continue to produce high levels of modern varieties (MVs) of crops. The data do not support the view that diminishing returns to varietal production have set in. Indeed, the rate of MV production as measured by releases has been steadily increasing for all crops in all regions, except for wheat and rice in Asia and Latin America where it has been roughly constant since 1985. In the 1990s MV production for all crops was more than double the rate in the 1970s, and four times the rate in the 1960s. For example, average annual wheat varieties released by national programmes rose from just over 40 between 1965 and 1970 to more than 80 between 1986 and 1990.
- IARCs and NARS have been the main producers of MVs in developing countries. Private firms produced some MVs, mainly as hybrids for maize, sorghum, millet and, more recently, rice. Moreover, private firm MV production has relied heavily on open pollinated 'platform' MVs generated by the IARCs and NARS programmes. Developed country organizations produced very few MVs for developing countries. NGOs generally did not produce MVs.
- IARC germplasm services provide a very important input to NARS crop germplasm improvement programmes. IARC content in released MVs was high for most crops. More than one-third of the approximately 8000 released crop varieties were crossed in an IARC programme. (For the Middle East and North Africa and for sub-

Saharan Africa, they accounted for more than half of all modern varieties released.) In addition, 17% of all NARS varieties relied on at least one IARC-crossed parent and another 23% relied on IARC ancestors.

- IARC programmes are both complementary to and competitive with NARS programmes. In examining the effect of IARC CGI programmes on NARS investments, Evenson found that for countries with small acreage planted to the crop or with low population densities, the competition effect was dominant, while for the largest countries and those where rural population densities are higher, the complementary effects dominated. When weighted by population, the complementary, i.e. enhancement, effects dominate. For all countries weighted by hectares planted, the net complementary effect of the IARCs produced roughly 15% more NARS CGI investment.
- The direct contribution of IARC programmes relative to the investment of resources is substantial. The proportion of total varieties produced by IARCs was well above their proportion of total resources invested in such production.
- With respect to adoption, the percentage of area planted to improved varieties was low for most crops (wheat in Asia is the exception) but has steadily grown such that presently improved varieties are dominant for most crops in most regions. IARC crosses are planted on roughly 36% of the area planted to MVs.
- With respect to production impacts, the conclusions from both the individual IARC case studies and from the three country studies show that without the IARCs the number of released varieties would have been 45–60% less (depending on assumptions).
- CGI contributions to annual productivity growth have been estimated by Evenson for all crops, by region and by decade. Growth from varietal improvement has been realized in all crops, but at very different rates by region. By the 1990s, all crops except beans were realizing high growth rates in productivity through varietal improvement. The average annual growth in productivity from CGI across all crops and regions between 1960 and 1998 was 0.718%, with the highest rates in Asia and the lowest in sub-Saharan Africa. Interestingly, crop productivity growth through breeding was higher in the 1980s and 1990s (averaging 0.830% per annum) than in the previous two decades (averaging 0.321% for the 1960s and 0.676% in the 1970s).
- The IARC contribution to the CGI gains has also been estimated via the counterfactual estimation. Depending on the assumptions used about substitution effects, and depending on the crop and the region, IARC's contribution as a share of total CGI annual growth varied between 40 and 45%.

- With respect to the economic and social consequences of CGI gains from IARC investments, Evenson and Rosegrant, using the IFPRI-based model, -IMPACT-, derive the following estimates based on their best estimates of what would have happened without the CGIAR input:
  (i) world food and feed grain prices (weighted by production) would have been 18–21% higher than they actually were, (and 35–66% higher in the absence of any CGI activity);[2]
  (ii) world food production would have been 4–5% lower – and not lower than that because of 1–2% higher production in the developed countries in response to higher prices, while developing countries would have produced 7–8% less;
  (iii) area planted to cropland would have been significantly higher, particularly for crops like rice. For all food crops, total acreage would have expanded by 1.5–2.7% (5-6 million ha in developed countries and 11–13 million ha in developing countries).
  (iv) food consumption per capita would have declined significantly for many groups. For all developing countries, the average reduction in caloric availability per capita would have been 4.5–5%, and up to 7% in the poorest regions. Furthermore, approximately 2–2.3% more children (13–15 million) – predominantly located in South Asia – would have been malnourished than otherwise, and infant mortality would have been higher;
  (v) imports of food in developing countries would have been about 5% higher.

Taken together, these are indeed important achievements from sustained investments in CGI research over a period of four decades. This first-of-its-kind comprehensive assessment of CGI programme impacts provides evidence of large scale success at the global level and in virtually every region and goes a long way towards dispelling the myth that the Green Revolution is over. The impact of reduced prices in terms of food security has been significant.

The findings of this study support the proposition that IARC investments have had positive impacts for all the study crops. These impacts have been large, partly because of higher leverage through IARC-NARS joint production. The placing of crop germplasm improvement at the core of IARC programmes appears to have been well justified.

[2] With respect to impacts on poverty alleviation, Evenson and Gollin conclude that the poor would have been hurt more by the higher prices in the absence of the CGIAR because they spend a higher share of their income on food.

At the same time, given the impression in some quarters that there have been significant environmental and other negative impacts from application of the Green Revolution technologies, SPIA commissioned other studies to look at such impacts. One of the outputs from this work has recently been published (Maredia and Pingali, 2002), while another is in the final editing stages (Nelson and Maredia, 2003). The conclusions of these authors is that yes, there have been some negative environmental impacts, but there also have been counterbalancing positive environmental impacts, particularly related to land savings. (More intensive production and greater output per hectare mean that less land would be required to produce a given output of food crops.)

SPIA wishes to congratulate Professors Evenson and Gollin and their colleagues in the IARCs for the important results and insights of this study on the impacts of CGI work in the CGIAR.

This has been a courageous and ambitious undertaking, fraught with many data constraints and methodological challenges. The study is based on an impressive amount of data and results, as shown in the accompanying tables and appendices. In fact the voluminous data accumulated for the study are far more than reasonably could be interpreted or commented on in any one book. While the authors have sought to highlight the key results and their interpretation in the brief narrative of each chapter, much more could still be said, debated and speculated upon. The reader is encouraged to consider the results of this study as a first approximation, an initial attempt in quantifying the benefits from CGI over the past four decades. We hope this will provide the impetus for a second generation of studies to confirm, to further explore, and to question some of the conclusions reached here, using new data, different methods and statistical tests, and different scales.

In the meantime, we believe that these findings represent a milestone in the assessment of the impacts of crop genetic improvement research and development, and that they will be of interest and use to many for a long time to come, but particularly to the NARS, the CGIAR members, centres, and Science Council, and to the broader community interested in the value and impacts of agricultural research.

As the authors conclude in the final chapter of the study, 'Consumers benefit most and poor consumers benefit most of all from agricultural research. Farmers are consumers too and for the world's smallest farm producer the total consumer gains are large.' From the producers' side, benefits also accrued. By adopting improved varieties, many farmers lowered costs of production and generated higher rates of return from their land, labour and capital. This, in turn, had positive impacts on income and helped reduce poverty in both land owning and labour producing households in some agricultural regions, but by no means all. An indirect spillover effect from modern variety adoption in other areas was declining crop prices. In the areas not touched by the Green Revolution, costs of production did not fall, and this, in turn, had an adverse effect on farmers' incomes in these regions. Crop germplasm improvement programmes have not yet delivered suitable crop varieties to them. Yet, for many this still represents the most promising way out of poverty. Thus, a key challenge now for the CGIAR and its NARS partners is to target CGI research investments to farmers who have thus far been bypassed by the Green Revolution, primarily in those resource-poor, marginal environments where modern varieties have not been adopted.

Acknowledgements

This book was inevitably the product of many individuals. Our main burden of obligation is clearly to all the authors who participated in the research reported here. We have appreciated the generosity of their contributions and their willingness to respond, often at short notice, to requests for information. We asked them to deliver the impossible – and we responded by editing and condensing their contributions, often drastically. For their forbearance and energy, we thank them.

In addition to the authors whose names are included on the individual chapter headings we acknowledge the many unnamed research assistants, enumerators, support staff, and clerical staff who have helped to collect and interpret the data, often under difficult circumstances. Most of these individuals carried out their efforts under the auspices of the international agricultural research centres of the Consultative Group on International Agricultural Research and their numerous national partners. These institutions deserve further credit for devoting time and resources to answering questions about impact.

We also thank, without naming, the numerous agricultural scientists who took time to answer their colleagues' impertinent questions about the impact of varietal improvement. We would like to believe that in many institutions, impact assessment work has opened up channels of communications between natural scientists and social scientists.

At a more specific level, this study owes a great deal to the members of the Standing Panel on Impact Assessment (SPIA) of the CGIAR's interim Science Council. In particular, we thank Hans Gregersen, Guido Gryseels and Tim Kelley for their contributions to the conceptualization and execution of the study, as well as Christina David, the focal point within SPIA and Peter Matlon, the focal point within the UNDP. The authors also wish to express their appreciation to the Cosponsors of the CGIAR (FAO, WB, UNDP) for generously supporting, through SPIA, this initiative.

We received valuable comments on this research from Dana Dalrymple, Don Duvick and Guy Manners. Cheryl Doss read and commented on several portions of the manuscript. Michael Morris provided helpful comments on several chapters. Three anonymous referees made detailed and important suggestions that greatly strengthened the manuscript. We also benefited from the comments of students at Williams College and Yale University.

Bob Evenson received outstanding research assistance from Ignez Tristao and Fabiana Tito.

At CABI *Publishing*, Tim Hardwick, Claire Gwilt and Rachel Robinson shepherded this volume along, addressing the inevitable delays and corrections with apparent patience and good nature.

We gratefully acknowledge institutional support from the Economic Growth Center at Yale University and from the Gaylord Donnelley Fellowship, given by the Yale Institute for Biospheric Studies, which supported Gollin during part of the preparation of the manuscript.

Finally, we would like to acknowledge the heroic efforts of Kathy Toensmeier in assembling the manuscript. This involved handling text and figures in at least a dozen different software formats, dealing with two editors (occasionally working at cross purposes) and more than a dozen sets of contributors. Without Kathy's determination, this book would never have found its way into print.

Acronyms and Abbreviations

ACA	alternative coconut ash
AICMIP	All India Coordinated Millet Improvement Project
AICSIP	All India Coordinated Sorghum Improvement Project
BGMV	bean golden mosaic virus
BRRI	Bangladesh Rice Research Institute
c.i.f.	cost, insurance, freight (included in price)
CGI	crop genetic improvement
CGIAR	Consultative Group on International Agricultural Research
CIAT	International Centre for Tropical Agriculture (Colombia)
CIMMYT	International Maize and Wheat Improvement Centre (Mexico)
CIP	International Potato Centre (Peru)
CIRAD	Centre de coopération internationale en recherche agronomique pour le développement
CLAYUCA	Consorcio Latinoamericano de la Yuca
CMS	cytoplasmic male sterility
COD	coefficient of diversity
COSCA	Collaborative Study for Cassava in Africa
CPRI	Central Potato Research Institute (India)
CRRI	Central Rice Research Institute (India)
DM	downy mildew
DRR	Directorate of Rice Research (India)
ECA	Economic Commission for Africa
ECABREN	Eastern and Central Africa Bean Research Network
EMBRAPA	Brazilian Agricultural Research Corporation

ES	economic surplus
FAO	Food and Agriculture Organization of the United Nations
FLAR	Latin American Fund for Irrigated Rice
GARB	gross annual research benefit
GEB	gross economic benefit
GRU	genetic resource unit
GTZ	German Technical Cooperation (Deutsche Gesellschaft für Technische Zusammenarbeit)
HYV	high-yielding variety
IAEG	Impact Assessment and Evaluation Group
IARC	international agricultural research centre
ICAR	Indian Council of Agricultural Research
ICARDA	International Centre for Agricultural Research in the Dry Areas (Syria)
ICRISAT	International Crops Research Institute for the Semi-Arid Tropics (India)
IDRC	International Development Research Council
IFAD	International Fund for Agricultural Development
IFPRI	International Food Policy Research Institute
IICA	Inter-American Institute for Agricultural Cooperation
IITA	International Institute of Tropical Agriculture (Nigeria)
IMPACT	International Model for Policy Analysis of Agricultural Commodities and Trade
INGER	International Network for Germplasm Evaluation and Research
INGER-Africa	International Network for Genetic Evaluation of Rice for Africa
IRAT	Institut de Recherches Agronomiques Tropicales
IRIS	International Rice Information System
IRR	internal rate of return
IRRI	International Rice Research Institute (Philippines)
IRTP	International Rice Testing Programme
ISVHAT	International Sorghum Varieties and Hybrid Adaptation Trials
LAC	Latin America and the Caribbean
LLS	late leaf spot
MENA	Middle East and North Africa
MSV	maize streak virus
MV	modern variety
NARS	national agricultural research system(s)
NCRI	National Cereals Research Institute (Nigeria)
NIC	newly industrialized country
NPV	net present value
OAU	Organization of African Unity

OPV	open-pollinated variety
PABRA	Pan African Bean Research Alliance
PBND	peanut bud necrosis disease
PCCMF	Central American Cooperative Network for Bean Improvement
PhilRice	The Philippines Rice Research Institute
PMV	peanut mottle virus
PPP	purchasing power parity
PROFRIJOL	El Programa Coperativo Regional de Frijol para Centro América, México y El Caribe
PROFRIZA	Proyecto Regional de Frijol para la Zona Andina
RCR	real cost reduction
SABRN	Southern Africa Bean Research Network
SADC	Southern Africa Development Community
SAFGRAD	Semi-Arid Food Grain Research and Development
SMIP	Sorghum and Millet Improvement Programme
SMY	scientist man-year
SPIA-TAC	Standing Panel on Impact Assessment of the Technical Advisory Committee
SSA	sub-Saharan Africa
SSD	single seed descent
SYE	staff-year equivalent
TFP	total factor productivity
TPS	true potato seed
TV	traditional variety
UNDP	United Nations Development Programme
USAID	United States Agency for International Development
USDA	United States Department of Agriculture
VT	varietal turnover
WANA	West Asia and North Africa
WARDA	West Africa Rice Development Association (Côte d'Ivoire)
WCA	West and Central Africa
WECAMAN	West and Central African Maize Network

Study Design and Scope 1

R.E. Evenson and D. Gollin

Prior to the 20th century, most crop genetic improvement was achieved by farmers through selection of seeds from superior plants. As human populations expanded into new regions, farmer selection produced increased genetic diversity in the form of distinct 'landraces', or traditional varieties, with different genetic characteristics within the cultivated crop species. In the first half of the 20th century, specialized crop breeding programmes were developed to exploit this farmer-created diversity to produce 'modern' crop varieties through systematic crossing and selection. These crop breeding programmes had by mid-century produced several generations of modern crop varieties in most cultivated species, suitable primarily for richer and more developed – and incidentally or not, temperate zone – countries.

Crop breeding programmes were much less developed in poor and tropical countries. The high degree of sensitivity of many crops to soil and climate characteristics meant that varieties suited to developed countries were seldom suited to poorer countries with different growing environments. Crop breeding programmes were thus required in many locations to produce varieties suitable for those locations. By the 1950s, few countries in the developing world possessed the research infrastructure required for effective plant variety breeding. As a result, most of the developing world lacked access to modern crop varietal technology. What was clearly needed was a model for multi-location crop breeding programmes that would utilize common breeding methods and strategies – as well as some common 'germplasm' in the form of parental breeding materials.

At the start of the 1960s, it was clear that private-sector firms were unlikely to make significant investments in crop improvement research targeted at the major crops grown in poor countries. Since there was no effective intellectual property protection of crop varieties at the time, few incentives existed for private company breeding programmes, except for 'hybrid' crops (i.e. varieties crossed from inbred parent lines to take advantage of heterosis effects). For hybrids, heterosis effects made first-generation seed attractive to farmers, who were willing to pay for the advantages offered by good seeds. In the 1960s, Plant Breeders' Rights were developed in order to provide incentives for private breeding programmes, and in the 1990s, conventional patent rights have been extended to crop varieties and biotechnology products. In 1960, however, most crop breeding programmes were in the public sector and were carried out at government agricultural experimental stations.

Thus, the international agencies concerned with promoting economic development after World War II were confronted with two realities. The first was that population growth was occurring at a rate that was historically unprecedented. Improvements in public and private health measures had brought about reductions in death rates in almost all developing countries. Even though birth rate declines followed death rate declines in most countries, the resultant demographic transition produced a population 'boom' in all developing countries. For some poor countries, this meant a tripling or more of population and of food demand over the second half of the 20th century.

The second reality was that most developed countries were already utilizing most of their land and water resources suited to crop production. Traditional crop improvement methods could not cope with the population-driven increases in demand.

The institutional response to these realities was to develop a system of international agricultural research centres (IARCs) funded through an international consortium of donors. This system eventually took on a formal structure as the Consultative Group for International Agricultural Research (CGIAR). Sixteen centres now form the CGIAR. Eight of these have mandates to develop technology for the major food crops in developing countries. Although some centres have a regional focus, and some are orientated towards the problems of specific regions or ecosystems, a number of institutions are directly mandated with crop-orientated research. In particular, they work with national agricultural research systems (NARS) to undertake and support crop breeding and genetic improvement.

The IARC strategy for crop genetic improvement encompasses the following:

- Developing, maintaining and evaluating basic crop germplasm collections (gene banks)

- Facilitating the exchange and use of germplasm collection materials with NARS programmes and with private seed firms
- Developing crossing and selection programmes to produce releasable varieties and/or advanced breeding lines for NARS breeding programmes (and private seed firms)
- Providing evaluations and information exchange to support the sharing and use of advanced breeding lines by NARS breeders (and private seed firms).

By the late 1960s, two IARC programmes – the International Rice Research Institute (IRRI), located in the Philippines, and the International Maize and Wheat Improvement Centre (CIMMYT), located in Mexico – were credited by the popular press with achieving a 'Green Revolution' in rice and wheat production. This Green Revolution was identified with the development of improved 'high-yielding' varieties of both rice and wheat and with the rapid adoption of these varieties by farmers in Asia and Latin America.

Today agricultural research takes place in a context profoundly different from the one that pertained 40 years ago. Astonishing new technologies have emerged, and scientific knowledge has advanced beyond any prediction. Most notably, the emergence of biotechnology – and the associated advances in our basic understanding of biological processes – have vastly changed the toolkit available to plant scientists. Along with these new technologies, changing legal views of intellectual property rights have contributed to significant shifts in the organization of agricultural research. In rich countries, private-sector firms have undertaken large investments in agricultural research based on biotechnology methods.

Against this backdrop, it is reasonable to step back and ask fundamental questions about the role of international crop research. Does varietal improvement still matter? Is public sector research required? Have national systems grown to the point where an international research centre is unnecessary? Have past investments in crop research led to improvements in productivity? Are continuing investments likely to remain worthwhile? Have the international research centres produced anything of value since, say, 1980 (i.e. after the Green Revolution)?[1]

Answering questions like these requires a careful methodological approach and lots of data. Fortunately, at the outset of the 21st century, we can draw on more than 40 years of experience with many crop improvement programmes in both IARCs and NARS. This volume represents an attempt to address some of these difficult questions. Specifically,

[1] In this book we define the Green Revolution in a broader context than the popular versions. We include all crops benefiting from conventional crop breeding programmes. We also include periods after 1980.

this volume grows out of a study commissioned by the Standing Panel on Impact Assessment of the Technical Advisory Committee (SPIA-TAC) of the CGIAR. The overall goal of this study was to document the impact of international research on crop genetic improvement in developing countries. The study focused on 11 major food crops: rice, wheat, maize, sorghum, millet, barley, beans, lentils, groundnut, cassava and potato.

This study had five formal objectives:

1. To document the *output* of crop genetic improvement programmes for IARCs, NARS and private firms, where output is measured in terms of the number of officially released crop varieties. This documentation is to include all periods and all regions in developing countries where the crop is important.
2. To evaluate the IARC contributions to crop genetic improvement output. This evaluation calls for varietal content measures identifying the institution responsible for crossing or selecting a released variety and its parents or other ancestors. It also requires statistical estimation of breeding production functions where germplasm (parental material) is explicitly treated as a factor of production.
3. To evaluate the farm production impact of crop genetic improvement products (varieties). This requires evidence of the adoption of varieties by farmers and of the production or productivity advantage of improved varieties over the varieties that they replaced. It also requires consistency between estimates of production advantage at the experimental plot, farm plot and aggregate production levels.
4. To evaluate the IARC programme effects on NARS and private-sector investments in crop genetic improvement programmes. This objective addresses the question of the 'NARS-strengthening' design element in IARC programmes.
5. To evaluate the economic consequences of crop genetic improvement programmes. This requires incorporation of the production advantage estimates from objective 3 into market models (both national and international) enabling the calculation of changes in equilibrium prices, production, trade, consumption and nutrition.

This volume consists of an introductory section (Chapters 2 and 3) and three main parts. Chapter 2 gives an overview which is designed to pick out some recurring themes and central messages from the subsequent chapters, and Chapter 3 offers a survey of methodological issues related to crop improvement studies.

Part II (Chapters 4–16) focuses on studies of individual crops and regions, with each chapter highlighting the experience of a single IARC with a particular crop. Some chapters cover several regions (e.g. Chapter 4 on wheat), while others deal with specific regions (e.g. Chapter 6 on rice in West Africa). The crop studies go into considerable detail on issues such as varietal production, adoption and advantage.

Partly to address the potential bias of the IARC crop-orientated analyses in Part II, this volume also includes three country studies in Part III (Chapters 17–20). These studies examine the impact of international research on productivity in India, Brazil and China. The three country studies directly address the need to have 'stories' of research impact that are consistent with national data. These studies essentially begin by measuring productivity increases in agriculture (TFP; total factor productivity). Using econometric techniques, the authors of these chapters then associate TFP gains with national and international crop genetic improvement (CGI) programmes. A disadvantage of the country studies is that they are necessarily unrepresentative. Although India, Brazil and China are good candidates for country studies because they have abundant data, they also have (arguably) the three strongest national agricultural research systems in the developing world. As a result, they are not necessarily typical. Studying these three countries cannot give us a true insight into the relationship between international research and productivity gain in smaller and poorer countries with less research infrastructure.

In order to address the concerns of these smaller countries, Part IV of the book offers three chapters that provide synthetic analysis based on cross-country data. This analysis looks at three issues: the impact of international research on the composition of the varieties grown in developing countries; the impact of international research on production; and the impact of international research on global economic outcomes, using the International Food Policy Research Institute (IFPRI) IMPACT model of the world agricultural economy. The IMPACT (International Model for Policy Analysis of Agricultural Commodities and Trade) model also makes it possible to consider the effect of international research on social indicators of interest, including poverty and hunger.

The original vision of the builders of the IARCs was dominated by food security concerns and the threat posed by the rapidly expanding population in developing countries. But food security has both global and local dimensions. At a global level, the studies in this volume show that varietal improvement programmes have contributed to what may be regarded as a success story. Food prices have fallen in all countries and consumers have benefited. However, for farmers, it is the local dimension of food security that has mattered most. When prices decline and costs do not, farmers are harmed. Varietal improvement programmes have not delivered modern crop varieties to all farmers and, for many farmers, access to modern varieties is a recent phenomenon. Both biological and political factors contribute to the uneven delivery of improved crop varieties to farmers. CGI programmes should be assessed against both global and local food security dimensions.

It should be noted that many of the chapters in this book – particularly the crop studies – are condensed from larger impact assessments undertaken by the same authors. Some of these assessments have been published as working papers or research centre documents, and readers in search of more detail are encouraged to contact the relevant authors or institutions directly.

Crop Genetic Improvement in Developing Countries: Overview and Summary

2

R.E. Evenson and D. Gollin

The second half of the 20th century brought economic development policies and investments into prominence. For the first time in modern history, international (multilateral) institutions were created for the specific purpose of achieving economic development objectives (e.g. the World Bank, regional banks and various agencies of the United Nations). Individual countries also established bilateral aid agencies (e.g. USAID, IDRC, GTZ). Expectations were high as colonial regimes were being ended. Many observers expected the subsequent decades to be characterized by per capita income 'convergence', in which the highest growth rates in per capita income would be achieved by countries with the lowest initial levels of per capita income.

By the 1960s, however, it was becoming increasingly clear that maintaining food production per capita was a challenging task. Improvements in health in the 1950s and 1960s were impressive in almost all developing countries. Infant mortality rates were declining, and life expectancy was increasing. Even though most countries experienced declines in birth rates shortly after declines in death rates (the lags were short, at least by historical standards), the dynamics of the demographic transition produced historically unprecedented increases in population in virtually all developing countries.

Agricultural policy makers in developing countries had experimented in the 1950s and 1960s with extension-led programmes predicated on the concept of the 'inefficient farmer'. The central idea was that technology was available to farmers, but that farmers' 'ignorance' – combined with the lack of community education and information programmes, as well as credit constraints and a high degree of risk aversion – were barriers to the full and effective use of technology. The agricultural aid programmes of the 1950s and 1960s also recognized the need for capacity building in universities, both for purposes of training extension and education specialists and for developing agricultural experimental station capabilities in national agricultural research systems (NARS).

By the late 1950s, evaluations of extension-led agricultural development programmes (including many integrated rural development programmes) were indicating relatively slow progress in productivity gains. The economist T.W. Schultz, in his classic work, *Transforming Traditional Agriculture*, argued that farmers with traditional technology (including farmer-selected crop varieties and livestock breeds) were actually efficient.[1] Schultz argued that development programmes would have to deliver new technology to poor farmers in developing countries in order to improve their situation. In Schultz's view, the location-specificity of crop and livestock technology meant that farmers in many parts of the world simply lacked access to modern technology; without the development of locally adapted technologies, they simply did not have a viable alternative to traditional practices.

These conditions led to the development of a system of international agricultural research centres (IARCs) that were eventually organized under the rubric of the Consultative Group for International Agricultural Research (CGIAR). The design features of most IARCs enabled them to specialize in one commodity or a small set of commodities and work with (and support) national agricultural research systems (NARS) in trying to use modern science to achieve productivity gains. Crop genetic improvement programmes were well suited to this design, although early IARC programmes also included other uses of science for agriculture, such as designing and engineering improved equipment. From the beginning, however, IARC programmes developed and maintained genetic resource collections (gene banks) and fostered free exchange of genetic resources between NARS programmes and IARC programmes. IARC programmes also supported researchers in the private sector, although these were few in number and importance in the developing world. Most IARCs developed strong breeding programmes, where advanced breeding lines and finished varieties were developed. These materials were made available to NARS programmes through gene banks and through international testing and exchange schemes.

By the late 1960s, the international centres appeared to be making significant progress. Improvements in crop productivity were most apparent in the two major cereal grains produced in developing coun-

[1] In Schultz's terms, they were 'poor but efficient', or equivalently, 'efficient but poor'.

tries, wheat and rice. In both crops, improvement was based on a new 'plant type'. Specifically, this plant type was shorter and earlier maturing, with less photoperiod sensitivity, than traditional tropical and subtropical varieties. The development of these plant types was not in any sense 'miraculous'.[2] However, these new plant types were popularized as miracles and represented as the foundation for a 'Green Revolution' in developing countries.

This popularized view of the Green Revolution was based on relatively patchy data showing rapid adoption of 'high-yielding' rice and wheat varieties in Asia and Latin America over the period from 1968 to the early 1980s. Until now, few data have been publicly available on other crops or regions, or on more recent time periods. For example, until recent years, little has been known about the development or diffusion of new rice and wheat varieties in the 1980s and 1990s. Similarly, few data have been available on rice and wheat varietal adoption in sub-Saharan Africa or in West Asia and North Africa. Finally, data on the development and diffusion of improved varieties in other crops have been relatively scarce. As a result, there has been little systematic work attempting to evaluate crop improvement in developing countries until this volume.

In spite of this absence of data, a large body of literature now exists discussing the Green Revolution. The literature includes studies claiming a miraculous transition to high productivity growth rates; it also includes studies that criticize the Green Revolution for many perceived failures. While a number of studies in this literature do make use of micro-data from particular locations and do bring important insights, it is striking how many authors have urged policy recommendations based on very limited evidence.

This volume, although incomplete in much of the country and crop detail, none the less provides a far more complete documentation of crop genetic improvement impacts than has previously been available from a single source. An attempt is made to cover both the production and diffusion of improved crop varieties, not just for wheat and rice, but also for nine other important food and feed crops in developing countries. Furthermore, we attempt to provide a comprehensive picture, at least at the regional level, not just for the 1960s and 1970s, but also for the 1980s and 1990s.

The extension of the analysis to the past two decades is critical. Contrary to popular belief, the productivity gains realized in these

[2] In the case of rice, it was clear that scientists knew from the outset exactly what they wanted to achieve, and they were able, in fact, to develop the most famous Green Revolution rice variety, IR8, within months of beginning breeding work.

decades have been large, and they have served a crucial purpose in helping to achieve global food security. Although it is not widely realized, the 1980s and 1990s were the decades of highest increments to population in almost all developing countries. Yet they were also decades of high productivity growth in crop agriculture. This volume will show that most (although not all) of these yield gains were produced by crop genetic improvement.

Taken together, the past four decades have been an era of rapid productivity and production gains in agriculture. In spite of historically unprecedented population increases – and in spite of limited natural resources – per capita food production in most developing countries has increased over these decades. Agricultural sectors in many countries have been transformed, not by industrialization, but by crop genetic improvement. The real price of food and feed grains is less than half its level of 50 years ago in international markets. Literally millions of people are alive today who would otherwise have died from hunger or from diseases related to inadequate nutrition. Tens of millions more people are eating more and better quality food than would have been possible if world food production per capita had remained at the levels of 1960.

To understand how remarkable this increase has been, consider some projections from only 20 years ago. In the widely cited *Global 2000 Report to the President*, a document summarizing a study initiated in 1977 at the behest of then US President Jimmy Carter and published in 1980, the most optimistic scenario suggested that per capita food production in less developed countries would increase by 19.5% from 1969–1971 to 2000. Per capita grain production was expected, in the most favourable projections, to reach 210 kg by the year 2000.

These projections, only 20 years old, and presented as optimistic upper bounds on likely trends, now appear astonishingly pessimistic. FAO's index of food production per capita for developing countries shows a 50% increase from 1969–1971 to 1998–2000. Per capita grain production in developing countries stands at 262 kg, fully 25% above the 'optimistic' case presented in *Global 2000*. Even excluding China, which has accounted for a large share of the developing world's increase in food and grain output, per capita grain production in 1999 was 226 kg, a 7.6% increase over the most optimistic scenario.

Food production is, of course, only a crude measure of success. Increases in production do not necessarily benefit all people equally. There remain important disparities in food consumption and food security across geographical regions, between rich and poor, between men and women. Moreover, today's world recognizes that the expansion of food production over the past 40 years has exacted some environmental costs.

Clearly it would be a mistake to attribute all the production increases to crop genetic improvement (CGI) research or to other research contributions. Farmers have expanded their use of fertilizers and pesticides. More land is irrigated, with greater efficiency than ever before. Mechanization has speeded up land preparation, allowing farmers to double-crop in some areas where previously they had only been able to grow a single crop. (In some areas, triple cropping has become the norm.) Farmers are better educated than ever before, and their knowledge gives them a better understanding of techniques and markets. Improvements in transportation infrastructure have altered the incentives for farmers who previously grew food primarily for subsistence consumption. Policy reforms in some countries have removed the heavy (if indirect) burdens of taxation that were often imposed on the agricultural sector. All these changes, and many others, have contributed to the expansion of world food production. Equally clearly, however, crop improvement has played a role. In almost all crops, and in almost all areas of the world, farmers are growing varieties that did not exist in 1960 or 1970.

Even if we accept that crop genetic improvement has played a role, the questions of policy interest are more specific. Were most of the benefits from plant breeding attained many years ago? How much impact has been generated by crop genetic improvement research in the past one or two decades? Has plant breeding been useful outside of a few highly favoured environments and crops? And how important have international agricultural research institutions been, as compared with national institutions in both rich and poor countries?

The Green Revolution: Mythology and Reality

For many casual observers, the successes and failures of international agricultural research can be summarized in a commonly held 'middle view' of the Green Revolution. Some key ingredients of this view are the following:

- Most of the increases in food production in developing countries were concentrated in the so-called Green Revolution crops: rice and wheat. Relatively little gain occurred in other crops.
- Most of the increases were associated with the development of crop varieties that performed well under irrigation and intensive use of fertilizers; relatively little happened in other growing environments.
- Most of the gains took place in Asia, and to a degree in Latin America. Little has been accomplished in sub-Saharan Africa.
- The major gains from the Green Revolution had largely been

realized by 1975 or 1980; since then, relatively little headway has been made.

- Yield gains that were achieved during the Green Revolution have begun to erode due to environmental degradation and other effects of unsustainable production.
- Scientists have tried and failed to develop high-yielding crop varieties for most marginal environments, where water, climate and soil constraints cannot be overcome through varietal improvement (or at least not through conventional breeding).

The simplistic view outlined here certainly has some underlying validity. But the succeeding chapters will also show some surprising divergences from this common view. Among the more surprising findings are the following:

- The Green Revolution is better understood as a 40-year history of steady productivity gains than as a one-time event. For all crops combined, the rate of production of improved varieties has been increasing in each decade.
- Technological advances have occurred in all crops, on all continents, and in all agroecological zones, although these advances have been uneven.
- The progress achieved in different areas is related to the effort expended on research, as well as to the pre-existing 'stock' of research done on similar crops and growing environments.
- Had international research on crop genetic improvement been halted in (say) 1980, the world would be demonstrably worse off.

The remainder of this chapter will briefly sketch out this 'alternative view' of the Green Revolution, drawing on the content of the book as well as some additional data. An implication of the alternative view is that while it is clearly desirable to pursue new technology paradigms or major steps in crop genetic improvement, the routine business of varietal improvement has had enormous value in the past. There have been large economic benefits from adapting modern varieties to specific locations. Similarly, there have been important gains from reducing the growth duration of crops, from improving resistance to biotic and abiotic stresses, and from similar 'small' achievements.

A Lengthy Record of Progress

For rice and wheat, the major advance of the 1960s was the introduction of semi-dwarfism into non-photoperiod-sensitive varieties of rice and wheat. This process had actually started somewhat earlier, with programmes such as the FAO *indica* × *japonica* crossing programme in rice, and the Rockefeller Foundation's efforts on wheat in Mexico, both of which dated back substantially before 1960. However, progress has continued fairly steadily since the 1960s. Some evidence suggests that the pace of crop improvement research has actually quickened, rather than slowed, in recent years.

One measure to consider is the release of new varieties. This is essentially a measure of research activity, rather than of impact, but it is suggestive. In essentially all of the crops for which we have data, varietal releases by national programmes rose steadily through the 1960s into the late 1980s and 1990s. In wheat, average annual releases in the 1965–1970 period were 40.8 varieties per year; in 1986–1990, the annual pace of release was approximately double, at 81.2. Annual releases of rice varieties tripled from 1965–1970 to 1986–1990, and remained at the high end of that level into the most recent years for which data are available. Maize releases have increased about fivefold between the 1965–1970 period and the present. The same pattern holds for sorghum. Even more pronounced increases are evident for crops that were relatively little researched, such as millet, barley and lentils.

A second measure of success is the rate of adoption of modern varieties. Adoption of modern varieties for all crops has continued steadily over time. As Evenson reports in Chapter 22 of this volume, for all developing countries, the adoption of modern varieties during the first 20 years of the Green Revolution – aggregated across all crops – reached 9% in 1970 and rose to 29% in 1980. In the subsequent 20 years, far more adoption has occurred than in the first two decades. By 1990, adoption of modern varieties had hit 46%, and by 1998, the most recent year for which data were available, adoption levels hit 63%. Moreover, in many areas and in many crops, first-generation modern varieties have been replaced by second- and third-generation modern varieties.

A third measure of success is yield increases. Yield increases cannot be attributed wholly to varietal improvement, but they provide another piece of evidence that the productivity impacts of crop genetic improvement research were not 'completed' in any sense by 1980.[3] For example, FAO data indicate that for all developing countries, wheat yields rose by 69% from 1980 to 2000; rice yields rose 42%; maize yields rose 40%; potato yields rose 38%; and even cassava yields rose 13%. In absolute terms (measured in kg ha^{-1}), yields for many crops rose more in the 1980–2000 period than in the 1961–1980 period. Average wheat yields in all developing countries rose by 789 kg ha^{-1} in 1961–1980 but by 1087 kg ha^{-1} in 1980–2000. Similarly, average rice

[3] As Chapter 3 will argue, yield increase is neither necessary nor sufficient as evidence of technological improvements.

yields in all developing countries rose by 914 kg ha^{-1} in 1961–1980 and by 1128 kg ha^{-1} in 1980–2000. In potatoes, average yields in all developing countries rose by 2488 kg ha^{-1} between 1961 and 1980, but in the period 1980–2000, they grew an additional 4193 kg ha^{-1}. (Because the yield *levels* were higher in 1980, the percentage change in yield actually fell slightly for wheat and rice. See Tables 2.1–2.3 for details.)

We would like to measure 'pure' productivity increases, rather than yield increases. A measure favoured by economists is 'total factor productivity' (TFP), which controls for increased use of inputs. We do not observe TFP for the aggregate data, although the country studies of Chapters 18–20 report TFP gains for India, China and Brazil. Where the data are available, we find evidence of continuing TFP gains in crop agriculture after the 1980s.

The question remains as to whether productivity gains reflect the ongoing influence of international research. This is a question of attribution, and methodologically we have few good ways of distinguishing the contributions of international research centres from those of other programmes. However, as Chapters 4–16 make clear, and as Evenson notes in Chapters 21 and 22, in essentially all crops and all regions, the continuing diffusion of modern varieties for developing countries has depended to a large degree on germplasm coming out of international research centres. The crop varieties being grown by farmers, along with those being used in national breeding programmes, appear to be based – in part or in whole – on genetic material supplied by international centres. Although strong national programmes have emerged in some countries, they appear to be complementing the efforts of the international centres, rather than substituting for them. Various measures of centre influence and contributions show a strong continued role for international research.

Progress Across Many Crops

A striking finding from the data is that substantial progress has been achieved across essentially all major crops. Rice and wheat were 'easy', in the sense that scientists began research with a 'blueprint' for varietal improvement and with a large stock of improved germplasm from temperate zones. In wheat, many good varieties were available in North America, Europe and Asia, with a key contribution coming from the variety Norin 10, a cultivar descended from Japanese semi-dwarfs. In rice, many good *japonica* varieties were available but lodging posed a problem for generating high yields from *indica* or *javanica* varieties suitable for the tropics. Again, semi-dwarfism (primarily taken from the variety Dee-Geo-Woo-Gen) provided the key for transferring high-yielding rice technology into varieties suitable for the tropics.

Table 2.1. Yield (kg ha^{-1}) of major crops for all developing countries (Source: FAOSTAT online data, 2 January 2002).

	1961	1965	1970	1975	1980	1985	1990	1995	2000
Barley	897	1,093	1,028	1,238	1,295	1,321	1,316	1,480	1,507
Cassava	7,398	7,929	8,480	8,596	9,123	9,811	10,014	9,743	10,267
Cereals, total	1,115	1,260	1,482	1,678	1,874	2,204	2,426	2,593	2,724
Lentils	517	576	535	580	546	716	767	794	835
Maize	1,128	1,252	1,494	1,696	1,969	2,178	2,447	2,738	2,766
Millet	578	525	737	684	650	723	769	705	739
Potatoes	8,492	8,833	9,645	10,489	10,980	11,875	12,475	14,041	15,173
Rice, paddy	1,756	1,930	2,276	2,406	2,670	3,182	3,468	3,593	3,798
Sorghum	685	698	859	1,028	1,058	1,089	1,084	1,062	1,136
Wheat	775	999	1,124	1,396	1,565	2,059	2,289	2,528	2,651

Table 2.2. Yield increases for major crops over selected time periods, for all developing countries (Source: FAOSTAT online data, February 2002).

	Yield increase 1961–70	Yield increase 1970–80	Yield increase 1980–90	Yield increase 1990–2000	Yield increase 1961–80	Yield increase 1980–2000
Barley	1.15	1.26	1.02	1.14	1.44	1.16
Cassava	1.15	1.08	1.10	1.03	1.23	1.13
Cereals, total	1.33	1.27	1.29	1.12	1.68	1.45
Lentils	1.03	1.02	1.41	1.09	1.05	1.53
Maize	1.33	1.32	1.24	1.13	1.75	1.40
Millet	1.27	0.88	1.18	0.96	1.12	1.14
Potatoes	1.14	1.14	1.14	1.22	1.29	1.38
Rice, paddy	1.30	1.17	1.30	1.10	1.52	1.42
Sorghum	1.26	1.23	1.03	1.05	1.55	1.07
Wheat	1.45	1.39	1.46	1.16	2.02	1.69

Table 2.3. Absolute increases in yield (kg ha^{-1}), major crops, for all developing countries (Source: FAOSTAT online data, 1 February 2001).

	Absolute increases in yield 1961–80	Absolute increases in yield 1980–2000
Barley	399	211
Cassava	1725	1145
Cereals, total	760	849
Lentils	28	290
Maize	841	797
Millet	71	89
Potatoes	2488	4193
Rice, paddy	914	1128
Sorghum	373	78
Wheat	789	1087

No such technological backlog – either of germplasm or of knowledge – was available for many other crops. There was little improved germplasm to be transferred for beans, lentils, barley, sorghum or millet – or even for potato. None was available for cassava. For maize, the more complicated mechanics of breeding and the high location-specificity of varietal technology made it difficult to adapt improved lines from the temperate zones to tropical climates and disease environments.

None the less, over the past 40 years, successes have been achieved for all of the major crops. For some crops, international breeding did not begin until the 1970s (sorghum, millet, barley, lentils, potatoes, cassava) or even until the 1980s (rice in West Africa). In most of these cases, research began with relatively small stocks of usable germplasm or knowledge. But advances have been striking.

Wheat

Wheat is the world's most extensively cultivated crop, with roughly 225 million ha under cultivation. Approximately 100 million ha are under cultivation in developing countries: 8 million ha in Latin America, 1.2 million ha in sub-Saharan Africa, 25 million ha in the Middle East and North Africa, and 65 million ha in Asia (of which 29 million ha are in China). Several wheat types are cultivated. Most are bread wheats, but significant quantities of durum wheats suited to pasta products are also grown. Wheats are classified as spring type or winter type, with northern temperate regions (i.e. Canada and the northern USA) producing traditional spring types. Winter types are produced in temperate and sub-tropical regions with mild winters, as in the southern growing regions of the USA. In tropical climates with relatively warm winters, spring types are planted, but in the autumn, as winter types.

Because of extensive investments in wheat research programmes in Europe and North America, the temperate zone spring and winter types had been considerably improved relative to the tropical spring types by the 1960s. The Rockefeller Foundation supported a wheat breeding programme in Mexico in the 1940s and 1950s under the direction of Norman Borlaug. This programme was eventually transformed into the wheat programme at CIMMYT where, after 20 years of dedicated breeding work, the Green Revolution semi-dwarf bread wheats were adapted for widespread use in Asia beginning in the mid 1960s.

The CIMMYT wheat programme has continued its programmes of germplasm management and diffusion of advanced materials to NARS breeders. In the 1990s, CIMMYT's wheat programme maintained a staff of 35 senior scientists (70 scientists) in a number of locations, with an annual budget of only $12 million. Approximately 1700 NARS scientists in many programmes are working on wheat improvement, with an annual budget of roughly $100 million in the 1990s. (In the 1960s, NARS programmes were one-third their present size.)

Figure 2.1 depicts the pattern of release of more than 2400 new wheat varieties since 1965 by major wheat producing regions. IARC content shares are also shown. For 1970, 1980, 1990 and 1998, adoption estimates are reported, and for 1990 and 1998, these show IARC content of farmer-adopted varieties.

These data show that varietal production increased in Latin America and Asia until the mid-1980s and that varietal release in the 1990s is somewhat lower than in the 1980s but still remains at a high annual level. These two 'mature' regions may be showing some evidence of 'exhaustion' of the genetic potential afforded by conventional breeding programmes. By contrast, the pattern of varietal release in sub-Saharan Africa and in the Middle East and North Africa is increasing with time.

The IARC content indicator measures are of interest because they reflect both the competitiveness of the IARC breeding programmes (as reflected in the IARC cross shares) and the germplasm contributions of IARC programmes (as reflected in the IARC ancestor shares). The IARC cross shares are lowest in Asia, where NARS programmes are strongest, and highest in sub-Saharan Africa and MENA (Middle East and North Africa), where the NARS programmes are less well developed. IARC ancestor shares are also high in all regions, reflecting high germplasm contributions.

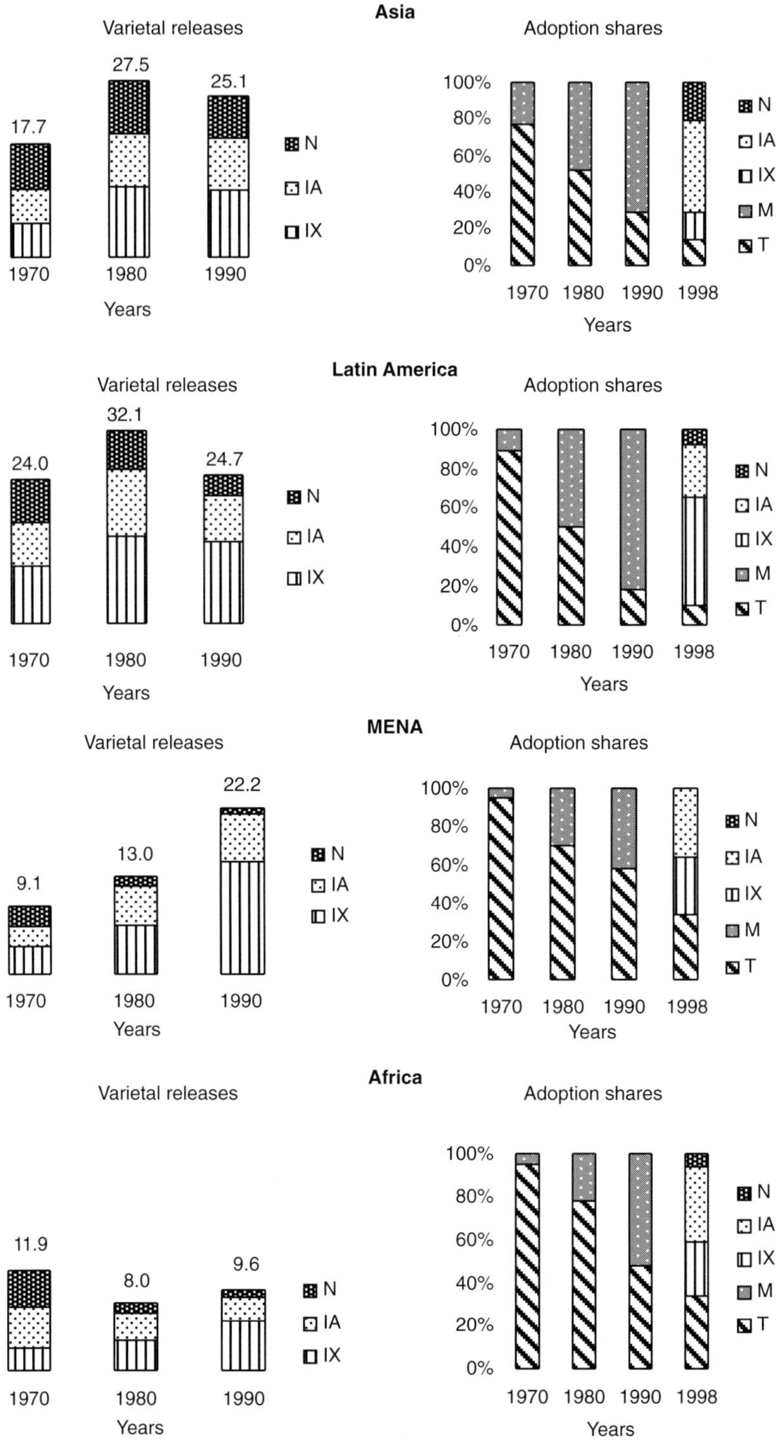
Asia
Varietal releases
17.7
27.5
25.1
N
IA
IX
1970
1980
1990
Years
Adoption shares
100%
80%
60%
40%
20%
0%
1970
1980
1990
1998
Years
N
IA
IX
M
T
Latin America
Varietal releases
24.0
32.1
24.7
N
IA
IX
1970
1980
1990
Years
Adoption shares
100%
80%
60%
40%
20%
0%
1970
1980
1990
1998
Years
N
IA
IX
M
T
MENA
Varietal releases
9.1
13.0
22.2
N
IA
IX
1970
1980
1990
Years
Adoption shares
100%
80%
60%
40%
20%
0%
1970
1980
1990
1998
Years
N
IA
IX
M
T
Africa
Varietal releases
11.9
8.0
9.6
N
IA
IX
1970
1980
1990
Years
Adoption shares
100%
80%
60%
40%
20%
0%
1970
1980
1990
1998
Years
N
IA
IX
M
T

Farmers have adopted modern varieties rapidly, and adoption rates have increased steadily since the Green Revolution period of the 1970s. The level of modern variety adoption in Asia was modest in 1970 in spite of the Green Revolution, and adoption levels were very low across Africa and the Middle East. By the 1990s, modern variety adoption levels were very high in Latin America and Asia. Germplasm generated by IARCs was disproportionately important in both regions, with IARC-developed varieties being adopted on a particularly large scale in Latin America. By contrast, in Asia, the varieties adopted most widely were developed by NARS but with IARC ancestry.

For both sub-Saharan Africa and MENA, modern variety adoption levels were high but lagged behind the Asian and Latin American levels by 10–15 years. IARC crossed varieties were less important in adopted varieties than in released varieties, while varieties developed in national programmes (with IARC backgrounds) were disproportionately represented in farmers' fields.

Rice

Rice is the most important crop in developing countries. Asian countries dominate production, with 133 million ha (India, with 43 million ha, and China, with 33 million ha, are the leading countries). Latin America and African countries each produce on roughly 8 million ha. Developed countries, including Japan, produce on only 5 million ha.

Rice is produced in several different environments. The dominant production environments are irrigated and rainfed 'paddy' environments. Rice is also produced in 'upland' and 'deepwater' environments, among others. Most upland production is in Africa and Latin America; most deepwater production is in Asia.

Figure 2.2 depicts the release and adoption pattern by region for roughly 1700 released rice varieties. Few, if any, of these varieties are suited to deepwater production environments. Varieties suited to upland production are also few in number, with a small concentration released in Brazil.

Three different IARCs have been involved in rice crop genetic improvement. For Asia, IRRI has played a major role in producing important rice varieties. The relationship between IRRI and Asian NARS is in some sense the most 'mature' of such relationships. After the 1970s, IRRI's role was increasingly that of a germplasm supplier,

Fig. 2.1 (opposite). Varietal releases and adoption shares: wheat. T = traditional variety; M = modern varieties; IX = IARC crosses; IA = NARS crosses, IARC ancestors; N = NARS crosses, NARS ancestors.

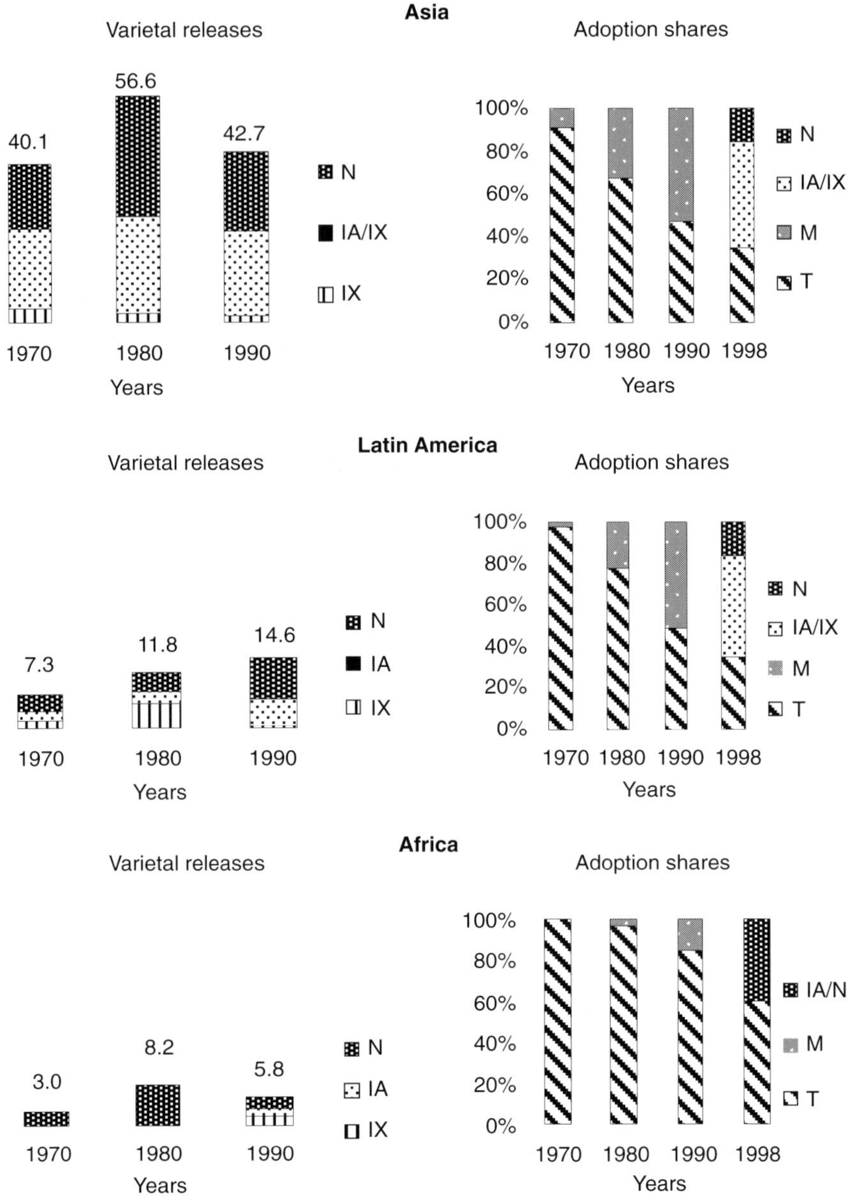

Fig. 2.2. Varietal releases and adoption shares: rice. T = traditional variety; M = modern varieties; IX = IARC crosses; IA = NARS crosses, IARC ancestors; N = NARS crosses, NARS ancestors.

producing parent material for NARS breeders. This role was facilitated by an international network for germplasm exchange that provides NARS breeders with ready access to breeding materials.

IRRI's success was first concentrated in irrigated rice environments and then extended to favourable rainfed environments. This success did not extend in any significant degree to upland rice environments nor to deepwater environments. The early IRRI Asian rice varieties were also not particularly well adapted to Latin America or to Africa.

CIAT (International Centre for Tropical Agriculture), in Colombia, established a rice-breeding programme a number of years after the IRRI programme was established. It was this programme that undertook the adaptive breeding required to bring the high-yielding semi-dwarf varieties from Asia to Latin America. In contrast to the release pattern in Asia, where releases levelled off in the 1980s and 1990s, annual releases have continued to increase in Latin America.

For African production conditions, neither IRRI nor CIAT had much success in generating varieties that met with widespread adoption. WARDA (West Africa Rice Development Association), the regional rice development centre, experienced considerable instability in the 1960s and 1970s, and was not effective until it was established as a centre capable of doing its own breeding, and not fully effective until it moved from an urban location in Liberia to an experimental station in Côte d'Ivoire. By the mid 1990s, WARDA's programme was beginning to show effectiveness.

The wheat and rice data show that the popular perception of the Green Revolution is at least incomplete, if not entirely naïve. In the case of wheat, CIMMYT has continued to play a major role in varietal production. The 1980s and 1990s were periods of higher annual varietal production than were the 1970s. The Green Revolution of the 1970s was extended beyond its 1970s boundaries in part through the efforts of CIMMYT to extend the area of suitability for the basic high-yielding plant type.

For rice, the picture is more complex. Varietal production in Asia was also higher in the 1980s and 1990s than in the 1970s. With respect to the more advanced Asian NARS, IRRI's role shifted to that of a germplasm supplier in the 1980s and early 1990s. However, in the later 1990s, the political opening of Cambodia, Vietnam and Laos, where national programmes were relatively undeveloped, put IRRI back in the position of breeding crosses for direct release. Outside Asia, the extension of the Green Revolution in rice to Latin America was greatly assisted by CIAT's programme. And, with a delay, WARDA is now assisting in the expansion of the rice Green Revolution to Africa.

Maize

Maize is grown in both temperate (mostly developed countries) regions and in tropical and subtropical regions (almost entirely in developing countries – 24 million ha in Africa, 28 million ha in Latin America, and 40 million ha in Asia). Major advances in maize varieties were achieved in temperate zone regions based on the 'hybrid' technique of breeding in the first part of the 20th century. Most of these hybrids were for maize fed to livestock. The transfer of hybrid maize technology to the tropics was very limited as of 1960. CIMMYT, the international centre mandated with maize production, chose to pursue improvements in both 'open-pollinated varieties' and hybrid varieties. A considerable quantity of maize produced in developing countries is consumed directly by humans.

Figure 2.3 depicts maize varietal releases for public and private breeders in Latin America and Africa. Note that IITA (International Institute for Tropical Agriculture) is the lead IARC concerned with maize crop genetic improvement for West and Central Africa.

Most of the public-sector NARS varietal releases in maize have been open-pollinated varieties, although in the 1990s hybrids became more important. Almost all private-sector varietal releases, by contrast, have been hybrids. By the 1990s, private-sector programmes were developing more varieties than public-sector programmes in Latin America. They are also becoming important in Asia and sub-Saharan Africa.

Early adoptions of modern maize varieties included both open-pollinated varieties and hybrids. By the 1990s, however, it was clear that farmers favoured hybrids over open-pollinated varieties. Private seed firms produced many (if not most) of these adopted hybrids. Thus, maize and to a limited extent, sorghum and millet – for which hybrids are rapidly being developed – represent cases where the private sector has produced genetic improvement for agriculture. This has not occurred in the other studied crops to any significant extent.

It is also clear from data on genetic resource content of varieties that CIMMYT and NARS breeding materials (germplasm) have been widely used by the private sector. The public sector thus created the 'platforms' on which the private sector was built. A similar phenomenon occurred in temperate zone maize breeding many years earlier.

In many respects, maize has been a Green Revolution crop in the same sense as rice and wheat. It is a major crop in developing countries. Maize is highly location-specific because of high photoperiod sensitivity. Many generations of breeding effort are required for improvements. The distinction between food and feed uses has turned out to be important. The major crop genetic improvement gains have been realized in hybrid varieties, chiefly for feed uses.

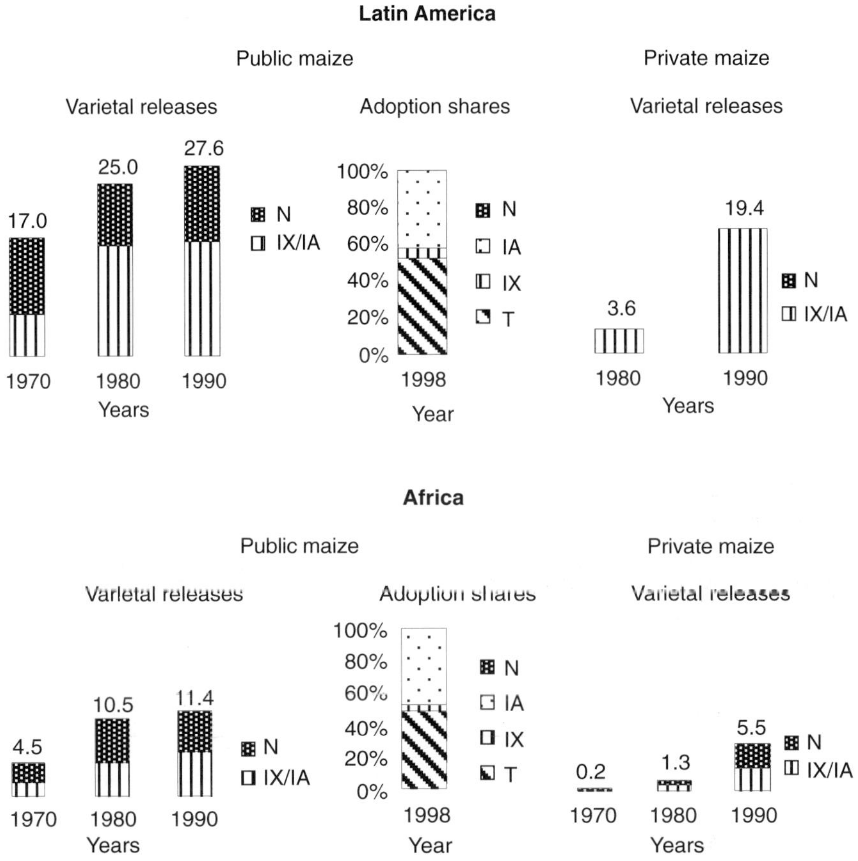

Fig. 2.3. Varietal releases and adoption shares: maize. T = traditional variety; M = modern varieties; IX = IARC crosses; IA = NARS crosses, IARC ancestors; N = NARS crosses, NARS ancestors.

Sorghum and pearl millet

Sorghum and pearl millet are grown extensively in semi-arid regions of Asia (chiefly India) and sub-Saharan Africa. Of the 45 million ha planted to sorghum, 23 million ha are in Africa and 14 million in Asia. For pearl millet, 38 million ha are planted worldwide, of which 20 million are in Africa and 16 million in Asia. ICRISAT (International Crops Research Institute for the Semi-Arid Tropics) has developed research programmes for both crops in both India and Africa. ICRISAT maintains germplasm collections and provides germplasm to approximately 750 sorghum scientists and 300 pearl millet scientists in NARS programmes.

India maintained breeding programmes for both sorghum and pearl millet prior to the ICRISAT programme and did release a number of modern varieties of sorghum and pearl millet in the 1960s and 1970s. Figure 2.4 depicts the release of varieties of sorghum by 5-year period after 1965. Annual varietal releases in both Asia (India) and Africa show upward trends, with annual releases in the 1990s being roughly double the releases in the 1970s. Varietal releases prior to 1980 were entirely NARS products. After 1980 the ICRISAT content became important, particularly in Africa where 70% of the releases in the 1990s were ICRISAT crosses.

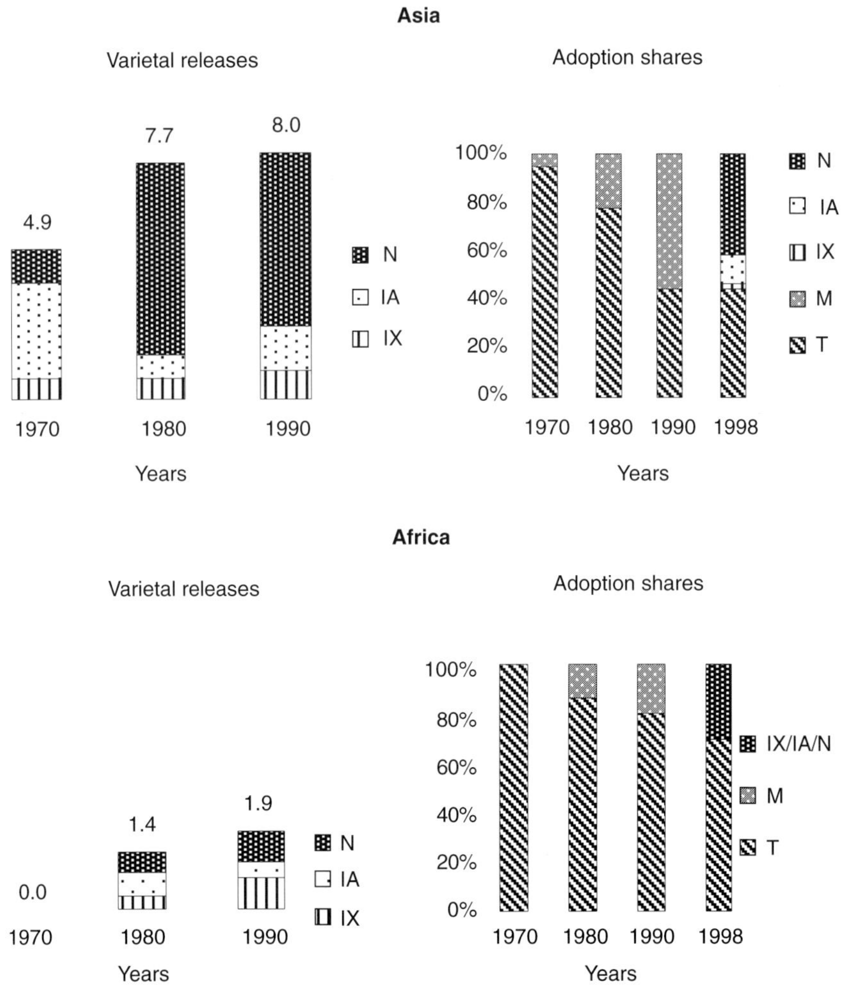

Fig. 2.4. Varietal releases and adoption shares: sorghum. T = traditional variety; M = modern varieties; IX = IARC crosses; IA = NARS crosses, IARC ancestors; N = NARS crosses, NARS ancestors.

During the 1990s, private firms began to develop and release hybrid varieties of sorghum. These releases are not included in the data in Fig. 2.4 but were significant during the 1990s (probably 30% of public releases). The relevant feature of this private-sector development is that ICRISAT material makes up a significant part of the genetic content of private varieties. Thus, private-sector firms are building programmes on public-sector ‘platforms’. It should be noted that a number of the public-sector varieties are also hybrids, requiring the seed production methods associated with hybrids. This also facilitated private-sector research programmes.

The adoption by farmers of improved sorghum varieties has differed markedly by regions. In Asia, adoption rates for modern sorghum varieties increased rapidly during the 1990s. Some of this increase was due to private-sector hybrids. Adoption rates have also increased in Africa, but with a delay. The modern varieties released in Africa prior to ICRISAT’s involvement achieved little adoption. By 1998, sorghum adoption in Africa had attained the same levels as those achieved in Asia 15 years earlier.

Figure 2.5 portrays varietal releases and adoption data for pearl mil let in both Asia and Africa. Here we note that with the establishment of the ICRISAT programme, varietal releases increased significantly. In fact, Africa effectively had no releases until the ICRISAT programme was established. ICRISAT parental material was especially important in both Asia and Africa.

The private sector has also begun to introduce hybrid pearl millet varieties in both Asia and Africa, and as with sorghum, these private sector varieties are built on platforms created by the NARS programmes and the ICRISAT programmes.

Adoption data are limited but do show patterns similar to those for sorghum. In 1998 Asian producers were utilizing modern varieties at a high level. These varieties also had significant ICRISAT content.

Barley

Barley is an important crop in a number of developed countries. It is also important in several developing countries with semi-arid or arid production conditions. These countries are chiefly in the Middle East–North Africa (MENA) region, where barley is grown to take advantage of limited winter rainfall. ICARDA initiated a barley programme in the late 1900s. Figure 2.6 shows varietal releases and adoption.

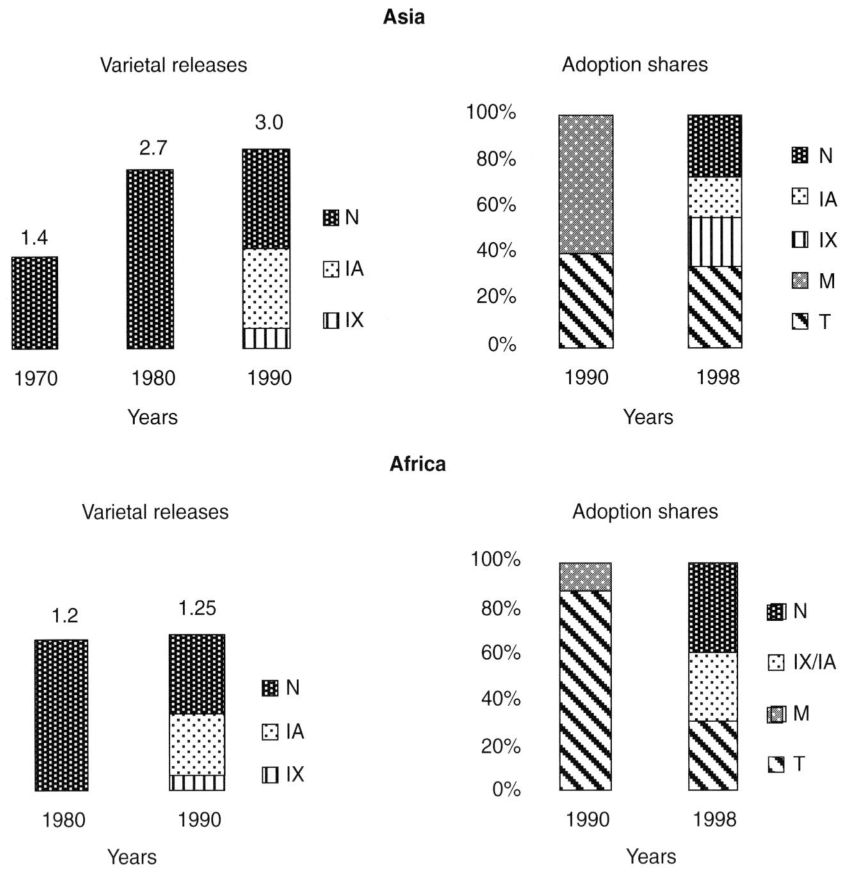

Fig. 2.5. Varietal releases and adoption shares: pearl millet. T = traditional variety; M = modern varieties; IX = IARC crosses; IA = NARS crosses, IARC ancestors; N = NARS crosses, NARS ancestors.

Beans

Dry beans are planted on 26 million ha worldwide, with approximately 6 million ha in Latin America and 35 million ha in Africa. China and India are also major producers. CIAT initiated a bean programme in South America in 1973, with its mandate extending to Central America in 1979 and to Africa in 1983. Seventeen Latin American NARS (88 breeders) had bean programmes in the 1970s. Only two programmes were active in Africa in the 1970s, but by 1998, 12 programmes in Africa (40 breeders) were operating.

CIAT in Colombia has been supporting bean improvement research for a number of years. Beans are an important source of protein in the

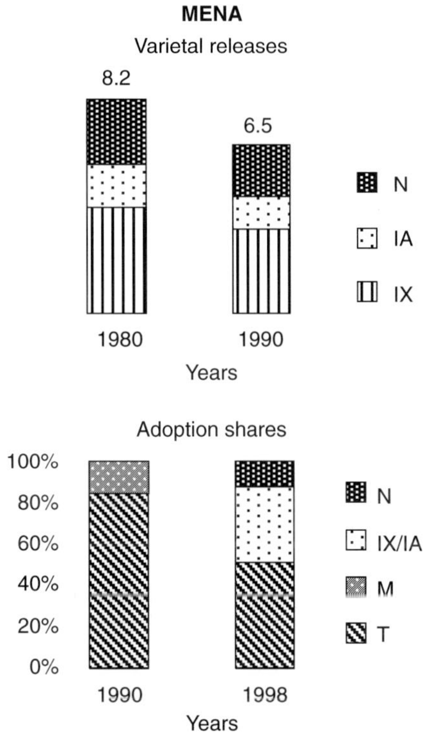

Fig. 2.6. Varietal releases and adoption shares: barley. T = traditional variety; M = modern varieties; IX = IARC crosses; IA = NARS crosses, IARC ancestors; N = NARS crosses, NARS ancestors.

diets of many consumers in Latin America, especially Brazil, and in Africa. Because of limited genetic improvements, beans have effectively been 'crowded out' of productive areas by crops with greater genetic improvement, especially maize and soybeans.

Varietal releases are shown in Fig. 2.7. These data show steady increases in releases in both Latin America and Africa, with high CIAT content (especially CIAT crosses) in both regions. Varietal releases in Africa were low in the 1970s but significant in the 1990s.

Adoption of modern bean varieties has been low in spite of high numbers of releases. This is partly due to high diversity in growing conditions and in taste patterns of consumers.

Lentils

Approximately 3.4 million ha of lentils are planted worldwide. India is the leading producer, with 1.2 million ha. Most of the remainder is pro-

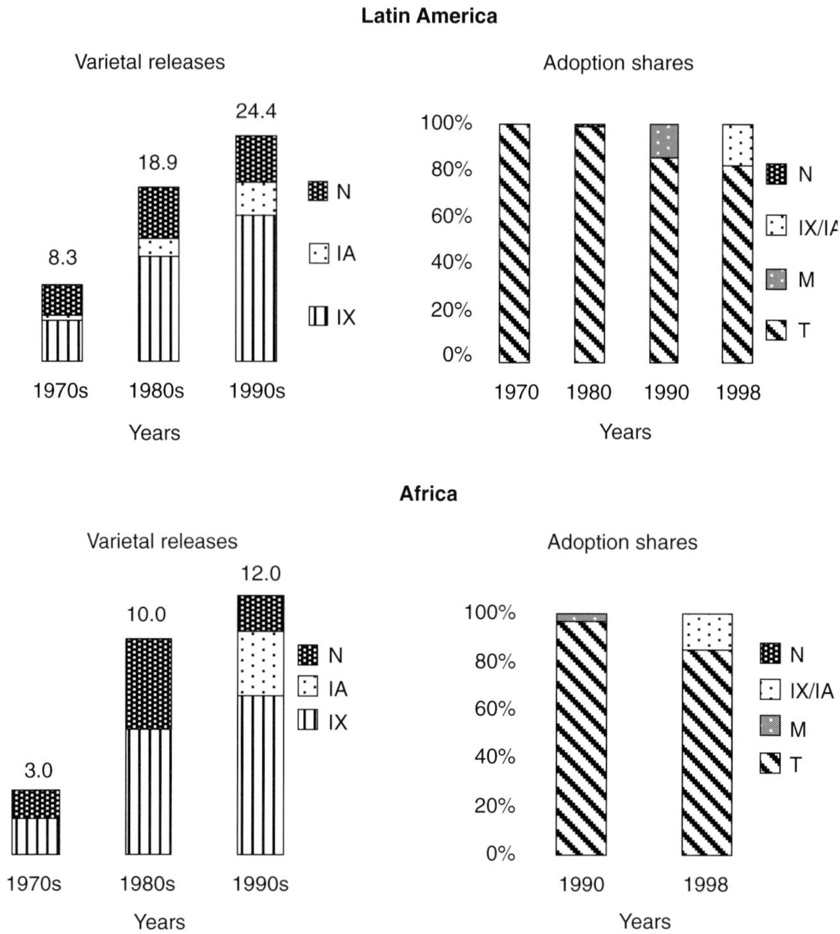

Fig. 2.7. Varietal releases and adoption shares: beans. T = traditional variety; M = modern varieties; IX = IARC crosses; IA = NARS crosses, IARC ancestors; N = NARS crosses, NARS ancestors.

duced in the dryland tropical regions of the Middle East and North Africa. ICARDA (International Centre for Agricultural Research in the Dry Areas) began lentil improvement in the late 1970s. Few improved varieties (except in India) were released until the 1980s and 1990s. In the 1990s most improved varieties were based on ICARDA crosses (Fig. 2.8).

Adoption of modern lentil varieties was negligible until the 1990s, but, in recent years, significant adoption has occurred. Virtually all modern varieties are crossed by ICARDA, attesting to a high impact in a crop where little was being accomplished prior to the IARC programme.

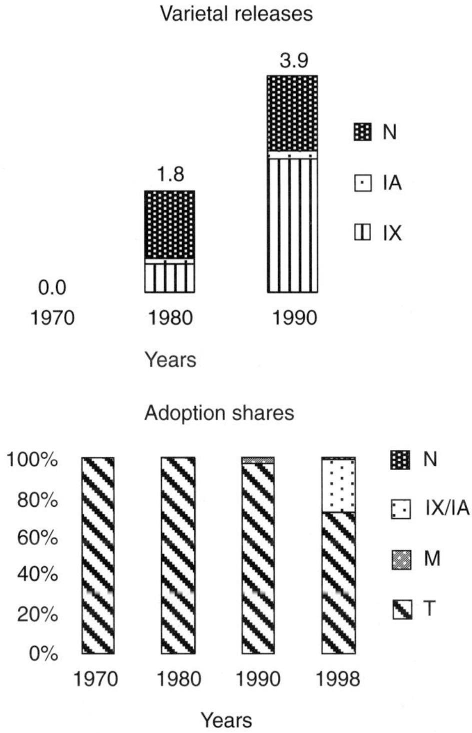

Fig. 2.8. Varietal releases and adoption shares: lentils. T = traditional variety; M = modern varieties; IX = IARC crosses; IA = NARS crosses, IARC ancestors; N = NARS crosses, NARS ancestors.

Groundnuts

Groundnuts are produced on 24 million ha, over 95% in developing countries. India, China, Nigeria and Sudan are the leading production regions. Few modern groundnut varieties suited to developing countries were released prior to 1980. Since then, varietal production in both Asia and Africa has increased steadily. By the late 1990s, adoption rates for modern varieties were very high in China, and high in India and parts of Africa (Fig. 2.9).

Potatoes

Potatoes are produced over a wide range of climate systems. Globally, 38 million ha are planted. Roughly one-quarter of this hectarage is in developing countries (Asia has 6 million ha, Africa 1 million, Latin

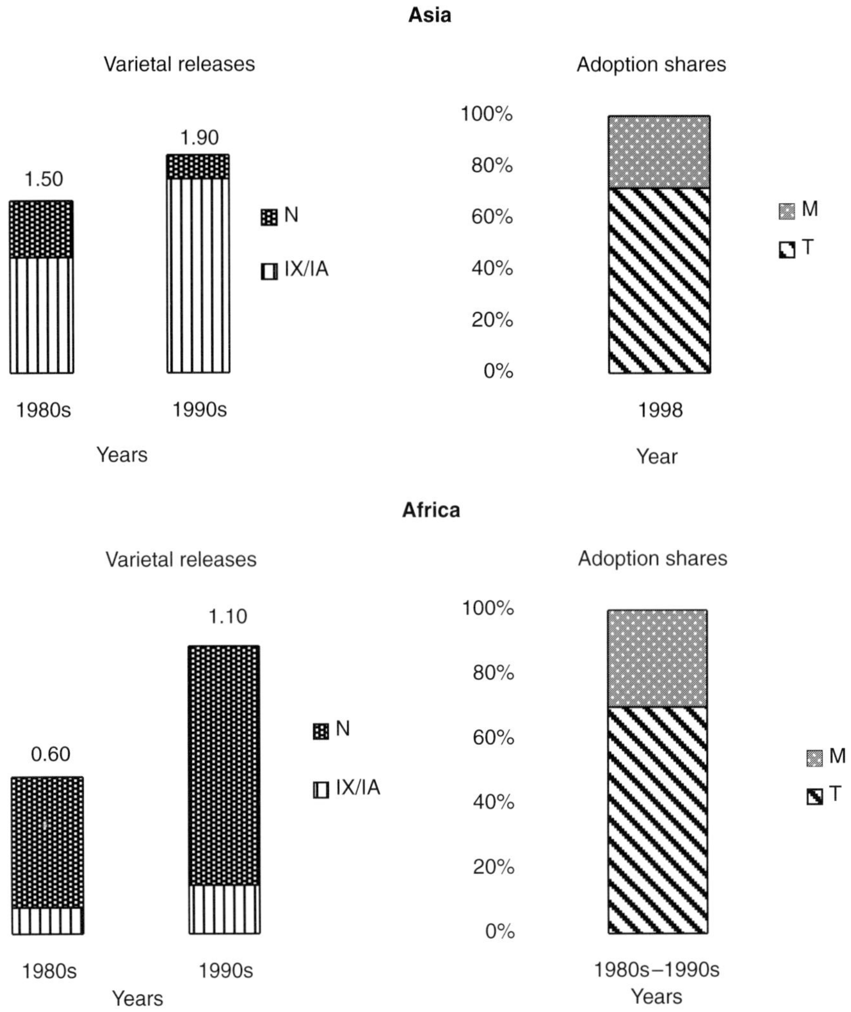

Fig. 2.9. Varietal releases and adoption shares: groundnut. T = traditional variety; M = modern varieties; IX = IARC crosses; IA = NARS crosses, IARC ancestors; N = NARS crosses, NARS ancestors.

America 1.5 million), and this hectarage is growing as new varieties are developed. Many potato landraces (farmers' varieties) continue to be planted for reasons of local preference. More than 30 developing countries have potato breeding programmes, with more than 300 breeders and related scientists working on genetic improvement in potatoes. CIP (International Potato Centre) provides genetic resource and breeding support in major regions.

Varietal production and adoption is shown for Latin America, Africa and Asia in Fig. 2.10; in all three regions rising levels of annual varietal release are noted. Rising levels of CIP contributions dominate only in Africa. Many of the NARS contributions are selections of landraces and much of the progress in potato productivity is due to improved handling and selection of seed by farmers.

The adoption of modern varieties is high in all three regions in the late 1990s. CIP content in these adopted cultivars is low, however, reflecting the importance of cultivar selection in local areas.

Cassava

Cassava is a major food crop in Africa, where 11 million of the world's 16 million ha are planted. Latin America plants 2.4 million ha and Asia 3.3 million ha. Cassava is thus produced almost exclusively in tropical developing countries. Two IARCs, CIAT in Latin America and IITA in Africa, support NARS programmes.

Varietal releases, as noted in Fig. 2.11, are of recent origin, with few improved varieties released before the 1980s. IARC content is high in releases, reflecting the fact that the IARCs have dominated the genetic improvement in this crop.

Modern varietal adoption rates remain low in the late 1990s, although significant adoption has taken place in Africa.

A Synthesis

Three synthesis chapters report comparisons of modern variety production and diffusion by crop, and also address several important policies. The first of these chapters (Chapter 21) is concerned with the production of modern varieties in all of the crops in the study. The chapter also addresses two important policy questions. The first of these is whether the IARC crop genetic improvement programmes had a germplasm impact on NARS programmes. This germplasm impact is based on the provision of advanced breeding materials to NARS programmes that made them more productive in producing NARS crossed varieties. The second question addressed in this chapter is whether the net impact of IARC programmes on NARS investments in crop genetic improvement was positive or negative. IARC programmes are both complementary to and competitive with NARS programmes. They complement NARS programmes by providing germplasm services. They compete with NARS programmes by developing finished crop varieties. Their net impact on NARS investments in crop genetic improvement

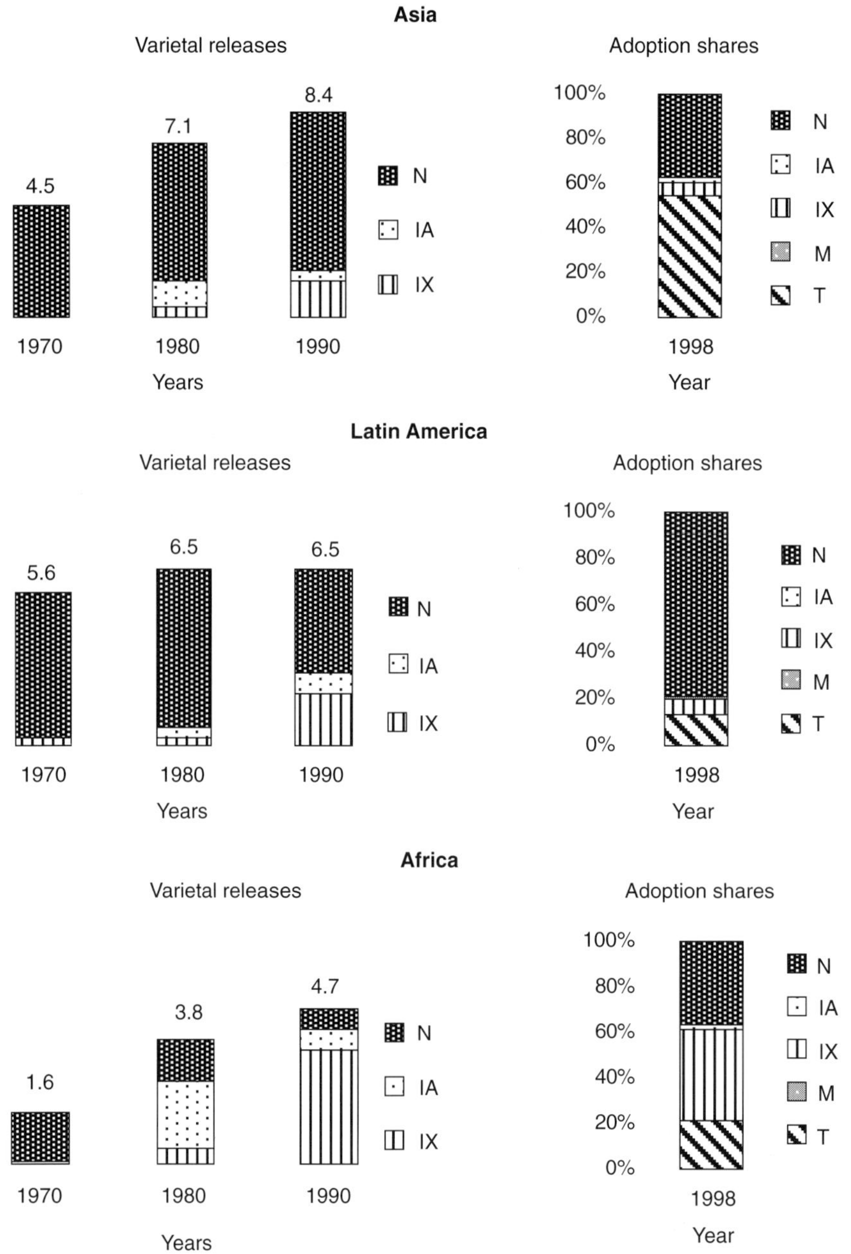

Fig. 2.10. Varietal releases and adoption shares: potatoes. T = traditional variety; M = modern varieties; IX = IARC crosses; IA = NARS crosses, IARC ancestors; N = NARS crosses, NARS ancestors.

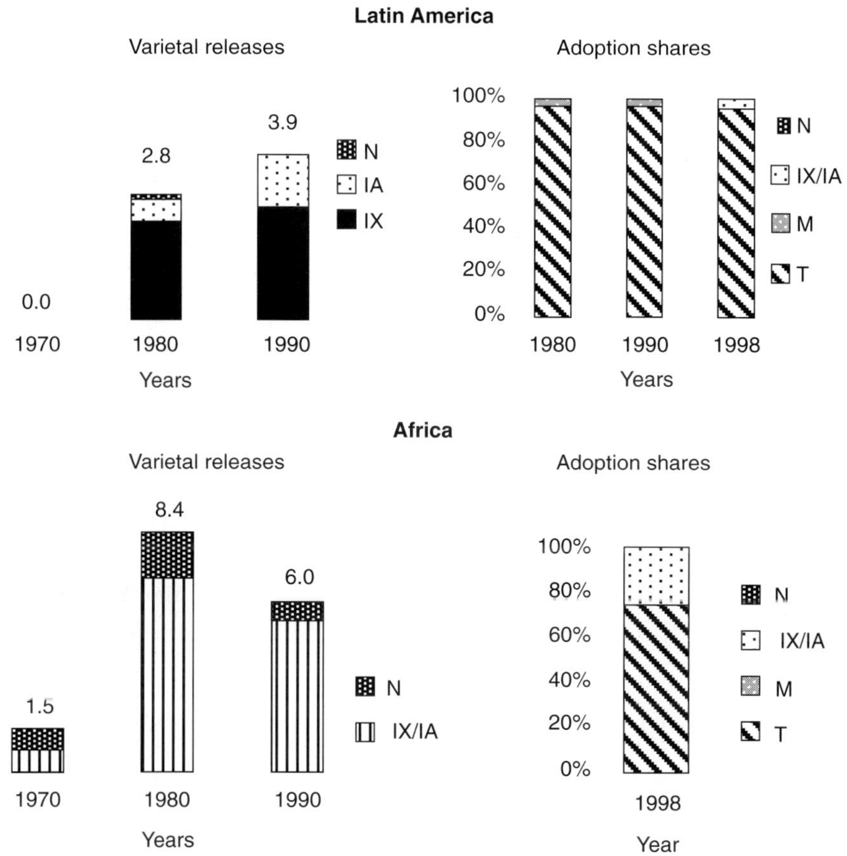

Fig. 2.11. Varietal releases and adoption shares: cassava. T = traditional variety; M = modern varieties; IX = IARC crosses; IA = NARS crosses, IARC ancestors; N = NARS crosses, NARS ancestors.

depends on the relative size of the complementary and competitive effects.

Varietal production by 5-year period, by crop, and by region for all crops is summarized in Chapter 21, along with measures of the contributions of international centres to the development of new varieties. We note first that for wheat and rice varietal production, peak rates of production were reached in the late 1980s and have been roughly constant for the past 15 years. However, rates of varietal production for both wheat and rice are considerably higher than they were in the 1970s.

By the 1990s, varietal releases in maize had surpassed those in rice. For all crops combined, rates of varietal production rose steadily over all periods. In the 1990s, modern varietal production was more than one-third higher than in the 1980s, more than double the rate in the 1970s, and four times the rate in the 1960s.

IARC content in modern varieties was high in most crops; 36% of all varietal releases (25% of unique varietal releases) were based on IARC crosses. For the Middle East and North Africa and for sub-Saharan Africa, more than half of all modern varieties were based on an IARC cross.

In addition, 26% of all modern varieties had an IARC-crossed parent or other ancestor. This attests to an IARC germplasm input in NARS crossed varieties. A strong IARC germplasm impact was identified in the statistical estimates reported in Chapter 21. These estimates suggested that germplasm contributions from international centres helped national programmes to stave off the 'diminishing returns' to breeding that might have been expected to set in had the national programmes been forced to work only with the pool of genetic resources that they had available at the beginning of the period. The estimates of Chapter 21 also showed significant germplasm impacts, as measured by cumulated varietal releases based on IARC crosses. From the 1970s to the 1990s, NARS crop genetic improvement resources roughly doubled. In the absence of IARC germplasm impacts, this would have led to a 70–80% increase in NARS varietal production. Because IARC crossed breeding lines combining large numbers of important traits were readily available to NARS breeders, however, NARS programmes were an additional 30–40% more productive. Actual varietal production by NARS in the 1990s was more than double the production rates of the 1970s.

The statistical study examining the net effect of IARC crop genetic improvement programmes on NARS investments found that the complementary germplasm effects of IARC programmes outweighed the potential 'crowding out' of national programmes by international ones. The analysis suggests that the availability of germplasm from the international centres induced national programmes to invest more than they would otherwise have done – so that NARS resources were about 15% higher than they would have been in the absence of IARC programmes.

Simply put, countries that perceived they could draw on useful flows of material from the international research system were likely to invest more in national research programmes than countries that did not have access to much useful material.

The second of the synthesis chapters, Chapter 22 reports an evaluation of the production gains associated with crop genetic improvement. These production gains are estimated utilizing data on farmer adoption of modern crop varieties from the crop studies (Chapters 4–16) and the country studies (Chapters 18–20). The country studies provided the major estimates of the productivity impacts of converting land under traditional (i.e. pre-1965) crop varieties to modern (post-1965) varieties. International crop TFP measures were also used to measure CGI impacts in production.

These computations are particularly instructive regarding regional contributions. The data on modern variety (MV) adoption differ significantly from data on MV production. This is particularly evident in crops where developed country stocks of genetic resources were not available for exploitation and for regions where little prior MV development had occurred. Crop genetic improvement programmes for cassava, groundnut, lentils, and beans, in particular, were characterized by periods where MV releases were being made with little MV adoption. This was also the case for sorghum and pearl millet, to some extent. For rice, on the other hand, MV adoption followed MV production closely in Asia and in Latin America. For sub-Saharan Africa, however, MV adoption has occurred only recently, even though MV production and release has been occurring for a number of years.

As shown below, there were striking differences in calculated crop genetic improvement contributions to annual productivity growth by crop, by decade and by region. These CGI contributions are in general consistent with actual productivity gains in these regions. They effectively explain why productivity gains in sub-Saharan Africa and in the Middle East/North Africa have differed greatly from comparable gains in Asia and Latin America. In the 1960s and 1970s, farmers in sub-Saharan Africa achieved almost no crop genetic improvement gains, at a time when Asian farmers were achieving high rates of these gains and of productivity gains generally. Farmers in the Middle East and North Africa also received few gains from crop genetic improvement. By the 1990s, this situation had changed, but there are still significant differences in the rate at which crop genetic improvement gains are being delivered to farmers in different regions.

Chapter 22 also reports calculations of the IARC contribution to crop genetic improvement gains. These calculations reflect the crop study data on IARC content in released and adopted varieties, as well as estimates of the germplasm impacts of IARC programmes on NARS programmes (from Chapter 21).

The final chapter in the volume reports estimates of the economic consequences of crop genetic improvement gains. The methodology used to make these estimates is first to define 'counterfactual' cases associated with crop genetic improvement, and then to compare prices, production, area, trade, and welfare in the equilibria that result from each of the cases. These equilibria involve multiple crops and multiple countries. For this purpose, the IMPACT model of the International Food Policy Research Institute (IFPRI) was utilized.

Two counterfactual cases were developed. The counterfactuals asked the following questions:

1. How would food prices, food production, food consumption and international food trade have differed in the year 2000 if the developing countries of the world were constrained to have had no crop genetic improvement after 1965, while developed countries realized the genetic gains that they actually achieved?
2. How would food prices, food production, food consumption and international food trade have differed in the year 2000 if the IARC system had not been built? We assume in this case that the developing countries would have achieved *some* gains based on their own national programmes (both public and private) but that they would not have benefited from targeted international research. As in the first case, developed countries were assumed to have experienced the same genetic gains that they achieved in reality.

The counterfactual scenarios showed that if developing countries had been constrained to have only farmer-selected crop genetic improvement after 1965, food prices would have been from 29 to 61% higher than they actually were. (Note that food prices have actually declined by more than 35% since 1965.) This may seem to be a modest price effect of something as drastic as the Green Revolution. But it must be remembered that developed countries are postulated to have had their actual productivity gains and that farmers in developed countries are price responsive. Accordingly, production would have risen in developed countries and exports to developing countries would have increased by 30%. (In reality, this might have been constrained by the ability of the developing countries to import food.) Production in developing countries would have been from 16 to 18% lower in this counterfactual. This reduction in production would have been greater, but farmers in developing countries are price responsive as well, and in the model they increase their production through increased use of land and other inputs.

Area cropped in both developed and developing countries would have been from 3 to 5% greater had genetic improvements not been realized. This expansion of crop land would have taken place on marginal and environmentally sensitive land, with attendant environmental consequences.

The most severe consequences of the counterfactual scenario, however, relate to food availability and consumption in developing countries. In the absence of the Green Revolution, food consumption per capita would have been from 10 to 13% lower in developing countries. Around 2–3% more children would have been malnourished (30–40 million children), and infant mortality rates would have been higher.

The second counterfactual attempts to simulate the economic implications of the IARC crop genetic improvement programmes. In the absence of these programmes, NARS programmes would have produced some advances. But the IARC programmes stimulated a bigger gain through modern variety breeding and germplasm impacts. Had the IARC programmes not been in place, world food and feed grain prices would have been from 18 to 21% higher than they actually were. Production would have been 4–5% lower, with developed country production offsetting part of the 7–8% reduction in developing countries. Cropped area would have been roughly 2% greater in both developed and developing countries. Exports from developed to developing countries would have been about 5% greater.

The political economy of 'food security' has two dimensions, global and local. From a global perspective, the Green Revolution as depicted in this volume is an extraordinary success. Food production per capita increased over a period of unprecedented population change. The prices of most food and feed commodities are lower in real terms today than at any other time in recent history. In addition, more people are better fed and enjoy better health nowadays.

However, there is another dimension to food security. This is the local dimension, and it is reflected in the fact that large numbers of people do not have an adequate diet and that large numbers of people have low incomes, earning $1 or $2 per day. After five decades of development programming, many developing countries remain in states of mass poverty.

It is essential, then, to assess the Green Revolution from this local perspective. Did the Green Revolution reduce mass poverty? Was it instrumental in creating conditions leading to increased incomes, particularly for labourers?

The counterfactual simulations reported in Chapter 23 only partially address these questions. However, the counterfactual concept is relevant to these questions, because countries were faced with the harsh realities of rapidly growing populations and limited land and water resources. The Green Revolution delivered modern varieties to farmers in an uneven fashion, and this unevenness had important welfare consequences for millions of poor farmers. Was there an alternative to the unevenness? Was there an alternative to the Green Revolution?

The chapters in this book address the uneven delivery issue to some extent. These chapters show that many biological factors did determine uneven delivery. For some production environments (e.g. deepwater rice) CGI programmes have produced little, simply because the genetic resources available to breeders were limited. For other cases, years of evaluation and selection to build effective breeding stocks suited to local production conditions were required. Managerial and political problems also hampered progress and contributed to unevenness as well.

Was there a real alternative to the Green Revolution? Did other development programmes contribute to the process of producing higher incomes? Certainly, there were many redistributive programmes and investments in health and schooling that contributed to welfare. But mass poverty is still a problem of very low incomes, and low incomes are related to low productivity. The Green Revolution did raise the productivity of land and water resources. It also raised the productivity of human resources and did lead to higher wages. This wage effect of the Green Revolution is difficult to detect in poor countries because of the absence of other productivity-enhancing programmes. In short, for the poorest countries, the Green Revolution, late and uneven as it was, was the 'only game in town'.

Finally, it should be noted that the Green Revolution did not damage the effectiveness of extension programmes or other programmes of importance in poor countries. On the contrary, the transformation of traditional agriculture was a transformation with consequences that went far beyond the performance of crop varieties.

References

Council on Environmental Quality and the Department of State (1980) *The Global 2000 Report to the President: Entering the Twenty-First Century.* US Government Printing Office, Washington, DC.

Schultz, T.W. (1964) *Transforming Traditional Agriculture.* Yale University Press, New Haven, Connecticut.

Objectives and Methodology: Crop Genetic Improvement Studies 3

R.E. EVENSON AND D. GOLLIN

The 13 chapters that make up Part II of this book all report data on the production of modern crop varieties (MVs), the diffusion of MVs to farms and the research impacts of MVs. Some of the chapters focus further on a particular region or on the impact of research done at a particular centre. In designing the study, we have sought to make these chapters parallel, as far as possible, but they are not entirely uniform. This chapter outlines the objectives and methodology used in the crop studies. It also points out some similarities and differences in the methods used in these chapters.

All the crop studies shared a common set of objectives. These were, in order of importance:

1. To document the production of modern varieties (as measured by releases), by period and by country.
2. To produce IARC (international agricultural research centre) content indicators in produced (released) varieties.
3. To document the adoption of modern varieties as of 1998 and for earlier years when possible.
4. To provide IARC content indicators in adopted varieties.
5. To estimate the production advantage of new varieties with respect to previously grown varieties, both for the case of areas shifted from traditional varieties to modern varieties (MV/TV) and for the case of areas shifted from older modern varieties to more recently developed modern varieties (MV/MV).

The methodology associated with the first four objectives is relatively straightforward, but requires extensive and detailed documentation. This is discussed below. The methodology associated with the fifth objective is more complicated. Production advantage is difficult to document at any level. Most crop studies are in a position to report a summary of 'micro' evidence on production advantage, based either on field trials or on-farm trials, but making micro-evidence consistent with macro (aggregated to district, regional or national level) evidence requires statistical sampling procedures that are generally not available in the context of crop studies. It is for this reason that country studies are part of the larger study. (See Chapter 17 for the methods and objectives of the country studies.)

A growing literature focuses on methodologies for the economic evaluation of agricultural research. Leading practitioners disagree on the appropriate ways to measure impact and to compute economic benefits. In this volume, we have largely tried to stay away from areas of methodological disagreement. Instead, these chapters are concerned with a logically prior issue: documenting research impacts. Before we ask whether research has been effective, it makes sense to ask how much research has been conducted. It also makes sense to ask whether there is any evidence that this research has actually found its way into use by producers. Only then does it make sense to measure economic impacts.

A number of the chapters present rough calculations of gross or net economic benefits from improved varieties. The authors have attempted to lay out as clearly as possible the assumptions that go into such calculations, but the reader should approach these figures with some caution. In particular, as should be obvious, these calculations are not directly comparable across chapters, nor are they based on any single methodological framework. The calculations are interesting, and we have retained them in the chapters because they provide useful information, especially crude estimates of the magnitudes of benefits, which can be compared with the magnitudes of costs.

Although the methodology of computing the benefits of research differs across centres, a major contribution of this volume is to provide essentially compatible data on the international centres' contributions to technology generation and diffusion. To our knowledge, this volume provides the most comprehensive and consistent reporting to date on crop genetic improvement (CGI) research products. As noted above, there were five objectives for each of the crop studies, and Chapters 4–16 offer broadly comparable data in response to each of these objectives.

Objective 1: Measuring varietal production

In recent years, modern crop varieties have been certified for release by national release agencies or boards. A breeding line or variety crossed and selected in an IARC breeding programme is submitted to one or more national release boards. Except for a few varieties released in the 1960s and 1970s by IARCs, these institutions do not release varieties directly. Instead, the materials may be screened, selected and released in a NARS (national agricultural research system) programme, or they may be used in a NARS breeding programme.

In general, national release boards have similar standards for release. Most attempt to impose minimum standards. In some cases, a country may engage in 'cosmetic' release to create the impression of breeding programme productivity, but this is relatively rare. For a few countries and crops, farmers make extensive use of varieties that have never been formally released. For other countries and crops, different research units issue different reports on the release status of varieties.[1] None the less, we take varietal releases as a relatively useful measure of research output.

Thus, for the purposes of this study, varietal production is measured by the number of releases. Since a given variety may be released in more than one country (especially true for IARC-crossed varieties), it would have been useful to record multiple releases. In practice, this was done only for rice varieties in Asia, where the multiple release ratio was 1.64 for IARC crosses and 1.06 for NARS crosses.

Objective 2: Measuring IARC content in released varieties

For most varieties in this study, genealogical information was available. This enabled the following classification by IARC content:

- an IARC cross (IX);
- a NARS cross with at least one IARC-crossed parent (IP);
- a NARS cross with at least one non-parent IARC-crossed ancestor (IA);
- a NARS cross with no IARC crossed ancestors (N).

These content measures (or indicators) were obtained for most of the varieties included in the study. They are more or less comparable across crops, although with maize breeding and sorghum breeding, ancestry may be somewhat more difficult to trace because of commonly used breeding techniques, such as population breeding and the use of composites. There are some difficulties in verifying the data from particular

[1] For example, with rice in India, there are large differences between the set of varieties listed as 'releases' by the Central Varietal Release Commission and those reported as releases by regional and local stations and breeders.

countries and, in a few cases, there are discrepancies in how different varieties are classified by different research institutions.

Some of the difficulties in classifying varieties reflect fundamental conceptual issues. For example, how should we classify a rice variety crossed in a NARS through a joint programme involving an IARC? What if the variety is developed at an IARC by a NARS breeder, using material that she has brought with her? What if a variety reflects the efforts of more than one NARS?

The crop studies in this volume have largely stayed away from some of the metrics that would attempt to weight IARC contributions by genetic diversity. A number of measures, such as the coefficient of parentage, would assign different weights to IARC ancestors depending on how far back these ancestors lie in the genealogy of the released variety. We have stayed away from such measures, preferring instead the transparency of the classification above. We acknowledge, however, that the measures used also have implicit weights attached to them. We further note that these IARC content measures are 'indicators' and should be interpreted as such.

Obviously there are difficulties in implementing the classification standard described above, or any other standard. None the less, Chapters 4–16 offer the best available information along these lines.

Objective 3: Measuring varietal adoption

Measuring varietal adoption is important, because, as the country studies show, there is considerable disparity in MV production (release) and adoption rates. Adoption measurement requires field surveys and a statistical sampling design. All governments measure crop production using surveys. Many of these surveys are formal systems with sampling weights, etc. A number of such systems record crop varieties and produce estimates of some aggregate (e.g. district, state, province) level of production and area by crop variety.

Some governments do not record varietal data, but do conduct surveys to monitor aggregate technology adoption rates. The early Green Revolution monitoring evidence popularized the use of measures of the percentage of area planted to MVs. Because of this, a number of such estimates are available. The authors of the crop studies in this volume were asked to achieve relatively complete and comprehensive estimates of MV adoptions by country for the most recent period and for earlier periods where possible.

There are several problems and issues that emerge here. First, figures on 'percentage of area planted to MVs' may not give good reflections of ground-level realities. Second, for some crops, such as maize and sorghum, farmers may report that they are using MVs, but if they are saving seeds, over a few years, the varieties actually planted may not perform like their MV ancestors. Third, the characterization of varieties as 'modern' or 'traditional' may not be particularly informative. Fourth, once MV adoption reaches a relatively high level, there is little additional information in the measure. Under these conditions, the MV adoption data cease to give useful information about the replacement of old varieties with new varieties.

To add complications, it is not clear *ex ante* that expanding the area under MVs is a necessary condition for research to have positive impacts. Under some circumstances, varietal improvement could actually reduce the fraction of area under MVs. For example, given suitable price elasticities of demand and elasticities of substitution between traditional and modern varieties, it might be possible for the introduction of an improved variety to *reduce* the area under MVs.[2]

None the less, for many crops, the area under MVs is continuing to expand, and there is useful information still to be gained from recording these data. We need to be careful in interpreting the data, however.

Objective 4: Measuring IARC content in adopted varieties

Most crop studies identified the major varieties actually adopted by farmers and computed IARC content measures for these varieties. In cases where earlier MV adoption data were estimated, it was usually not possible to impute IARC content. Chapter 22 reports synthesis comparisons of IARC content in produced (released) MVs and adopted MVs.

To the extent that the adopted varieties are known and that their genealogies can be traced, it is straightforward to measure the IARC content.

Objective 5: Estimation of MV/TV and MV/MV impacts on production

Most IARC and NARS crop genetic improvement programmes maintain yield evaluation programmes. These are used for varietal selection and

[2] Suppose, for example, that the grain from modern varieties is not readily substitutable for the grain from traditional varieties (not at all substitutable, in the extreme case). Suppose also that the demand for modern grain is inelastic (perfectly inelastic, in the extreme case). Then improving the yield of the modern variety will lead in equilibrium to a reduction in the area planted to modern varieties.

for testing and comparison. There are international yield trial systems and observational nurseries as well.

Field trials may be conducted on experimental station premises, where experimental protocols can be implemented, or on farmers' fields under farmer management. Numerous varietal trials enable comparisons of yield or production associated with varieties. This information is valuable to breeders.

Few, if any, crop genetic improvement programmes have made an effort to bring a statistical design structure to their field trial systems, which might enable these micro- and field-level data to be 'blown up' and made consistent with macro (e.g. district) data. This is a serious limitation, because it inhibits the use of the detail in yield studies and of the experimental control aspects of these studies.

As noted above, country studies where district, state and province data are directly employed to estimate varietal inputs are reported in Chapters 17–20 of this volume. (Chapter 17 discusses country study methods.) A brief discussion of the evaluation principles, however, will be useful in the context of varietal impact estimates in the crop studies. The most useful framework for discussing these effects is the total factor productivity (TFP) framework. TFP measures are designed to capture changes in production that are over and above those directly attributable to changes in input or factor use. That is, they are measures of increases in output per unit of total inputs.

When a new crop variety with superior biological performance displaces an older crop variety, this typically will produce TFP gains if the crop variety characteristics are not treated as specific inputs. However, during the period over which the varietal displacement took place, other activities may well have produced TFP gains. This places the evaluator in the position of specifying a statistical multiple source model (as in the country studies of Chapters 18–20) and estimating varietal impacts using such a model.

The alternative to specifying a multiple source model is to rely on evidence from an experimental situation. There is high potential value to field evidence because of this experimental control feature. However, if one cannot have confidence in the micro–macro consistency, one has to resort to the multiple source methodology (described in Chapter 17). In practice, the multiple source methodology produced a high degree of consistency between the country study estimates.

Within the context of the multiple source methodology, it is useful to note that a change in varieties is likely to change the productivity of other factors of production (including natural resources) and that a change in another factor of production may change the productivity of a variety. For example, it is well known (and built into the multiple source model of Chapter 19) that the first modern varieties of rice and wheat introduced into Asia had the effect of raising the productivity of water and of solar radiation. As irrigation investments were made in response, the productive value of the varieties themselves was raised. In at least some crops, these production interactions are quite important (see Chapter 15 on potatoes).

It is also useful to note that most crops have quality distinctions, and that changes in the relative demand for different qualities occur over time. This presents the evaluator with additional complexity. Demand shifts, for example, may lead to price increases for high-quality but low-yielding varieties. These price increases may lead farmers to reduce production of high-yielding low-quality varieties in favour of low-yielding high-quality varieties. They may also induce the plant breeder to shift breeding emphasis towards quality objectives. In such cases, the evaluator cannot simply use quantity of production as a measure of research impact. Price premia and discounts must also be considered.[3]

Similarly, varietal improvements that lead to changes in the agronomic characteristics of varieties are not readily evaluated by looking at quantity of production. Consider, for example, the impact of introducing a short-duration modern rice variety. In some locations, this variety may make possible double cropping (e.g. with wheat) on land previously single cropped in rice. The introduction of the new variety may actually reduce the yield per hectare of rice and possibly also of wheat. However, farmers are much better off, and more food is being produced in total.

Similarly, if a disease-resistant maize variety is introduced to a region, it may be planted in areas where maize was previously uneconomic. Maize yields in this region may actually be lower than in other regions, so *average* maize yields will go down. But the value of production will have risen.

These examples are provided simply to make the case that crop-by-crop production and yield gains may be inadequate measures of the impact of crop improvement research. Ideally, we would look not at crop-by-crop impacts, but at changes in entire production systems, taking into account changes in the total value of agricultural production. In some cases, we might need to look for benefits outside the agricultural sector, i.e. at economy-wide growth.

Such analysis was beyond the scope or capacity of the crop studies included in this volume, however. Many authors have reported yield advantage, measured in some way. There are many different approaches to measuring this yield advantage. Some chapters used farm-level

[3] Again, the more complicated TFP measurement methodology is capable of handling this problem, but the data requirements may be insurmountable.

studies, and others used the best guesses of scientists, or experimental-station studies. In all cases, the authors have used the best available data on yield advantages and have tried to document the assumptions being made.

Since different authors had access to different types of data, and since they took different approaches to calculating yield advantages, it should be obvious that the economic computations that follow are largely non-comparable across crops. It would be misleading to compare the gross (or net) economic benefits that different authors have calculated and to use these as the basis for comparisons of effectiveness across centres. These calculations are useful in giving an indication of the magnitude of research benefits, but they are imprecise and sensitive to changes in assumptions.

More complicated approaches are addressed elsewhere in this volume. In Chapter 22, an effort is made to synthesize available estimates (most from the country studies in this volume) and to arrive at a consensus estimate of MV/TV effects that is then used to compute crop genetic improvement productivity impacts. These, in turn, are subjected to a statistical comparison with actual macro-data.

Taken together, the chapters of Part II offer a rich and detailed picture of 40 years of crop improvement research. They provide the most detailed evidence yet that research has led to the production of improved varieties; that these varieties draw heavily on IARC ancestry; that these varieties are used by farmers; that the varieties selected by farmers include the ones with IARC ancestry; and that there are substantial economic benefits associated with varietal improvement. This is a powerful and compelling story of research effectiveness. Although there are admittedly difficulties in assessing impact at the level of a single crop, and although production advantage is an imperfect measure of impact, the crop studies taken together suggest that the benefits associated with crop improvement are very great. This is consistent with the lessons emerging from Parts III and IV.

Wheat 4

P.W. Heisey, M.A. Lantican and H.J. Dubin

In this chapter we report some results from the International Maize and Wheat Improvement Centre (CIMMYT) wheat impacts study, which focused on wheat genetic improvement research at CIMMYT and in national agricultural research system (NARS) wheat improvement programmes. Since the mid-1970s, CIMMYT has also collaborated with the International Centre for Research in the Dry Areas (ICARDA) on wheat improvement research for spring bread and durum wheat in the West Asia/North Africa (WANA) region; since 1990 this collaboration has been extended to facultative and winter wheat improvement for this region. The data reported here represent countries producing over 98% of the wheat grown in developing countries.

IARC and NARS Investments in Wheat Genetic Improvement

In this section, we analyse investments in wheat genetic improvement both by international agricultural research centres (IARCs) and NARS. International wheat improvement research is collaborative and depends on international testing by a network formed by CIMMYT and national research systems worldwide (Maredia and Byerlee, 1999). In the WANA region, CIMMYT also collaborates with ICARDA on wheat genetic improvement. We begin with IARC investments and then turn to the NARS.

From its inception, CIMMYT's primary research focus has been on the genetic improvement of two cereal crops, wheat and maize. As a result, CIMMYT's entire budget could be considered to be devoted to genetic improvement of these two crops. On the other hand, certain CIMMYT research products over the years, such as farming systems research, natural resources research, and some economic analysis, for example, might not be thought to be directly related to crop genetic improvement.

In the following analysis, we take two approaches to measuring investment in wheat genetic improvement research at CIMMYT. In both approaches, we assume that the entire wheat programme staff, including representatives of disciplines such as pathology, agronomy, physiology and others, as well as plant breeding, is focused on the genetic improvement effort. Our first assumption is that CIMMYT's entire budget can be charged to crop genetic improvement. In this approach, we allocate the total budget – including money spent on other programmes and on administration – between wheat and maize according to the proportion that the wheat programme budget comprises of the total budgets of the two crops programmes.[1] The alternative assumption is that the total CIMMYT budget is allocated to wheat genetic improvement according to the proportion that wheat programme senior staff represent relative to all CIMMYT senior staff, including staff in the other research programmes, external relations and administration. Total investments in wheat genetic improvement at CIMMYT are presented in Table 4.1. All expenditures are in 1990 US dollars. The first assumption probably overestimates the true investments in wheat genetic resource improvement; the second assumption is probably an underestimate.

By the first assumption, real CIMMYT investment in wheat genetic improvement rose steadily until the late 1980s, after which it fell significantly. By the second measure, real investment began to fall slightly earlier, from the mid-1980s. By both assumptions, real CIMMYT investment in wheat genetic improvement in recent years is now roughly back at the level of the 1970s. By the high assumption, CIMMYT today invests about $12 million (1990 dollars) annually in wheat genetic improvement; by the low assumption, the investment is about $7–8 million per year.

We use the larger measure, based on reported CIMMYT wheat programme budgets, to calculate real expenditure on wheat improvement per wheat scientist. This measure has fluctuated, but the general trend has been downward since about 1980.

[1] Currently there are five research programmes at CIMMYT, Wheat, Maize, Economics, Applied Biotechnology and Natural Resources.

Table 4.1. CIMMYT wheat research expenditures and research staff.

	1966	1970	1975	1980	1985	1990	1995	1999
Wheat budget 1[a] ('000 1990 $)	949	8,066	10,445	14,747	16,421	16,182	11,802	12,364
Wheat budget 2[b] ('000 1990 $)		5,777	7,771	10,179	11,256	9,897	7,041	7,847
Senior wheat scientists[c]		19	33	30	37	36	33	36
Senior wheat scientists+senior research support[d]		26	40	40	54	58	59	71
$/scientist ('000 1990 $)[e]		425	317	492	444	450	358	343

[a] Total CIMMYT budget (in 1990 dollars) is apportioned to the CIMMYT wheat programme according to the rule 'wheat programme budget/(wheat programme budget+maize programme budget)'.

[b] Total CIMMYT budget (in 1990 dollars) is apportioned to the CIMMYT wheat programme according to the rule 'wheat programme senior scientists/total CIMMYT senior staff'.

[c] Wheat programme senior staff. All disciplines, plant breeding as well as other disciplines, are assumed to contribute directly to the crop genetic improvement effort.

[d] Wheat programme senior staff plus other CIMMYT non-crop programme senior staff (such as biotechnology, natural resources, economics, external relations), apportioned to the wheat programme according to the rule 'wheat programme senior staff/(wheat programme senior staff+maize programme senior staff)'.

[e] Budget figure from first row divided by number of senior staff, wheat programme.

Source: *CIMMYT Review*; *CIMMYT Annual Reports*; Pardey *et al.* (1999).

Allocating ICARDA expenditures to wheat genetic improvement is more difficult. Based on ICARDA reports of staffing and research programmes, as well as estimates of joint CIMMYT/ICARDA investments in 1990 (Byerlee and Moya, 1993), we estimate that in the 1990s, ICARDA may have invested about $1 million (1990 dollars) in wheat genetic improvement annually.

Byerlee and Traxler (1995) estimated purchasing power parity (PPP) expenditures, in 1990 dollars, by NARS on spring bread wheat genetic improvement. Their estimates did not include China. Their estimates were based on comparing numbers of scientists working on spring bread wheat genetic improvement to numbers of agricultural scientists in general, and then applying this percentage to PPP expenditures on all agricultural research. The latter data were taken from Pardey and Roseboom (1989). We extended these estimates to all wheat, outside of China, by using the ratio 'all wheat releases/spring bread wheat releases' for different periods to adjust investment figures upward.[2] For China, we applied the same methods used by Byerlee and Traxler (1995) to research data provided by Fan and Pardey (1992).

Real investments in NARS wheat genetic improvement research grew steadily from the mid-1960s to about 1990 (Table 4.2). In 1990, NARS invested about $100 million on genetic improvement research.[3]

It is difficult to measure NARS investments in wheat genetic improvement past about 1990. Most publicly available data on developing NARS agricultural research investments end in the late 1980s or early 1990s. The consensus is that worldwide, in both developing and industrialized countries, public investment in agricultural research has stagnated or grown slowly over the 1990s. This certainly seems true in Latin America and sub-Saharan Africa. Projections of 1980s trends in China (Fan and Pardey, 1992) or India (Evenson *et al.*, 1999) would suggest some continued growth in NARS investments.

Table 4.2. Wheat genetic improvement research expenditures by NARS, 1990 dollars.[a]

	1965	1970	1975	1980	1985	1990
World	29.9	41.1	56.2	74.1	86.9	97.5
Asia	12.0	15.8	22.4	32.4	40.1	45.9
China	6.6	7.5	10.0	14.7	20.0	22.9
India	4.2	7.0	10.1	13.6	14.7	16.0
Other Asia	1.1	1.3	2.3	4.1	5.4	7.1
Latin America	5.4	8.8	12.4	16.2	16.3	16.6
Sub-Saharan Africa[b]	1.7	2.5	3.8	4.3	3.4	3.7
WANA	10.8	14.1	17.6	21.1	27.1	31.2

[a] For countries excluding China, wheat improvement research expenditures were calculated from data provided by Byerlee and Traxler (1995) for spring bread wheat by adjusting by the proportion 'total releases/spring bread wheat releases' for the relevant periods. For China, the same methods used by Byerlee and Traxler were applied to research expenditure data reported by Fan and Pardey (1992) and data on numbers of wheat genetic improvement researchers from the CIMMYT Wheat Impacts database.
[b] Excludes South Africa.
Source: Author's calculations based on data in Byerlee and Traxler (1995); Bohn *et al.* (1999); Fan and Pardey (1992); Evenson *et al.* (1999); and CIMMYT Wheat Impacts database.

[2] This measure was considered preferable to others, for example 'total improved wheat area/total improved spring bread wheat area', because it could be applied in a more time-specific fashion. Furthermore, we felt that the measure we used would be less likely to result in upward biases, particularly in the WANA region.

[3] Recall that CIMMYT invested about US$16 million annually (high estimate) in wheat genetic improvement during the same period.

Another indicator for NARS, the numbers of scientists working on wheat genetic improvement, does appear to have grown over the 1990s (Table 4.3). The research intensity measure, scientists per wheat area, also appears to have grown over the same period.[4,5] Only in sub-

Table 4.3. Scientist person-years devoted to wheat genetic improvement, developing countries, 1992 and 1997.

1992	Wheat area ('000 ha)	Scientist person-years	Scientists per million hectares
World	98,396,962	1,233	12.5
Asia	63,935,597	719	11.2
China	30,560,894	410	13.4
India	24,005,969	200	8.3
Other Asia	9,368,734	109	11.6
Latin America	8,330,807	132	15.8
Sub-Saharan Africa[a]	1,233,044	62	50.3
WANA	24,897,513	320	12.9

1997	Wheat area ('000 ha)	Scientist person-years	Scientists per million hectares
World	103,354,008	1,709	16.5
Asia	64,645,591	996	15.4
China	29,490,841	673	22.8
India	25,607,333	200	7.8
Other Asia	9,547,417	123	12.9
Latin America	9,463,297	170	18.0
Sub-Saharan Africa[b]	3,288,170	104	31.5
WANA	25,956,950	439	16.9

[a] Excludes South Africa.
[b] Includes South Africa.
Source: Authors' calculations from data in Bohn *et al.* (1999); FAO Agrostat (http://apps.fao.org/cgi-bin/nph-db.pl?subset=agriculture), 3-year averages; and CIMMYT Wheat Impacts data files.

[4] Note that these estimates, obtained by questionnaire responses counting actual scientists, are lower than publications-derived estimates presented by Evenson (2000), by a factor of ten. Scientist count estimates are likely to be underestimates, as they tend to miss scientists working on wheat improvement in universities, for example, but we feel they are likely to be more accurate than estimates based on publications.

[5] Another research intensity measure is scientists per unit of wheat production. For both measures, there is a clear inverse relationship between wheat production or wheat area, on the one axis, and research intensity on the other (Bohn *et al.*, 1999; Heisey *et al.*, 1999).

Saharan Africa does this measure appear to have fallen substantially. This result is partly due to the inclusion of South Africa in 1997, but not in 1992, but would have held even if South Africa had been excluded in 1997.

Anecdotal evidence suggests that in smaller research programmes, declining support per scientist has combined with relatively stagnant wheat improvement budgets, while some larger programmes have continued to make strong investments in wheat genetic improvement. There is little hard data, however. If wheat genetic improvement research budgets had increased over the 1990s consistent with late 1980s trends in aggregate research budgets, increased investments in the developing world outside of Latin America and sub-Saharan Africa would imply total investments are over $140 million (1990 dollars) today.[6] So, at present the range of expenditures in NARS wheat genetic improvement might fall somewhere between 100 and 150 million 1990 dollars.

Wheat Variety Releases in Developing Countries, 1966–1970

CIMMYT does not release varieties directly to farmers. NARS may test CIMMYT germplasm for direct release, or they may adapt or incorporate CIMMYT material into their own research to develop final varieties (Byerlee and Moya, 1993). The number of wheat varieties released annually by NARS doubled between 1966 and the mid-1980s, when it reached the level (about 80 releases per year) where it has essentially remained ever since (Table 4.4).[7] Over the past 15 years, a greater number of releases in the WANA region has counteracted somewhat lower rates of release in China and India. Lower rates of release in these two large producers with stronger and more mature wheat programmes probably is indicative of more precise varietal targeting, and not of any declines in investment. Furthermore, larger producers tend to release fewer wheat varieties per unit of area than smaller producers. This general observation is clouded by many variations, however. For example,

[6] By most accounts total public agricultural research expenditures in Latin America and sub-Saharan Africa stagnated over the 1980s and early 1990s.

[7] Some caution must be exercised in interpreting these figures, because it is likely that, with one exception, earlier release information is underestimated relative to later release information. This general trend probably does not hold true at the end of the period for which estimates are made. Here, too, there is probably a downward bias to the estimates.

Table 4.4. Average annual wheat varietal releases by region, 1966–1997.

	1966–70	1971–75	1976–80	1981–85	1986–90	1991–97
World	40.8	54.2	63.0	76.2	81.2	79.3
Asia	8.6	16.4	18.0	30.0	21.2	22.4
China	2.6	5.8	7.8	14.0	12.2	10.4
India	3.8	7.0	6.4	11.6	6.0	7.6
Other Asia	2.2	3.6	3.8	4.4	3.0	4.4
Latin America	17.6	20.2	27.0	30.6	33.4	26.3
Sub-Saharan Africa[a]	10.2	9.6	7.8	8.0	8.2	9.6
WANA	4.4	8.0	10.2	7.6	18.4	21.0

Source: CIMMYT Wheat Impacts data files.
[a]Includes South Africa.

Latin America and sub-Saharan Africa release far more varieties per unit of wheat area than other developing countries (Byerlee and Moya, 1993; Heisey *et al.*, 1999). Part, but not all, of this finding can be explained by the small amount of wheat produced by many countries in these regions.

Spring bread wheat releases dominate the varieties released in developing countries, just as spring bread wheat is the dominant wheat type grown in the developing world. Although spring bread wheat as a percentage of total wheat releases has fallen somewhat, from over 80% around 1970 to over 70% in the 1990s, this percentage is still higher than spring bread wheat's percentage of total wheat area, which is about 66%. Spring durum wheat releases comprised about 6% of all releases in the 1960s, and make up about 10% today. Spring durum wheat constitutes a little over 6% of all wheat area. Facultative and winter wheat releases have been at their highest level, 16%, in the 1990s, but this is still lower than the 27% of total wheat area represented by these types.

CIMMYT Content in Released Wheat Varieties

In the late 1960s, about one-third of all the wheat varieties released by developing countries were CIMMYT crosses, and another sixth had at least one CIMMYT parent. By the 1990s, these fractions had risen to about half CIMMYT crosses and another quarter that had a CIMMYT parent. At both the beginning and end of the time period covered in this study, an additional 7–8% could be traced to at least one CIMMYT ancestor.

These proportions varied by wheat type. Worldwide, around 90% of all spring bread wheat releases had at least one CIMMYT ancestor by the 1990s, and the percentage for spring durum wheat was even higher – nearly all spring durum wheat releases had a CIMMYT ancestor. In fact, direct CIMMYT crosses as a proportion of all spring durum wheat releases rose from one-quarter in the late 1960s to over three-quarters by the 1990s. In contrast, almost no facultative and winter releases were direct CIMMYT crosses over much of the period covered by this study. Facultative/winter varieties with some CIMMYT ancestry (cross, parent, or any ancestor) have constituted about 35–40% of all releases since the 1970s. In the mid-1980s, CIMMYT opened a collaborative winter-wheat breeding programme in Turkey, and in 1990 this effort merged with ICARDA's highland wheat programme. In the 1990s, for the first time a notable percentage (15%) of winter wheat releases were based on direct CIMMYT crosses (Table 4.5).

There are also regional differences in CIMMYT content in wheat releases. In the 1990s, for example, virtually all the spring bread wheat releases in WANA, sub-Saharan Africa, and Asia, outside of China and India had some CIMMYT content. Furthermore, CIMMYT crosses featured particularly heavily (62–73%) in releases in these regions. In India and Latin America, about 90% of the spring bread wheat releases had some CIMMYT content in the 1990s, and about half the releases were CIMMYT crosses. In contrast, although 60% of the spring bread wheat crosses in China had some CIMMYT content in the 1990s (and the percentage was even higher in some of the earlier periods), no direct CIMMYT spring bread wheat crosses have been released in recent years.

There are fewer regional variations in IARC content in spring durum wheat releases, where nearly all recent releases have CIMMYT or CIMMYT/ICARDA content, and three-quarters of these releases are CIMMYT crosses. Regional variation reappears more strongly in facultative/winter bread wheat releases. In China, home to over 60% of the wheat area planted to winter types, 20–25% of recent releases have some CIMMYT content, but nearly all of these releases have CIMMYT ancestors further removed than the parent generation. In the other important winter-wheat regions of the developing world, in WANA, CIMMYT content has risen to about 60% of the releases in the 1990s, with 29% coming from CIMMYT crosses, compared with none from CIMMYT crosses before 1990. In smaller winter-wheat areas, CIMMYT content was almost universal in Latin America in the 1990s, but minimal in South Africa, the only country in sub-Saharan Africa growing facultative or winter types (see Appendix Table A4.6).

Adoption of Semi-dwarf Wheat Varieties

Since the initial introduction of semi-dwarf wheat varieties in the 1960s, adoption has grown steadily, although at different paces in different countries. In 1970, semi-dwarf varieties were unimportant except in Asia

Table 4.5. CIMMYT content in released wheat varieties by period (proportions).

	1966–70			1971–75			1976–80			1981–85			1986–90			1991–97		
	Cross	Parent	Ancestor	Cross	Parent	Ancestor	Cross	Parent	Ancestor	Cross	Parent	Ancestor	Cross	Parent	Ancestor	Cross	Parent	Ancestor
Spring bread	0.38	0.20	0.06	0.48	0.24	0.06	0.41	0.38	0.06	0.46	0.32	0.09	0.52	0.29	0.10	0.53	0.29	0.08
Spring durum	0.25	0.00	0.08	0.65	0.00	0.00	0.77	0.08	0.00	0.68	0.18	0.03	0.75	0.20	0.02	0.77	0.20	0.02
Facultative/ winter bread	0.00	0.04	0.12	0.09	0.27	0.09	0.00	0.26	0.12	0.00	0.22	0.10	0.02	0.18	0.15	0.15	0.14	0.12

Source: CIMMYT Wheat Impacts database; CIMMYT Wheat Pedigree Management System.

outside of China. By the late 1990s, semi-dwarfs covered over 80% of all wheat area, with adoption rates of 90% or greater in Latin America and Asia outside of China. Fully 60% of all wheat area in developing countries was planted to wheat varieties with some CIMMYT content; excluding China, this figure would rise to 75% (Table 4.6).[8]

The contribution of breeding programmes can be measured in many ways. The broadest definition of CIMMYT contribution, which we have applied in the previous paragraph, is an 'any ancestor' rule. As we have seen in the section on releases, there are other categories, such as varieties with CIMMYT parents, and crosses made by CIMMYT. The last category, 'crosses made by CIMMYT', would constitute perhaps the narrowest definition of CIMMYT contribution. Pardey *et al.* (1996) summarize some of the measures used by other researchers to apportion contributions of different breeding programmes to pedigree-bred varieties, and propose several new measures. In Table 4.7, we apply the 'any ancestor' rule, the geometric rule developed by Pardey *et al.* (1996), and the 'CIMMYT cross' rule to our database on wheat planted in developing countries, for 1990 and 1997. The geometric rule gives a weight of 1/2 to the cross itself, 1/8 to the crosses used as parents, 1/32 to the crosses used as grandparents, and so on. In the earliest generation considered, weights are doubled to make all weights sum to 1. In our calculations, we used five generations.

Table 4.6. Percentage total wheat area planted to semi-dwarf wheat varieties by period (percentage of wheat area planted to varieties with CIMMYT content).

	1970	1977	1983	1990	1997
World	14	33	50	67 (47)	81 (60)
Asia	19	42	55	74 (46)	86 (59)
China	0?	10?	31?	60? (3)	79 (22)
India	36	74	76	87 (86)	92 (90)
Other Asia	51	74	87	91 (91)	94 (91)
Latin America	11	24	68	82 (77)	90 (83)
Sub-Saharan Africa	5	22	32	52 (51)	66 (57)
WANA	5	18	31	42 (36)	66 (54)

Source: Dalrymple (1978, 1986); Anderson *et al.* (1988); Byerlee and Moya (1993); and CIMMYT Wheat Impacts data files.

[8] If areas covered by landraces and unknown varieties were excluded from the calculation, in 1997 varieties with some CIMMYT content would have covered 64% of all improved wheat area (including improved tall varieties), or 80% of all improved wheat area outside of China.

Table 4.7. CIMMYT contribution to all wheat area planted in developing countries, 1990 and 1997; a comparison of 'any ancestor', geometric rule, and CIMMYT cross.[a] (Figures expressed as percentages of total area.)

'Any ancestor' rule		Geometric rule		CIMMYT cross	
1990					
Excluding China	Including China	Excluding China	Including China	Excluding China	Including China
67.3	46.5?	38.4	NA	34.4	24.2?
1997					
Excluding China	Including China	Excluding China	Including China	Excluding China	Including China
77.4	62.6	39.4	28.7	29.4	21.0

[a] Wheat areas planted to landraces and unknown improved varieties are included in the calculations. For these areas, CIMMYT contribution is set to zero.
Source: Authors' calculations based on CIMMYT Wheat Impacts data files; CIMMYT Wheat Pedigree Management System; and Byerlee and Moya (1993).

During the 1990s, area planted to wheat varieties with some CIMMYT content increased at the same time that the area planted to direct CIMMYT crosses decreased. Apportionment of CIMMYT content by the geometric rule was almost constant between 1990 and 1997. These results conceal some clear regional differences, not all of which are evident in Table 4.7. As the figures indicate, both for historical reasons and the substantial differences in growing environments from much of the rest of the developing world,[9] China's wheat genetic improvement programme has made less use of CIMMYT germplasm than programmes elsewhere in the developing world. A significant amount of Chinese wheat area today (some 6–7 million ha around 1997) is sown to wheat varieties with some CIMMYT ancestry, but for the most part these are not CIMMYT crosses nor varieties with a CIMMYT parent.[10]

[9] These include large areas sown to winter and facultative habit wheat, some area sown to high-latitude spring wheat, and special disease problems even in the areas sown to low-latitude spring wheat (Byerlee and Moya, 1993).

[10] See He and Rajaram (1997) for a further discussion of CIMMYT–China collaboration in wheat breeding.

Outside of China, all three indicators of CIMMYT content presented here – any ancestry, geometric rule or CIMMYT cross – increased between 1990 and 1997 in the wheat areas of WANA, Latin America and sub-Saharan Africa. In South Asia, however, the area planted to wheat with some CIMMYT content increased between 1990 and 1997, but areas apportioned according to the geometric rule or CIMMYT cross fell significantly. This was enough to drive the aggregate estimate for CIMMYT crosses down for the developing world excluding China. At the same time, as Table 4.6 indicates, the percentage of area planted to wheat varieties with some CIMMYT ancestry is higher in South Asia than it is in any other world region. In other words, CIMMYT germplasm is present in nearly all the wheat grown today in South Asia, but particularly in India, where a substantial area is sown to varieties with large original CIMMYT content, with several generations of subsequent crossing and selection by NARS scientists. Wheat scientists in South Asia have also incorporated earlier improved tall varieties into their germplasm base.

Economic Impacts of IARC Wheat Genetic Improvement Research

Yield and economic impacts of HYV wheats

High-yielding variety (HYV) wheats are clearly associated with increased yields. This has been demonstrated under many circumstances using various methods. Byerlee and Moya (1993) present evidence that wheat breeders have made significant yield progress since the initial development of HYV wheat. We extended the survey of Byerlee and Moya and reviewed 30 published and unpublished studies recording yield progress in trial data, from a variety of bread-wheat growing environments in the developing world. All but two studies showed significant gains in yield potential, even when the data were restricted to HYVs, giving further credence to the argument that HYV turnover in wheat, as well as initial HYV adoption, can lead to significant yield gains. A major scientific review (Evans and Fischer, 1999) also concludes that the genetic yield potential of wheat has continued to increase.

Using many years of international trial data in many different locations, Maredia *et al.* (1999) showed that across most major spring-wheat growing environments in the developing world, yields of CIMMYT crosses were greater than or equal to yields of varieties originating from that environment. Weighted by relative areas in the different environments and adjusting trial yield data to yields under farmers' conditions, these yield advantages would translate into an average yield advantage of about 200 kg ha^{-1} in farmers' fields. Note that this is a yield advantage of CIMMYT crosses over other improved wheat varieties, some of which may also have some CIMMYT content.

Improved yield performance has been associated with higher yield potential, and in a survey of more than 70 wheat breeders in developing countries, Rejesus *et al.* (1996) found that yield potential was one major reason for the use of CIMMYT wheats in these programmes. Superior stress resistance, however, is also another major contribution associated with CIMMYT wheat. One major long-term breeding effort has focused on resistance to wheat diseases, particularly the rusts. Over the period 1966–1988, considerably more of the increase in yields in CIMMYT-derived cultivars may have been due to superior leaf rust resistance than to increase in physiological yield potential (Sayre *et al.*, 1998). The respondents to the survey of breeding programmes also indicated that disease resistance was another major reason for incorporating CIMMYT germplasm (Rejesus *et al.*, 1996). CIMMYT wheats have also improved nitrogen-use efficiency (Ortiz-Monasterio *et al.*, 1997), tolerance to heat (M. Reynolds, unpublished data) and tolerance to drought (R. Trethowan, unpublished data). A summary of research findings (Smale *et al.*, 2001) demonstrates that these increases in yield potential and stress tolerance have occurred at the same time that genealogical diversity and molecular genetic diversity have also increased. Econometric evidence has also confirmed the yield and productivity contributions of HYV wheat.

It is not surprising, then, that studies of the economic returns of wheat research have generally showed positive rates of return. Evenson (2001) recently reviewed a large number of studies of the rate of return for agricultural research. Out of some 15 studies of wheat research in developing countries, only two found near zero or negative returns. These studies focused on all wheat research, but it is evident that wheat genetic improvement has been a major contributor to the positive rates of returns found in most studies.

Evaluating the IARC contribution to the economic benefits from HYV wheat

Byerlee and Traxler (1995) have provided the most comprehensive attempt to date to evaluate the economic impact of the joint CIMMYT–NARS wheat genetic improvement effort, focusing on spring bread wheat. They estimated an *ex post* rate of return for wheat breeding research for developing countries, with the highest returns in South Asia, and in irrigated and high rainfall environments. By 1990, they argued that more than two-thirds of the benefits from wheat improvement research were coming from varietal turnover rather than initial HYV adoption. They projected that future rates of return would be 35% or greater. In monetary terms, Byerlee and Traxler (1995) estimated that the total economic surplus in developing countries was about $2.5 billion annually, for a total research cost that never exceeded $70 million annually.[11] They assumed a lag of 17 years from initial investment to peak benefits.

It is possible, however, to couple straightforward projections of the major indicators used by Byerlee and Traxler (1995) with simple assumptions about supply and demand elasticities, to make some rough comparisons. These back-of-the-envelope calculations suggest that by the late 1990s, the Byerlee–Traxler assumptions would suggest that without the CIMMYT–NARS wheat improvement research investment, developing country wheat yields might have been 8–9% lower than they would have been in its presence; developing country wheat production might perhaps have been 24 million metric tonnes lower; and international wheat prices may have been around 7% higher than they would have been with international wheat genetic improvement research.

In terms of methodology, Byerlee and Traxler (1995) used relatively simple price assumptions to capture the effects of large regions' positions as net wheat importers or relatively self-sufficient producers. They did not actually consider the price effects of changing levels of wheat supply. Their estimates are of benefits from international wheat crop improvement research in total, not simply benefits that might be attributed to CIMMYT. Furthermore, it is possible that the yield assumptions they used no longer track wheat yield changes in farmers' fields, especially since aggregate statistics show that country wheat yields are still growing but no longer at the phenomenal rates seen from the Green Revolution through the mid-1980s. On the other hand, Byerlee and Traxler (1995) are only analysing research in spring bread wheat for four major environments in which this type of wheat is grown. With sufficient data, on all spring-bread wheat area, as well as spring durum and facultative/winter wheat, the analysis would have resulted in larger yield, output and price effects than reported.[12, 13]

[11] As noted, Byerlee and Traxler (1995) were estimating the costs of spring bread wheat genetic research only.

[12] It should be remembered, however, that the four spring bread wheat environments chosen were those in which CIMMYT wheat improvement research had had the largest impacts.

[13] Spillover benefits to industrialized countries are also ignored.

Whatever methods we used to estimate the economic impacts of IARC wheat genetic improvement research, it is quite clear that without this research:

- annual wheat production in developing countries today would be significantly lower than it is;
- total wheat production in developing countries over the past 30 or more years would have been much lower than it actually has been;
- wheat imports by developing countries today would be notably larger than they are in reality;
- real world wheat prices today would be significantly higher than their actual levels;
- the area of land planted to wheat in developing countries would be slightly higher than it actually is.

Economic benefits of crop genetic improvement research – further considerations

In our opinion, there are several significant issues that have only been partially addressed by studies estimating the economic costs and benefits of international crop genetic improvement research. Economists' discussions usually focus on several key assumptions: the trading positions (exporting, importing or self-sufficient) of the countries or regions in the analysis; the elasticities of supply and demand; and the nature and measurement of the cost reduction, or supply shift, attributable to the research in question.

However, there are other important questions that receive considerably less attention. If the study estimates returns from research, time lags are nearly universally assumed, as it is somewhat difficult to estimate appropriate time lags empirically. A related question is the development of an appropriate counterfactual scenario. If a given research programme did not exist, would no alternative programme have come into existence? It is likely, for example, that in the absence of CIMMYT, a more limited form of international exchange of wheat germplasm would have developed; genes for stature, disease resistance and other important traits would have eventually been used in wheat grown in the developing world, and so on. To the best of our knowledge, Evenson's recent attempt (Evenson, 2000) to estimate NARS varietal production in the absence of IARC crop germplasm improvement investment is one of the only efforts to delineate the counterfactual empirically. An alternative approach would be to estimate the total impacts of international crop genetic improvement research, as do Byerlee and Traxler (1995), and then partition those impacts to the

IARCs and NARS, perhaps using the various means of partitioning developed by Pardey *et al.* (1996). This latter approach does not, however, consider the catalytic contribution of IARC crop germplasm improvement to NARS research.

Partitioning benefits among different breeding institutions is only one example of the larger issue of aggregation. Are research benefits for a crop consistent with empirically observed aggregate supply and demand, and changes accounted for by other changes in input use? To what extent are research benefits attributable to genetic improvement, or to other research; for example, crop management research?

Finally, *ex post* studies do not estimate the marginal benefits to be obtained from continued investment in international wheat genetic improvement research by CIMMYT, ICARDA and the NARS. Byerlee and Traxler (1995), as noted, are an exception to this observation. For example, they suggested that at 1990 investment levels, research aimed only at maintaining yield levels would of itself generate high internal rates of return.

Summary and Conclusions

CIMMYT invests between $7 million and $12 million (1990 dollars) in wheat genetic improvement research today, and ICARDA perhaps another $1 million. This amount is significantly lower than it was in the late 1980s. NARS wheat genetic improvement investment may have been about $100 million (1990 dollars) in 1990. There is little evidence that NARS investment has fallen, and it may be somewhat larger today, particularly because of increased investment in a few large producers.

Wheat varietal releases by NARS increased sharply from the mid-1960s to the mid-1980s; and since that time annual releases over all developing countries have averaged about 80. In recent years, a somewhat lower apparent rate of release in China and India has been counteracted by higher releases in the WANA region. Although in recent years the proportion of facultative/winter releases has grown, it is still lower than the proportion of wheat area planted to these growth-habit types in developing countries. By the 1990s, the proportion of wheat releases that had some CIMMYT content had risen to over 80% from less than 60% in the late 1960s. In the 1990s, nearly all spring durum wheat releases had at least one CIMMYT ancestor, 90% of all spring bread wheat releases had at least one CIMMYT ancestor, and 40% of all facultative or winter bread wheat releases had at least one CIMMYT ancestor. The degree of CIMMYT content in releases has varied over time, over countries and regions, and over growth habit types.

By the late 1990s, HYV wheat covered over 80% of all wheat area, with particularly high adoption rates in Asia outside of China, and Latin America. At this time, 60% of all wheat area was planted to varieties with some CIMMYT content. This proportion would rise if only scientifically bred varieties were considered, or if attention were restricted to countries other than China. The proportions of spring bread wheat and spring durum wheat area planted to CIMMYT-content varieties were particularly high. The proportion of facultative/winter-wheat area sown to CIMMYT-related varieties was much lower. Compared with 1990, by the late 1990s more wheat area in some large producers, particularly India, was devoted to local crosses rather than to direct CIMMYT crosses. Wheat varieties in India and the rest of South Asia are none the less largely based on CIMMYT material.

HYV wheats have been associated with significant increases in wheat yield, in recent years as much or more through their superior stress resistance as through their superior yield potential. Genetic improvement has contributed to increases in aggregate wheat yields throughout the developing world, greater wheat production, reduced wheat imports by developing countries, and lower world wheat prices. Economic returns on international investment in wheat genetic improvement have been high, and returns on further investment are likely to be high as well.

Despite this record of outstanding achievement, the future of international wheat genetic improvement research remains cloudy. There are at least three uncertainties hanging over future impacts of international wheat genetic improvement. First is the efficiency of the system as it is constituted today. Second are emerging biophysical challenges in world wheat production. Third are evolving institutions for wheat improvement research in an environment characterized by greater intellectual property protection and increasing private sector investment.

One weakness identified in the current system is the slow rate of varietal replacement in many wheat growing areas (Heisey, 1990; Heisey *et al.*, 1999). Although varietal replacement from one generation of HYVs to the next has been shown to bring benefits, in many cases it is likely that economic returns on investment in wheat genetic improvement would be increased significantly if the lag time between varietal release and substantial adoption was reduced. Another major issue is the degree to which wheat genetic improvement research programmes, particularly small ones, should attempt ambitious crossing programmes or rely more on spill-ins from the international system (Maredia and Byerlee, 1999).

As noted, there is no convincing evidence that wheat genetic improvement, which to date has been achieved almost entirely through so-called conventional breeding, has reached yield ceilings.[14] None the less, yield gains in farmers' wheat fields have slowed over the past 15 years or so. The slowing of gains in farmers' yields in advanced irrigated wheat-growing areas such as the Punjabs of India and Pakistan, or northwest Mexico, is particularly notable. It is likely that crop management issues and resource degradation play important roles in these areas. In the intermediate term, both genetic solutions and crop management research may be needed to keep developing country wheat production increasing at the rates necessary to prevent imports from rising too high and to maintain the long-term decline in real world wheat prices.

Increased private-sector investment in crop improvement, changes in intellectual property rights regimes, and potential technical changes such as genetic engineering have begun to make fundamental changes to the nature of plant breeding and germplasm exchange. Although the force of these changes has been felt primarily in industrialized countries, it is clear that the impact will be global. Germplasm exchange, one of the cornerstones of the international wheat genetic improvement research effort, has already become more circumscribed than in the past. The proportion of wheat genetic improvement investment performed by the private sector varies from one industrialized country to the next. In general, it is considerably less than investment in other crops such as maize. None the less, private-sector wheat research investment is growing in industrialized countries, and it is likely that this will spread to high-potential areas in the developing world, particularly if economically feasible wheat hybrids are developed. As the decline in the CIMMYT wheat research budget over the last 10–15 years shows, the level of funding of international public-sector wheat research is by no means certain. Can the private sector cover some of the gap? In some cases, it may be able to do so, but it is unlikely that private and social optima in the allocation of wheat genetic improvement research resources will be equivalent. There is a strong, continuing need to find stable sources of funding for public wheat improvement research for the developing world.

In short, the international wheat genetic improvement system linking CIMMYT with the NARS, and with other international institutions

[14] Among crop scientists, there are, however, debates over deeper issues, such as whether there is potential to improve crop biomass, in addition to partitioning more biomass to grain yield, or whether most of the yield gains represent gains in stress-free yield potential or better resistance to stress.

such as ICARDA, has proved to be an outstanding success in delivering improved wheat varieties and tangible economic benefits throughout the wheat-growing developing world. In the future, however, the system is likely to change significantly. Both political will and financial resources will be necessary in order to maintain a strong public-sector component in international wheat research.

References

Anderson, J.R., Herdt, R.W. and Scobie, G.M. (1988) *Science and Food: the CGIAR and Its Partners.* Consultative Group on International Agricultural Research, Washington, DC.

Bohn, A., Byerlee, D. and Maredia, M.K. (1999) Investment in wheat improvement research in developing countries. In: Maredia, M.K. and Byerlee, D. (eds) *The Global Wheat Improvement System: Prospects for Enhancing Efficiency in the Presence of Spillovers.* CIMMYT Research Report No. 5. CIMMYT, Mexico, pp. 36–45.

Byerlee, D. and Moya, P. (1993) *Impacts of International Wheat Breeding Research in the Developing World, 1966–1990.* CIMMYT, Mexico.

Byerlee, D. and Traxler, G. (1995) National and international wheat improvement research in the post-Green Revolution period: evolution and impacts. *American Journal of Agricultural Economics* 77, 268–278.

Dalrymple, D.G. (1978) *Development and Spread of High-yielding Varieties of Wheat and Rice in Less Developed Nations.* Foreign Agricultural Economic Report No. 95. United States Department of Agriculture, Economic Research Service, Washington, DC.

Dalrymple, D.G. (1986) *Development and Spread of High-yielding Wheat Varieties in Developing Countries.* United States Agency for International Development, Washington, DC.

Evans, L.T. and Fischer, R.A. (1999) Yield potential: its definition, measurement, and significance. *Crop Science* 39, 1544–1551.

Evenson, R.E. (2000) Crop germplasm improvement: a general perspective. Paper presented at the annual meeting of the American Association for the Advancement of Science, Washington, DC, 21 February 2000.

Evenson, R.E. (2001) Economic impact studies of agricultural research and extension. In: Gardner, B.L. and Rausser, G.C. (eds) *Handbook of Agricultural Economics.* Elsevier Science, Amsterdam, pp. 575–628.

Evenson, R.E., Pray, C.E. and Rosegrant, M.W. (1999) *Agricultural Research and Productivity Growth in India.* IFPRI Research Report No. 109. IFPRI, Washington, DC.

Fan, S. and Pardey, P.G. (1992) *Agricultural Research in China: Its Institutional Development and Impact.* ISNAR, The Hague, The Netherlands.

He Zhonghu and Rajaram, S. (eds) (1997) *China/CIMMYT Collaboration on Wheat Breeding and*

Germplasm Exchange: Results of 10 Years of Shuttle Breeding (1984–94). Proceedings of a conference held in Beijing, China, 4–5 July 1995. Wheat Program Special Report No. 46. CIMMYT, Mexico.

Heisey, P.W. (ed.) (1990) *Accelerating the Transfer of Wheat Breeding Gains to Farmers: A Study of the Dynamics of Varietal Replacement in Pakistan.* CIMMYT Research Report No. 1. CIMMYT, Mexico.

Heisey, P.W., Lantican, M.A. and Dubin, H.J. (1999) Assessing the benefits of international wheat breeding research: an overview of the global wheat impacts study. In: Pingali, P.L. (ed.) *CIMMYT 1998–99 World Wheat Facts and Trends. Global Wheat Research in a Changing World: Challenges and Achievements.* CIMMYT, Mexico.

Maredia, M.K. and Byerlee, D. (eds) (1999) *The Global Wheat Improvement System: Prospects for Enhancing Efficiency in the Presence of Spillovers.* CIMMYT Research Report No. 5. CIMMYT, Mexico.

Maredia, M.K., Ward, R. and Byerlee, D. (1999) Assessing potential international transferability of wheat varieties. In: Maredia, M.K. and Byerlee, D. (eds) *The Global Wheat Improvement System: Prospects for Enhancing Efficiency in the Presence of Spillovers.* CIMMYT Research Report No. 5. CIMMYT, Mexico, pp. 22–35.

Ortíz-Monasterio, J.I., Sayre, K.D., Rajaram, S. and McMahon, M. (1997) Genetic progress in wheat yield and nitrogen use efficiency under four nitrogen rates. *Crop Science* 37, 898–904.

Pardey, P.G. and Roseboom, J. (1989) *ISNAR Agricultural Research Indicator Series: a Global Data Base on National Agricultural Research Systems.* Cambridge University Press, Cambridge. Published for the International Service for National Agricultural Research (ISNAR).

Pardey, P.G., Alston, J.M., Christian, J.E. and Fan, S. (1996) *Summary of a Productive Partnership: The Benefits from US Participation in the CGIAR.* EPTD Discussion Paper No. 18. IFPRI, Washington, DC.

Rejesus, R.M., Smale, M. and van Ginkel, M. (1996) Wheat breeders' perspectives on genetic diversity and germplasm use: findings from an international survey. *Plant Varieties and Seeds* 9, 129–147.

Sayre, K.D., Singh, R.P., Huerta-Espino, J. and Rajaram, S. (1998) Genetic progress in reducing losses to leaf rust in CIMMYT-derived Mexican spring wheat cultivars. *Crop Science* 38, 654–659.

Smale, M., Reynolds, M., Warburton, M., Skovmand, B., Trethowan, R., Singh, R., Ortíz-Monasterio, J.I., Crossa, J., Khairallah, M. and Almanza. I. (2001) *Dimensions of Diversity in CIMMYT Bread Wheat from 1965 to 2000.* CIMMYT, Mexico.

Appendix 4.1

Table A4.1. Average annual spring bread wheat varietal releases by region, 1966–1997.

	1966–70	1971–75	1976–80	1981–85	1986–90	1991–97
World	33.4	45.2	49.4	59.4	61.3	58.3
Asia	6.4	14.8	14.8	23.8	16.8	17.1
China	0.6	5.2	5.2	9.2	8.2	5.7
India	3.6	6.0	5.8	10.4	5.6	7.0
Other Asia	2.2	3.6	3.8	4.2	3.0	4.4
Latin America	16.0	16.8	22.6	24.8	28.5	22.7
Sub-Saharan Africa	8.8	8.8	6.4	6.6	6.0	5.7
WANA	2.2	4.8	5.6	4.2	10.4	12.7

Table A4.2. Average annual spring durum wheat varietal releases by region, 1966–1997.

	1966–70	1971–75	1976–80	1981–85	1986–90	1991–97
World	2.4	4.6	5.2	6.8	8.8	8.0
Asia	0.2	1.0	0.6	1.4	0.4	0.6
China	0.0	0.0	0.0	0.0	0.0	0.0
India	0.2	1.0	0.6	1.2	0.4	0.6
Other Asia	0.0	0.0	0.0	0.2	0.0	0.0
Latin America	0.6	0.8	0.8	1.6	1.6	2.1
Sub-Saharan Africa	0.4	0.0	0.6	1.0	0.2	1.0
WANA	1.2	2.8	3.2	2.8	6.6	4.3

Table A4.3. Average annual facultative/winter wheat varietal releases by region, 1966–1997.

	1966–70	1971–75	1976–80	1981–85	1986–90	1991–97
World	5.0	4.4	8.4	10.0	10.8	13.0
Asia	2.0	0.6	2.6	4.8	4.0	4.7
China	2.0	0.6	2.6	4.8	4.0	4.7
India	0.0	0.0	0.0	0.0	0.0	0.0
Other Asia	0.0	0.0	0.0	0.0	0.0	0.0
Latin America	1.0	2.6	3.6	4.2	3.4	1.4
Sub-Saharan Africa	1.0	0.8	0.8	0.4	2.0	2.9
WANA	1.0	0.4	1.4	0.6	1.4	4.0

Table A4.4. CIMMYT content in released spring bread wheat varieties by region and period (proportions).

	1966–70			1971–75			1976–80			1981–85			1986–90			1991–97		
	Cross	Parent	Ancestor	Cross	Parent	Ancestor	Cross	Parent	Ancestor	Cross	Parent	Ancestor	Cross	Parent	Ancestor	Cross	Parent	Ancestor
World	0.38	0.20	0.06	0.48	0.24	0.06	0.41	0.38	0.06	0.46	0.32	0.09	0.52	0.29	0.10	0.53	0.29	0.08
Asia	0.44	0.25	0.03	0.46	0.27	0.11	0.23	0.61	0.08	0.30	0.45	0.18	0.24	0.38	0.36	0.38	0.31	0.14
China	0.00	0.00	0.00	0.38	0.15	0.27	0.08	0.58	0.15	0.17	0.63	0.15	0.10	0.44	0.44	0.00	0.38	0.20
India	0.39	0.22	0.06	0.37	0.40	0.00	0.24	0.69	0.03	0.25	0.40	0.23	0.14	0.46	0.36	0.49	0.29	0.16
Other Asia	0.64	0.36	0.00	0.72	0.22	0.06	0.42	0.53	0.05	0.71	0.19	0.10	0.80	0.07	0.13	0.68	0.26	0.03
Latin America	0.39	0.14	0.09	0.54	0.17	0.00	0.50	0.22	0.05	0.60	0.20	0.01	0.55	0.31	0.00	0.50	0.31	0.08
SS Africa	0.27	0.32	0.04	0.32	0.39	0.11	0.38	0.47	0.09	0.46	0.30	0.12	0.70	0.23	0.03	0.62	0.25	0.10
WANA	0.54	0.09	0.00	0.67	0.12	0.04	0.57	0.36	0.00	0.52	0.29	0.05	0.81	0.10	0.00	0.73	0.25	0.00

Table A4.5. CIMMYT content in released spring durum wheat varieties by region and period (proportions).

	1966–70			1971–75			1976–80			1981–85			1986–90			1991–97		
	Cross	Parent	Ancestor	Cross	Parent	Ancestor	Cross	Parent	Ancestor	Cross	Parent	Ancestor	Cross	Parent	Ancestor	Cross	Parent	Ancestor
World	0.25	0.00	0.08	0.65	0.00	0.00	0.77	0.08	0.00	0.68	0.18	0.03	0.75	0.20	0.02	0.77	0.20	0.02
Asia	0.00	0.00	0.00	0.40	0.00	0.00	0.67	0.00	0.00	0.57	0.29	0.00	0.00	0.50	0.50	0.75	0.25	0.00
China	na	na	na	na	na	na	na	na	na	na	na	na	na	na	na	na	na	na
India	0.00	0.00	0.00	0.40	0.00	0.00	0.67	0.00	0.00	0.50	0.33	0.00	0.00	0.50	0.50	0.75	0.25	0.00
Other Asia	na	na	na	na	na	na	na	na	na	1.0	0.00	0.00	na	na	na	na	na	na
Latin America	1.00	0.00	0.00	1.00	0.00	0.00	0.75	0.25	0.00	0.875	0.125	0.00	1.00	0.00	0.00	0.93	0.07	0.00
SS Africa	0.00	0.00	0.00	na	na	na	0.33	0.33	0.00	0.40	0.20	0.00	1.00	0.00	0.00	0.43	0.57	0.00
WANA	0.00	0.00	0.17	0.64	0.00	0.00	0.88	0.00	0.00	0.71	0.14	0.07	0.73	0.24	0.00	0.77	0.17	0.03

Table A4.6. CIMMYT content in released facultative/winter wheat varieties by region and period (proportions).

	1966–70			1971–75			1976–80			1981–85			1986–90			1991–97		
	Cross	Parent	Ancestor	Cross	Parent	Ancestor	Cross	Parent	Ancestor	Cross	Parent	Ancestor	Cross	Parent	Ancestor	Cross	Parent	Ancestor
World	0.00	0.04	0.12	0.09	0.27	0.09	0.00	0.26	0.12	0.00	0.22	0.10	0.02	0.18	0.15	0.15	0.14	0.12
Asia	0.00	0.00	0.10	0.00	0.00	0.33	0.00	0.00	0.15	0.00	0.00	0.17	0.00	0.00	0.25	0.00	0.03	0.21
China	0.00	0.00	0.10	0.00	0.00	0.33	0.00	0.00	0.15	0.00	0.00	0.17	0.00	0.00	0.25	0.00	0.03	0.21
India	na	na	na	na	na	na	na	na	na	na	na	na	na	na	na	na	na	na
Other Asia	na	na	na	na	na	na	na	na	na	na	na	na	na	na	na	na	na	na
Latin America	0.00	0.00	0.20	0.15	0.31	0.00	0.00	0.44	0.00	0.00	0.48	0.05	0.00	0.41	0.06	0.50	0.40	0.10
SS Africa	0.00	0.00	0.20	0.00	0.00	0.25	0.00	0.00	0.25	0.00	0.00	0.00	0.10	0.10	0.10	0.00	0.10	0.00
WANA	0.00	0.20	0.00	0.00	1.00	0.00	0.00	0.43	0.29	0.00	0.33	0.00	0.00	0.29	0.14	0.29	0.18	0.11

5 International Research and Genetic Improvement in Rice: Evidence from Asia and Latin America

M. Hossain, D. Gollin, V. Cabanilla, E. Cabrera, N. Johnson, G.S. Khush and G. McLaren

Rice is the developing world's most important food crop. For 2.6 billion people around the globe – and primarily in developing countries – rice is a major staple food. Rice is harvested on about 155 million ha, 96% of which is in developing countries.

Asia accounts for more than 90% of the world's rice area and a comparable fraction of production (see Appendix 5.1). In a number of countries of Asia, rice accounts for more than half of all human food energy – and in the poorest countries, such as Bangladesh, Myanmar, Cambodia, Laos, Vietnam and eastern India, it contributes over two-thirds of the human energy intake (see Appendix 5.2). Rice is also important in the diets and cultures of a number of countries in Africa (Madagascar, Sierra Leone, Senegal, Côte d'Ivoire, Guinea and Liberia) and in Latin America, where the crop is the staple food for many groups and individuals.

This chapter assesses the contributions of international research centres to rice productivity gains in the developing countries of Asia and Latin America. Chapter 6 offers a separate account of international research and its effectiveness in West Africa, which is consequently omitted from this chapter. The main institutional actor considered in this chapter is the International Rice Research Institute (IRRI), located in the Philippines, which holds a global mandate for rice improvement research. In collaboration with national and international research centres, IRRI works 'to improve the well-being of present and future generations of rice farmers and consumers, particularly those with low incomes' (IRRI, 1996). In addition to IRRI, however, several other international institutions have played important roles in rice breeding. Most notably, in Latin America, the International Centre for Tropical Agriculture (CIAT), with headquarters in Colombia, has carried out an important programme of rice genetic improvement.

This chapter focuses specifically on 12 countries in South and Southeast Asia: Bangladesh, Cambodia, India, Indonesia, Laos, Malaysia, Myanmar, Pakistan, the Philippines, Sri Lanka, Thailand and Vietnam. Omitted are a number of countries in East Asia (where rice is enormously important) and in West and Central Asia and North Africa. China, the largest rice producing country in the world, is covered in a separate chapter in this volume. Japan and South Korea are developed countries with substantial rice research capacity and declining demand for rice. Central and West Asia and North Africa have been excluded because of lack of adequate data. The countries in these regions account for only 1% of total rice land.

The time period covered is 1965 to 1999. Although IRRI was in existence for several years before 1965, we have few data for the years 1960–1964, and the diffusion of modern varieties (MVs) was negligible. The data for years after 1995 are incomplete. Where we have reported data for the years 1995–1999, we would like to caution the reader that the data are partial and preliminary.

This study draws on several sources of data. In addition to published data from national research institutes, national statistical organizations and international organizations (including IRRI), we drew on information about breeding activity and varietal releases provided by scientists throughout the region. IRRI has collected similar data for a long time, and some have been published previously (e.g. Hargrove *et al.*, 1980). Many of these data are maintained in the International Rice Information System (IRIS) at IRRI.

To update the data, we sent a questionnaire to 28 breeding stations in South and Southeast Asia to collect information on the numbers of varieties released, the crosses involved in the released varieties, the breeders' objectives, the pool of genetic material used in breeding, and the most popular varieties adopted by farmers in the area served by the breeding stations. For a random sample of recent crosses, the breeders were asked to enumerate the genetic traits they hoped to obtain from each parent. To follow up on this survey, one of the authors visited a number of the breeding stations to conduct in-depth interviews with the plant breeders, while other plant breeders were interviewed during their visits to IRRI.

Background

International rice research dates back about 50 years, to the early work of the United Nations Food and Agriculture Organization (FAO), which undertook a concerted effort to develop *indica* × *japonica* crosses in the 1950s. Most Asian countries have national breeding programmes that date back further still, with some tracing their roots to colonial institutions founded as long as 100 years ago. Research in both China and Japan can be traced back still farther; the importance of rice improvement was widely recognized in these countries long before the science of genetics was formalized. Dalrymple (1986) offers a summary of the history of rice breeding.

IRRI was established in 1960 and breeding began shortly thereafter. Drawing on the experience of the Rockefeller Foundation's wheat breeding programme in Mexico, IRRI scientists sought to develop a set of short stiff-straw rice varieties that would respond well to fertilizer applications. At the time, most tropical rices were tall and tended to fall over (lodge) when fertilizer was applied. As a result, IRRI breeders quickly moved to introduce semi-dwarfism into *indica* rices. Within the first year of breeding, a number of highly successful crosses were made.

In particular, IR5 and IR8 were released to national programmes relatively quickly. The latter of these became associated with the Green Revolution in rice and was widely cultivated in Asia and in parts of Africa and Latin America by the early 1970s. Throughout the early 1970s, a succession of additional IRRI varieties was released, adding traits such as disease- and pest-resistance, cooking quality and wider adaptability. As national programmes grew in strength, IRRI abandoned the practice of releasing varieties directly and instead shifted to a strategy of supplying germplasm and elite breeding lines to national programmes for evaluation, selection and use.

Over the succeeding three decades, an enormous amount of breeding effort went into the creation of new rice varieties (Hargrove, 1979; Hargrove and Cabanilla, 1988; Evenson and Gollin, 1997). Our data include information on 2677 rice varieties released around the world, including approximately 2000 dating from after the inception of IRRI.[1] In the post-IR8 period (since 1970), the 12 countries in South and Southeast Asia mentioned above released 1507 varieties.[2]

The remainder of this chapter attempts to document the output of national and international breeding programmes along with evidence on the role the international research system – as embodied by IRRI and

[1] Exact numbers depend on which of two sources of data we use for India's releases. Publications of the All India Coordinated Rice Improvement Programme and the Central Variety Release Committee include a narrower set of varieties than those listed by regional and local breeders.

[2] Using the narrower measure of Indian releases, the figure is 1198.

CIAT – has played in the development of improved varieties.[3] We will assess how the IARC role has changed over time and how it differs across countries and regions.

In conducting this analysis, we must recognize the complexity and multiplicity of IRRI's and CIAT's contributions to rice varietal improvement. Whereas the first generation of modern varieties was based to a large degree on international research, subsequent varieties have involved collaboration among national and international centres (Evenson and Gollin, 1997). Indeed, it makes little sense to discuss IARC impact in isolation from the work of the NARS.[4]

This study will examine IARC contributions to the development, release and adoption of new varieties; to the improved productivity of those varieties; and to improvements in the characteristics of those varieties. The questions addressed in the study fall into four categories requiring distinct types of data.

First, we examine data on the inputs into research – both financial investments and human capital. This information allows us to assess the absolute size of research investments on rice crop improvement. It also gives us some sense of the relative size of investments in international rice research, relative to investments in national programmes.

Second, we consider evidence on varieties released by national programmes. By analysing the patterns of varietal release, and in particular by analysing the genealogies of released varieties, we can arrive at a thorough understanding of IARC contributions to national programmes.

Third, we can look at data on the diffusion of improved rice varieties. Until recently, such data have been limited to aggregated figures on the area planted to 'modern varieties' by country or by region. In this study, however, we make use of detailed data on the area planted to specific varieties of rice.

[3] In this chapter, we will not directly consider the impact of the FAO breeding programme, nor will we focus on the use of materials developed by other international institutions, such as IRAT in West Africa. Without denying the importance of these institutions, we focus on IRRI because it has been the pre-eminent international institution involved in Asian rice research over the past 40 years, and on CIAT because of its central role in Latin America.

[4] It is difficult to overemphasize the importance of the collaborations between IARCs and the various NARS. Perhaps it is most accurate to say that there is not a clear delineation between IRRI and the NARS. Scientists move back and forth across institutions, seeds and genetic resources are exchanged with relative freedom, and knowledge is readily shared. IRRI's impact is thus closely bound to the impact of the NARS.

Fourth, we can examine data on the yield and productivity effects of improved varieties and on the characteristics of varieties released and planted by national programmes. Such data are incomplete, but they help to illustrate how international researchers and their NARS collaborators have added new and valuable traits to the pool of germplasm available to farmers.

Research Inputs and Investments

Both national and international research institutions have contributed to the development and release of new rice varieties. In considering the impact of research programmes, it is useful to have some measures of the investments made by these institutions. Have investments in rice research risen or fallen over the past several decades? What patterns can be detected in research output per unit of input?

Unfortunately, data are scarce, particularly for earlier time periods. We also lack data on research investments by international researchers outside the CGIAR. For the most part, however, in the period under investigation, it seems reasonable to focus on research investments by IRRI and its NARS collaborators.

Investments in South and Southeast Asia

For the NARS in our study, we have little data on the number of scientists identified in rice improvement over the past four decades. Two sources give us snapshots of NARS staffing at different moments in time. For 1983, we have data on a number of countries, taken from a directory of rice workers. For 1999, we have partial coverage from a directory of rice scientists collected as part of an IRRI survey of the rice research capacity of selected Asian NARS. Table 5.1 presents this information.

For those countries with data in both time periods, there is an apparent increase in the number of scientists (those with MS or PhD degrees) involved in rice research. The increase over the last two decades was small for South Asia but substantial for Southeast Asia. For Vietnam, the number increased from 21 to 80, and for the Philippines from 108 to 205 over the 1983–1999 period. For 1999, the numbers show employment of 15 rice scientists per million ha of rice area in South Asia, and 21 for Southeast Asia. These coefficients, applied to the total rice area in South and Southeast Asia, give a crude estimate of the number of rice scientists employed in national pro-

Table 5.1. Number of scientists engaged in rice research in selected Asian countries 1983 and 1999.

	No. of scientists		No. of scientists per million ha of rice land	
Country	1983	1999	1983	1999
Bangladesh	134	163	13	16
Cambodia	na	35	na	18
India	546	625	13	15
Indonesia	66	na	7	na
Laos	na	31	na	43
Nepal	15	na	11	na
Pakistan	24	na	12	na
Philippines	108	205	34	53
Sri Lanka	14	17	18	25
Thailand	138	133	14	14
Vietnam	21	80	4	11

na, not applicable.
Source: IRRI, 1983. *The International Directory of Rice Workers*, International Rice Research Institute, Los Baños, Philippines. *Annual Reports*, Bangladesh Rice Research Institute, and personal communications with the Director of Research of State Universities and ICAR institutes in India. The numbers do not include those in charge of management and support services.

grammes at 1720 (880 for South Asia and 840 for Southeast Asia). This number is almost seven times the number of scientists (with MS or PhD degrees) working at IRRI in 1999 (Table 5.2). At IRRI, the number of scientists increased by 53% during the 1980s but declined in the 1990s due to a number of staff reduction programmes and budget cuts. Indeed, there has been a substantial decline in the number of scientists in the areas of plant protection and crop and natural resource management. This was partially offset by an increase in the numbers of scientists in the fields of breeding, biotechnology and physiology, and geographical information systems.

Data on the expenditures and financial investments supporting NARS researchers are even more limited. For India, data from two major rice research institutes – the Central Rice Research Institute at Cuttack (CRRI), and the Directorate of Rice Research at Hyderabad (DRR) – show a total budget of US$3.23 million supporting 164 rice scientists, indicating an expenditure of US$19.7 thousand per scientist. If we assume that scientists in other institutions were funded at comparable levels, it suggests that India spent about US$12.3 million on rice research in 1998.

Table 5.2. The number of rice scientists in the International Rice Research Institute, Philippines.

Disciplines	1979	1989	1999
Plant breeding, biotechnology, genetics resources, and physiology	39	59	71
Entomology and plant pathology	22	47	38
Agronomy and agroecology	20	52	42
Soil and water sciences	17	34	31
Experimental farming and engineering	14	15	10
Social sciences	18	25	21
Training and information centre	11	15	13
Country offices	32	17	25
Total	173	264	251

Note: Includes nationally recruited staff (NRS) with MSc degrees and above, post-doctoral fellows and project scientists, but excludes visiting scientists and consultants.
Source: IRRI, *Annual Report 1979*; *Program Report 1989*; and *Program Report 1999.*

In similar fashion, we can estimate expenditures in other NARS. For example, the government of Bangladesh allocated US$3.4 million in 1998 to the Bangladesh Rice Research Institute (BRRI) for 164 rice scientists (excluding the management and the support staff), implying an annual budget of US$20.9 thousand per scientist. The Philippines Rice Research Institute (PhilRice) received an allocation of US$3.54 million for 162 rice scientists (US$21.9 thousand per scientist). Extrapolating these numbers for countries lacking such data, we estimate that the NARS in South and Southeast Asia currently spend about US$36.2 million per year for rice research (US$17.9 million for South Asia, and US$18.3 million for Southeast Asia). The investment is almost equivalent to the US$34 million that IRRI spent in 1999 (IRRI, 2001). Taking NARS and IRRI investments together, the total annual investment in rice research in low-income Asia (excluding China) at present appears to be about US$70 million.

Not the entire amount is spent on research for genetic improvement. IRRI's medium term plan for 1994–1998 (IRRI, 1993) showed an allocation of 37% of its resources for genetic enhancement and breeding (including collection, conservation and evaluation of germplasm). For 2000, the share was estimated at 36.6% (IRRI, 2001). It is difficult to compile similar data for NARS. The data obtained from CRRI and DRR in India show that 36% of the scientists were engaged in research related to genetic improvement of rice, 27% for crop protection (including host plant resistance), 29% for crop and natural resource management, and 8% for socioeconomic and policy research. For the Philippines, the other country for which detailed information was available, the numbers were 44% for genetic improvement and seed production, 26% for crop protection, 22% for crop and natural resource management, and 8% for socioeconomic and policy studies. We may thus assume that 40% of the rice research investment in South and Southeast Asia (US$28 million) is allocated annually for activities related to production of improved rice varieties.

Investments by CIAT and Latin American NARS

Rice breeding in Latin America has historically occupied a smaller role than in Asia. During the past century, however, rice has emerged as the most important food grain in much of tropical Latin America. Because rice is a convenience food, relative to many other starch foods, which require extensive processing, rice consumption has risen steadily with urbanization. In tropical parts of Latin America, rice now accounts for more calorie consumption than either wheat, maize, cassava or potatoes.

About 20–25 rice breeders were active in the region for most of the period from 1970 to 1988, almost all operating in NARS institutions. The most active breeding programmes have been in Brazil, Colombia, Mexico and Peru.

CIAT maintained a small but active breeding programme through the 1970s, with noteworthy expansion of effort in the 1980s and 1990s. CIAT has also participated extensively in germplasm exchanges with IRRI and other institutions, including those in France and the USA. Through most of the 1970s, CIAT was staffed with three principal scientists (PhD level) working on rice breeding. That number began to increase in the early 1980s. By the early 1990s there were six scientists – out of a rice programme staff of eight scientists – working on rice genetic improvement. This number declined to around four in the latter part of the decade.

Varietal Improvement and IARC Contributions to Genetic Improvement

National programme releases in South and Southeast Asia

In the past 40 years, the national agricultural research systems of South and Southeast Asia have released 2040 varieties of rice on which we have data.[5] The rest of the world has released about 600 more varieties for which we have data. Table 5.3 shows a breakdown of the data for South and Southeast Asia by country and by time period.

One notable feature of the data is that there appears to have been a continuing flow of varietal releases over time, with the region's NARS turning out about 50 new varieties per year from the mid-1970s until the mid-1990s. For the most recent period, it is difficult to tell whether lower numbers of releases reflect a decline in the rate of production of new varieties or simply a lack of information about the latest releases.

Table 5.3. Number of improved rice varieties released in South and Southeast Asia by time period.

	No. of varieties						No. of
						Rice land	varieties per
Country	Pre-1970	1971–80	1981–90	1991–99	Total	(million ha)	million ha
Bangladesh	13	29	28	30	100	10.5	10
Cambodia	4	0	6	32	42	2.0	21
India	208	211	347	170	936	44.8	21
Indonesia	30	26	68	26	150	11.6	13
Laos	8	1	1	11	21	0.72	29
Malaysia	14	15	17	6	52	0.67	78
Myanmar	34	38	83	8	163	5.5	30
Pakistan	21	7	6	2	36	2.4	15
Philippines	14	38	20	56	128	3.9	33
Sri Lanka	22	18	22	13	75	0.83	90
Thailand	72	15	21	14	122	10.0	12
Vietnam	93	11	44	67	215	7.6	28
Total	533	409	663	435	2040	100.5	20

Note: The numbers include varieties with at least partial genealogical records. For India, the numbers include all varieties listed as released. The data from the All India Coordinated Rice Improvement Programme show only 454 varieties as officially released until 1995. The number of varieties released in India may be an underestimate of the level of production, as many varieties distributed under the 'mini-kit' programme are cultivated by farmers without being released.
Source: IRRI.

5 Although we have attempted to limit our data set to varieties actually released to farmers, the data may include some elite lines used in breeding programmes. The data also include some traditional varieties. There may also be commonly grown varieties that are not included here as releases. For example, some Indian varieties are in common use in Bangladesh, but they have not been officially released by the national system in that country. For consistency, we have attempted to limit our data set to the varieties formally released by national programmes.

Individual countries have, however, witnessed different patterns. For example, Vietnam's research system increased its releases of new varieties dramatically from the 1980s onward. Laos released a number of varieties in the 1990s after a long period with few releases. Cambodia's research system also picked up in the 1990s. In all of these countries, the increased rate of releases corresponded to increases in the human capital for the rice research system, as well as to increased cooperation with the international scientific community. IRRI contributed directly to the development of rice research capacity and germplasm improvement research through bilateral country programmes in Cambodia and Laos, and extensive research and training support for Vietnamese scientists (Raab *et al.*, 1998).

There are of course substantial differences in release patterns across countries. By either measure available to us, India has released by far the most varieties – unsurprising given that it has the largest rice area in the world (45 million ha). Vietnam, Indonesia and Myanmar are the other leaders in numbers of released varieties. In part, these differences reflect differences in the size and scope of research efforts; in part, they reflect differing standards of release and differing coverage of data. None the less, the data suggest an active and productive research effort throughout the region.

NARS uses of IRRI materials

By tracing the genealogies of released varieties, we can examine IRRI's contributions to varietal improvement in a number of different forms.

IRRI-developed varieties

Of the 2040 released varieties in the data for which we could trace ancestry, 219 were known to be IRRI lines released directly in other countries (without further breeding).[6] This figure represents about 11% of all releases. Table 5.4 shows the breakdown of these by country. For the region as a whole, the direct release of IRRI lines reached a peak in the 1976–1980 period and declined to a low level in the 1990s. Within the aggregate data, however, there are important differences across countries. About 25% of releases in the Philippines were direct IRRI releases; for Vietnam, about 20% of releases were bred at IRRI. Myanmar released many IRRI lines in the 1980s. By con-

[6] In some cases, it is difficult to tell whether the variety itself was developed at IRRI. The genealogical data on which the analysis is based does not always permit a clear distinction between IRRI-developed varieties and those with other IRRI ancestry. Under the circumstances, the figures given here should be viewed as conservative estimates.

Table 5.4. Contribution of IRRI to released varieties in South and Southeast Asia, by country.

Country	IRRI crosses released as varieties	Varieties with an IRRI parent	Varieties with IRRI materials in previous ancestors	Released varieties linked with IRRI materials
	(Numbers are percentage of total releases)			
Bangladesh	11.0	46.0	8.0	65.0
Cambodia	23.8	7.2	0.0	31.0
India	5.2	33.1	9.5	47.8
Indonesia	10.0	42.0	16.0	68.0
Laos	4.8	38.1	0.0	42.9
Malaysia	11.5	28.9	7.7	48.1
Myanmar	23.9	20.2	0.7	44.8
Pakistan	22.2	25.0	0.0	47.2
Philippines	26.6	38.2	4.7	69.5
Sri Lanka	2.7	30.7	21.3	54.7
Thailand	0.0	10.7	4.9	15.6
Vietnam	20.5	28.8	3.7	53.0
Total	10.7	31.1	7.9	49.7

trast, Thailand did not release any IRRI-developed varieties, according to our data, and India and Sri Lanka had fewer than 10% releases of this type.

As a percentage of the total released varieties, IRRI-developed varieties appear to have reached their highest level in the 1970s, when about 18% of all releases in the region were IRRI-developed lines (Table 5.5). This corresponds to the period in which the first modern varieties were developed that displayed effective resistance to a number of important diseases and pests. In subsequent years, the fraction of IRRI-developed varieties has fallen substantially; in the 1990s, only about 3% of varieties can be identified as direct IRRI crosses.

IRRI materials as parents

Excluding the IRRI-developed varieties, 31% of all varieties originated from one or more parents developed at IRRI (Table 5.4). Thus, including IRRI-developed lines, about 42% of the released varieties in the data originate from one or more parents developed at IRRI. The numbers of IRRI-parent releases reached a peak in the 1970s, when 60% of varieties released in South and Southeast Asia had IRRI parents, including those that were directly developed at IRRI. Since then, the proportion has remained stable at about 40% of all varieties released.

Table 5.5. Contributions of IRRI to released varieties in South and Southeast Asia by time period.

Time period	IRRI crosses released as varieties	Released varieties with IRRI parents	Released varieties with IRRI ancestry
Pre-1970	11.6	15.6	16.0
1971–75	16.9	59.1	61.0
1976–80	17.7	60.4	64.7
1981–85	11.9	42.8	54.7
1986–90	10.7	40.2	49.6
1991–95	3.3	35.9	49.3
1995–99	3.1	45.8	54.2
Total	10.7	41.8	49.7

Bangladesh, Indonesia, Laos and the Philippines had the highest proportion of releases in this category, with as many as 45% of locally bred varieties using at least one IRRI parent. By contrast, around 10% of Thai and Cambodian varieties had a known IRRI parent.

In a number of major rice producing countries, the use of IRRI parents has fallen perceptibly since the early years of the Green Revolution (Table 5.5). In Sri Lanka, for example, over 70% of varieties released during the 1971–1975 period had at least one IRRI parent. By the 1990s, however, only about 20% of releases had an IRRI parent, and none of the country's releases were crossed at IRRI. Taken together, this suggests a change in the respective roles of IRRI and the national programme.

IRRI provision of other ancestors

As NARS capacity has grown, IRRI has increasingly provided national programmes with elite lines for use in breeding. These are used as parents of released varieties, but sometimes they appear as grandparents or more remote ancestors. Many varieties thus have IRRI ancestry, but not at the parental level. Excluding IRRI crosses and varieties with IRRI parents, this pool accounted for 213 released varieties in the data. This is 7.9% of all varieties in the data (Table 5.5). Combined with the previous two categories of materials, this brings the total proportion of varieties with some IRRI ancestry to 50%.

As would be expected, the varieties that fall in this category are relatively recent. There are almost none prior to 1976, but in the period since 1980, this category of IRRI-derived material has risen to more than 18% of all releases.

The overall contribution of IRRI (including all three categories) to the improved germplasm released by NARS can be seen from Table 5.5. The share increased from 16% in the 1960s to over 60% in the 1970s.

Thereafter it remained at a level of 50%. Across countries the contribution was the highest in the Philippines, Indonesia and Bangladesh and lowest in Thailand, Cambodia and Laos, where farmers are still growing mainly traditional varieties.

Types of IRRI materials in use

Are national programmes today using IRRI materials that were already available a decade or two ago? To what extent are newer IRRI materials being used? To answer these questions, we identified the date of crossing for all IRRI material appearing in the genealogy of a released variety. We then asked how many of the released varieties in our data set make use of materials crossed at IRRI after 1990, or after 1980, etc.

IRRI's first semi-dwarf varieties (IR5 and IR8) were crossed in the years 1962–1964, although they did not reach a usable stage for breeding or multiplication until 1966. These earliest semi-dwarfs (which also included less widely known breeding lines, such as IR127 and IR262) were extremely influential in national breeding programmes. Our data indicate that 18% of the varieties were based on these lines, with no subsequent IRRI crosses in their genealogies. Most of these releases came in the years 1971–1985, but a few releases were based on this germplasm throughout the 1990s.

A second generation of IRRI materials was developed from crosses made in the period from 1965 to 1971. This included the widely used varieties IR20, IR22 and IR24, along with the less widely used IR26, IR28 and IR29.[7] Some of these varieties displayed useful disease resistance, with several drawing on the hardy Indian variety TKM6. In our data set, 10% of the released varieties trace to these second-generation IRRI materials, with no subsequent IRRI lines in their genealogies. The pattern of use is similar to that for the first-generation materials; these materials were most widely used in the late 1970s and diminished in importance in each subsequent period.

In the succeeding years, disease and pest resistance continued as a major goal of IRRI's breeding. A major innovation was the incorporation of *Oryza nivara* into the IRRI breeding pool, conferring resistance to the grassy stunt virus. Varieties based on this germplasm were developed from 1972 to 1976, including the sister selections IR32, IR38 and IR40. Another set of varieties with multiple disease resistance was represented by IR36

[7] These varieties were in fact released by IRRI. Subsequently, IRRI abandoned the practice of releasing varieties itself, but the Philippine Seed Board released IR36 and a number of subsequent IRRI lines, using the IR designation.

and its sister variety IR42. This third generation of material provided the foundation for a large proportion of the varieties in the data – about 18% overall and more than 20% of the varieties released in the early 1990s.

The following generation of IRRI breeding materials consisted of crosses made at IRRI in the years from 1977 to 1980. An additional 77 varieties (3.8%) in the data set include in their pedigrees crosses made at IRRI in this period (i.e. IR15329 to IR33898). All but three of these were released after 1985, accounting for about 7% of all releases in this period. The total number of varieties based on this germplasm is not large, but it is significant. If IRRI's breeding had been halted in 1976, 4% of the varieties in the data set could not have been created. The importance of these varieties is far greater, however. One of the crosses made in this period resulted in IR64, which is at present the most widely cultivated rice variety in the world.

Finally, for the region as a whole, 30 varieties (1.5%) include in their pedigrees crosses made at IRRI in the years 1981 to the present (i.e. crosses from IR33899 to IR75000). Note that we would not expect to see many varieties in the data set with such relatively recent IRRI crosses. It would be normal to see time lags of 10 years or more between the time a cross is made and the time when a resulting variety is released by a national programme.[8]

The aggregated data for all 12 countries in our data set are summarized in Fig. 5.1. In this figure, the successive waves of IRRI germplasm can be seen in the released varieties of different periods. One striking feature of the data is the continuing importance of the first- and second-generation IRRI germplasm. The more recent vintages of IRRI germplasm appear to be following essentially the same pattern of use as early vintages. The newer material is being used; but it takes at least 10 years for it to show up in any substantial numbers of released varieties.

Because the time lags involved are large, a reasonable way to assess the usefulness of IRRI's recent crosses is to look at the current use of these materials in NARS breeding programmes. Cabanilla *et al.* (2000) carried out an analysis of the breeding materials used in a number of national programmes. Based on a random sampling of crosses made at 28 different rice experimental stations in South and Southeast Asia, they found that many breeding programmes made extensive use of recently developed IRRI lines. The implication is that IRRI's recent breeding effort

[8] Once a cross is made at IRRI, the resulting offspring must be grown through as many as eight generations before they become sufficiently homozygous (genetically uniform). This process is likely to take 3–5 years. If the variety is then used as a parent, an additional 3–5 years is required.

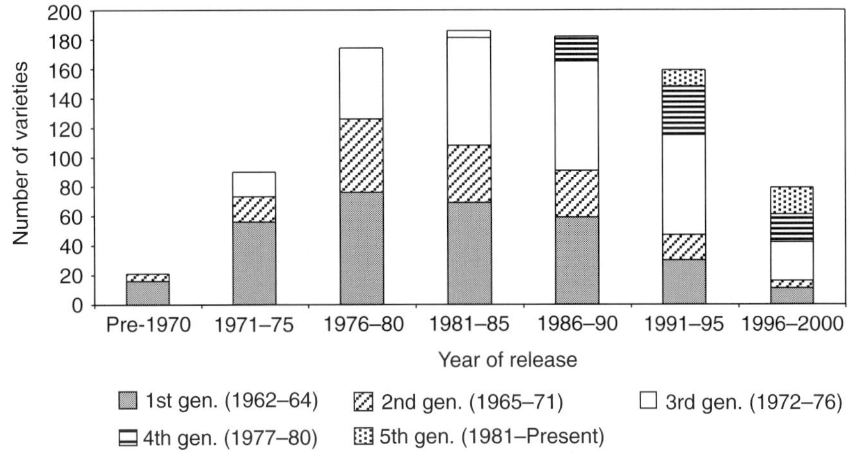

Fig. 5.1. Released varieties with IRRI ancestors of different vintages.

has already made a difference in the pool of released varieties, and it seems set to make further contributions in the years ahead.

IRRI as a source of traits

IRRI's materials were initially attractive to national breeding programmes primarily as a source of a single trait: the semi-dwarfing gene. However, IRRI varieties soon incorporated 'bundles' of other useful traits and characteristics. IRRI's strength has been in providing valuable bundles to national programmes, which in many cases have added locally desirable traits to produce final packages attractive to farmers.

Other studies – particularly by breeders and scientists – have focused on the changing characteristics of IRRI varieties over time (Khush, 1995; Khush and Hossain, 2000). To assess IRRI's contribution to the overall bundling process, we measured the number of distinct landraces in the genealogies of released varieties, and the extent to which this number is dependent on IRRI breeding.[9]

[9] A landrace is a crop variety that has been selected through traditional farmers' practices over an extended period of time. We use the term loosely here, without distinguishing between varieties selected by farmers and varieties selected by early plant breeders and scientists. We use the term to refer to any ancestor that cannot be traced further, i.e. any ancestor that is not a product of hybridization. Operationally, we also treat mutant selections as landraces, along with some varieties of unknown origin that we cannot trace further. Thus, we use the term to refer to the same concept that Cabanilla *et al.* (2000) characterize as 'ultimate ancestors'.

The number of landraces in the genealogy of a released variety offers a useful measure of breeding intensity or complexity. A simple selection from a traditional variety traces to one landrace. A hybrid between two landraces – such as the early Indonesian variety Bengawan, which crossed Cina with Latisail, would have two landraces in its genealogy.[10] A larger number of landraces then denotes more breeding effort. Some varieties have 20 or more distinct landraces in their genealogies.[11]

For the whole data set, the average number of landraces per released variety is just over 6.3 (see Table 5.6). Figure 5.2 shows the trend in this number over time. From a low of about two landraces per released variety in the pre-1970 period, the average number has risen to about 8.2 landraces per released variety in the latest period. This implies a substantial increase in the genealogical complexity of succeeding generations of varieties. In turn, this implies greater bundling of desirable traits. Of the 6.3 landraces in the genealogy of an average released variety, IRRI has contributed about 4.6, meaning that (on average) the national programmes are typically combining an IRRI variety with one or two landraces (Table 5.6). This pattern has grown more pronounced over time. In the late 1990s, an average released variety had 8.2 distinct ancestors, of which IRRI served as the source of 6.6.

IRRI's contribution of landrace packages has grown over time. While the average number of landraces per released variety has risen from two to 11, the number of these landraces that are obtained independently of IRRI has remained fairly constant, varying between one and two. Some countries – notably Sri Lanka – have tended to release varieties with more

[10] Note that the term 'hybrid' is used in rice breeding to refer to any cross made from traditional plant breeding techniques. Varieties based on such crosses produce fertile seeds and are homozygous – genetically quite uniform. By contrast, popular discussions of 'hybrid varieties' refer to a somewhat different phenomenon, by which the first generation offspring from inbred parent lines can be made genetically uniform. These offspring display a 'heterosis' effect that results in increased yield. The seeds produced by the offspring, however, may exhibit high rates of sterility and heterozygosity – a high degree of genetic variation – making it impractical for farmers to save and replant seeds. Until the past decade, it has been unfeasible to produce the second type of hybrids in rice. Recent technological advances, such as the development of a number of useful lines displaying cytoplasmic male sterility, have made hybrid rice more practical, and in China there is currently a considerable area devoted to hybrid rice. In our data, however, there are fewer than five hybrid rices, in this sense.

[11] In computing this measure, we have elected not to count duplicated landraces, although these do none the less reflect breeding effort. Other measures include the total number of ancestors (including all parents, grandparents, etc.). These numbers tend to be highly correlated.

Table 5.6. Average numbers of identifiable ancestors, for those varieties with at least partial genealogical records, by time period.

Country	Time period								Landraces introduced through IRRI materials
	Pre-1970	1971–75	1976–80	1981–85	1986–90	1991–95	1996–2000	Total ancestors	
Bangladesh	2.9	7.5	8.1	8.9	7.3	7.7	2.4	6.7	4.9
Cambodia	0.0	0.0	0.0	1.0	17.0	5.3	5.9	6.0	5.8
India	2.3	4.0	4.9	4.9	5.6	6.3	7.9	5.4	3.1
Indonesia	2.6	9.3	10.0	11.0	13.0	6.9	16.0	9.3	6.9
Laos	0.0	0.0	1.0	16.0	0.0	10.0	13.0	7.1	5.3
Malaysia	2.6	3.5	5.0	6.7	8.8	10.0	27.0	6.8	4.6
Myanmar	3.0	4.4	4.3	6.1	7.5	13.0	8.6	5.5	6.1
Pakistan	1.5	2.8	4.0	4.7	0.0	18.0	10.0	4.1	2.6
Philippines	3.4	7.4	9.5	13.0	14.0	15.0	16.0	12.0	10.8
Sri Lanka	3.8	4.7	7.2	7.3	6.5	10.0	13.0	6.9	2.8
Thailand	1.1	5.3	3.1	6.4	3.4	5.7	16.0	2.8	0.8
Vietnam	3.0	5.5	12.0	9.6	8.6	7.5	12.0	7.1	7.3
Total	2.9	8.4	8.2	9.8	9.0	7.8	8.2	6.3	4.6

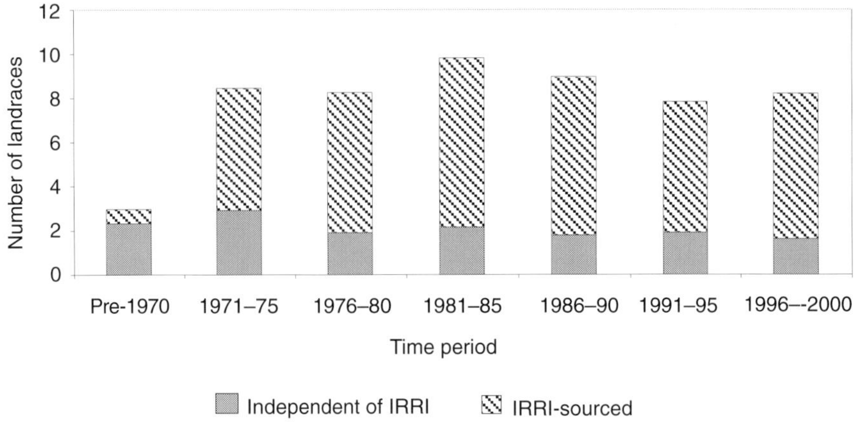

Fig. 5.2. Landrace content of released varieties.

landraces that occur independently of IRRI breeding. Others, such as Laos and Cambodia, have released few varieties that incorporate landraces independent of IRRI's breeding. This reflects the relatively recent nature of breeding efforts in those countries, combined with the extent of their cooperation with IRRI. Since IRRI scientists have been actively involved in those countries, it is possible that landrace material from those countries is entering their released varieties via IRRI's breeding efforts.

To conclude, IRRI has put together packages of traits that the national programmes find useful. The IRRI rices are occasionally useful in their existing form and are released directly by NARS. But more often, the IRRI rices lack one or more locally important traits, such as resistance to a locally common problem, or a particular quality characteristic. In these cases, the NARS use IRRI materials as building blocks, combining them with local materials or with other modern varieties to get desirable bundles of traits.

CIAT and Latin American varietal improvement

CIAT and the Latin American Fund for Irrigated Rice (FLAR) have data covering 299 rice varieties released from 23 national programmes in Latin America and the Caribbean. By far the most active national programme in the region has been Brazil, which released 95 varieties. Other major programmes include those in Mexico, Peru and Colombia.

Of all the released varieties, fully 40% were crossed at CIAT, with varying degrees of selection by collaborating NARS institutions. These included 34 CIAT lines released as varieties in Brazil, 17 released in Colombia after further selection by the national programme, and nine in Guatemala. The proportion of Latin American varieties crossed at CIAT is far higher than for IRRI in South and Southeast Asia. There appears, however, to be relatively modest use of CIAT crosses as parents or other ancestors of released varieties. Only 13 varieties in the region – about 6.5% – were developed from CIAT parents or ancestors. Possibly the research time lags are such that CIAT material has not yet entered the pool of released varieties through this channel.

An additional 108 varieties released by Latin American and Caribbean NARS (36% of the total) were acquired through international networks – primarily the International Network for Germplasm Evaluation and Research (INGER), coordinated by IRRI. These varieties include a number of Asian varieties released in Latin America and the Caribbean. Chaudhary *et al.* (1999) document at least 13 instances of IRRI-developed varieties released in the region, with another 12 varieties from Asian NARS released in Latin America and the Caribbean. Overall, only 19% of the varieties released in the region were entirely NARS-developed.

Adoption and Use of Improved Varieties

The previous section focuses on varietal releases. These are a measure of research activity, but in many ways they are a poor measure of the productivity of research (Evenson, 1998b). Release decisions are in some sense arbitrary; they reflect a judgement on the part of varietal release committees or scientists about the usefulness of their output. The breeders' judgement may not be shared by farmers, who evaluate varieties based on their own criteria of usefulness and their relative importance (Paris *et al.*, 2000). Thus, a research system could generate lots of 'spurious' releases that are little used. Moreover, different countries adhere to different standards for release. As a result, it is not possible to make many meaningful comparisons of release patterns across countries.

To get a better idea of research productivity, we must look beyond varietal releases to data on varietal adoption or use.

Diffusion of improved varieties

By the late 1990s, nearly 75% of the rice area in Asia was planted to improved or so-called high-yielding varieties (HYVs) or modern varieties (MVs). For the 12 South and Southeast Asian countries in our data, the figure was 71%. The situation varied considerably across countries, as well as for different regions or geographical units within countries (see Appendix 5.1). An estimate of the area covered by modern varieties at different points of time over the last three decades is shown in Table 5.7.

Table 5.7. The trend in the adoption of modern rice varieties in South and Southeast Asia.

Year	South Asia			Southeast Asia		
	Harvested area (million ha)	Area under MVs (million ha)	Rate of adoption (%)	Harvested area (million ha)	Area under MVs (million ha)	Rate of adoption (%)
1966	47.4	1.0	2.1	31.2	1.6	5.1
1976	52.0	16.0	30.8	33.0	8.1	24.5
1981	55.3	24.1	43.6	35.2	12.5	35.6
1986	56.0	29.0	51.8	35.3	17.7	50.0
1991	56.9	34.9	61.3	35.6	20.8	58.4
1999	60.0	42.5	70.8	42.1	29.7	70.5

Source: IRRI, World Rice Statistics database.

The trend in the rate of adoption for three countries for which we have uninterrupted time series data can be seen from Fig. 5.3. The rate of adoption varied across countries, often depending on the development of irrigation infrastructure. The adoption of modern varieties in East Asia, where irrigation infrastructure was already highly developed, was essentially complete by the 1970s. Adoption was also very high in the Philippines in the 1970s, due to large irrigation infrastructure projects implemented in the 1950s and 1960s. In eastern India and Bangladesh, the rate of adoption was low initially, but picked up later, starting in the late 1970s when the governments started implementing flood control, drainage and irrigation projects and provided incentives for farmers to invest in groundwater irrigation via tube wells. However, in Punjab, Haryana and Tamil Nadu – states of India where irrigation infrastructure was already developed – the rate of adoption of MVs was very high in the 1970s. In Thailand, Myanmar, Cambodia and Laos, as well as the plateau uplands of eastern India, where rice is cultivated mostly under rainfed conditions, farmers are still growing traditional varieties on large tracts of land.

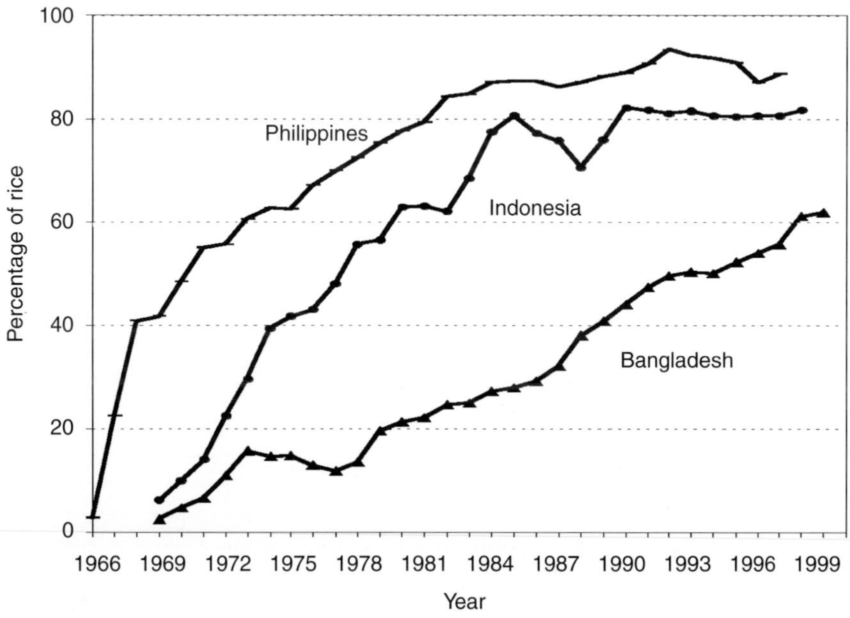

Fig. 5.3. Trend in the rate of adoption of MVs, Bangladesh, the Philippines and Indonesia. *Note:* Data on Indonesia from 1990–98 are own estimates as government data are not available. Source: Bangladesh: BBS; Philippines: Philrice–BAS; Indonesia: Biro Pusat Statistik.

The other factor behind low adoption rates in some areas is the lack of desirable quality traits (such as aroma) in modern rices. In Thailand, for example, the adoption of modern rices has remained low because scientists have not succeeded in developing high-yielding varieties comparable in quality to the jasmine rice for which Thailand has an established export market. In the Chhatishgarh and Chottonagpur plateau in eastern India, the duration of the rainy season is short and the monsoon is erratic. Consequently, there is little demand for HYVs that are intolerant to drought and take more than 110 days to mature. In regions with deep flooding, such as in the river basins of Bangladesh and Cambodia, as well as parts of Uttar Pradesh, Assam and Bihar in India, farmers cannot grow semi-dwarf HYVs because of the risks of flooding and submergence. Farmers still grow many traditional varieties in such environments, saving and replanting their own seeds.

Although popular perception characterizes the 'Green Revolution' as a phenomenon of the late 1960s and early 1970s, the expansion of MV area has shown a relatively steady trend. For the region as a whole, the period 1980–1995 saw an increase in MV adoption that was as large, in absolute terms, as that which occurred in the period from 1965 to 1975. Because the increase comes on a much larger base, the percentage increase is much smaller, but the absolute rates of diffusion have remained extraordinarily high.

This is also true in Latin America, where 40% of total rice area is under modern varieties. The figures range as high as 87% in Colombia and 73% in Peru. However, in Brazil, the region's largest producer, only 25% of the rice area is under modern varieties. This reflects the importance of upland rice in Brazil and the lack of suitable high-yielding drought-tolerant varieties for uplands. On irrigated rice and favourable upland areas in Brazil, almost 95% of area is planted to modern varieties. In less-favourable upland environments, however, traditional varieties predominate.

Area planted by variety: patterns and summary

Until recently, it has been difficult to obtain data on the actual use of specific varieties by farmers. Typically, if countries kept track of diffusion at all, they recorded area planted to 'modern varieties', as a single category. Indonesia alone has long collected data on area planted by variety, based on crop cut surveys. In 1991–1992, however, IRRI asked national programme collaborators to share information on the area planted to specific modern varieties. For South and Southeast Asia, data were obtained from Bangladesh, Malaysia, Myanmar, Thailand and Vietnam.

Appendix 5.3 summarizes the data, providing the names of the five most popular varieties grown in the 12 countries in our study. These varieties are important; typically at the village level, the five leading varieties account for between 30 and 80% of area planted.[12] Summing together 11 of the 12 countries in the data (excluding only India), we find that these leading varieties account for almost 45% of total rice area. Of the 55 varieties observed in this sample, 18 were IRRI crosses, 11 derived from IRRI parents, and seven with other IRRI ancestry, 16 had no IRRI ancestry (12 of these are in fact traditional varieties). Although the IRRI crosses made up only about one-third of the varieties, they covered 40% of the area planted to these leading varieties. Taken together, the 36 leading varieties with some IRRI ancestry covered over two-thirds of the area planted to these leading varieties – and in fact covered about 30% of the total rice area in the 11 countries under consideration. This is a strong indication that varieties originating at IRRI are being widely used in farmers' fields. In fact, it appears that released varieties with IRRI ancestry are planted on disproportionately large shares of total area.

According to IRRI estimates, the most popular variety in the region is IR64, which is grown in large areas of Indonesia, the Philippines, Vietnam, and the Indian states of Andhra Pradesh and Orissa. IR64 may have spread to nearly 13 million ha of rice land in Asia. Other IRRI varieties that remain popular are IR36, IR42, IR66 and IR8 – which was released in 1966 but continues to cover large areas in several Indian states and in Bangladesh.[13] 'Mahsuri', a variety developed by the FAO *indica* × *japonica* crossing programme in the 1950s, and now cultivated under many names, remains one of the most popular varieties grown in the rainfed lowlands of several Indian states, Nepal, Bangladesh and Myanmar.

Hargrove (1979) and Hargrove and Cabanilla (1988) noted from a similar study that about 60% of the widely grown varieties were locally developed, while 30% were introduced from IRRI and 10% were from other countries. Our data suggest that the situation has changed little since then. About one-third of the widely grown varieties were introduced from IRRI, and most of the rest were developed locally, with varying degrees of IRRI germplasm.

[12] These leading varieties typically account for a larger fraction of total area in favourable irrigated environments. They may be relatively less important in less favourable environments, where a more diverse array of varieties is often cultivated.

[13] In previous years, IR8, IR20, IR36, IR42, IR50 and IR66 were popular in many countries.

Indonesia

For a more detailed picture, we can draw on data from Indonesia, where an unusually detailed data set includes time series observations on area planted by variety. This allows us to trace the entire life cycle of varieties as they have moved in and out of use by farmers. The following paragraphs summarize some of the information derived from this data.

Area planted to IRRI varieties

In the Indonesian data, IRRI-developed varieties made up only 20% of all varieties released but accounted for over 70% of the area planted. In 1972, IRRI crosses accounted for over 40% of Indonesia's rice area. IR5 was the most important of these crosses, with IR8 and IR20 also grown. Over the next several years, the use of IRRI-developed crosses dropped rapidly, but the use of IRRI parent material increased. However, in the years following 1974, IRRI-developed varieties rose to cover as much as 80% of Indonesia's area in the early 1980s, and the figure still remained over 70% by 1990. Much of this reflects the enormous popularity of IR36 and subsequently IR64 (Fig. 5.4). In 1993 IR64 was grown on 6.3 million ha, about 57% of the rice area.

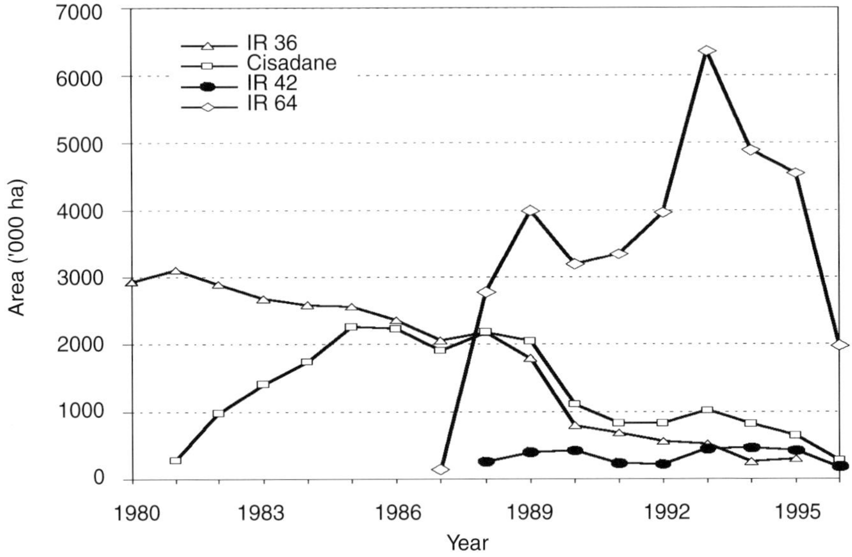

Fig. 5.4. Area ('000 ha) planted to most popular rice varieties in Indonesia, 1980–1996.

IRRI parent materials rose rapidly in importance from essentially none of Indonesia's area in 1972 to as much as 35% by the mid-1970s. Throughout the 1980s, IRRI parent varieties have typically covered between 20 and 30% of the rice area. By comparison, these varieties represent 36% of Indonesian releases. Materials with no IRRI ancestry accounted for almost 60% of the rice area in 1972, but by 1990 they accounted for just 11% of the total. The decline in actual area, as opposed to shares, was not as acute; none the less, the area planted to materials with no IRRI ancestry fell by about 60% over the period.

Impact on Production

It is difficult to quantify the production impacts of international rice research (Evenson, 1998b). As noted in Chapter 3, it is conceptually and empirically difficult to disentangle the impact of research from the increased use of inputs. It is also difficult to attribute research impacts to particular programmes or institutions.

None the less, the story of the Green Revolution in rice over the past 40 years stands as an enormous success. According to FAO data, the developing countries of Asia had 1.6 billion people in 1961. Over the following four decades, to 2000, the population more than doubled, to 3.4 billion people. During the same period, rice area expanded from 107 million ha to 139 million ha, an increase of only 30%. However, rice production grew by 170%, from 199 million t in 1961 to 540 million t in 2000. About 83% of the production increase was attributable to growth in yield, from 1.85 t ha^{-1} to 3.94 t ha^{-1} within the last four decades.

In Latin America, irrigated rice yields increased from 3.3 t ha^{-1} in the mid-1960s to 4.6 t ha^{-1} in 1995. When combined with expansion in rice area, this resulted in a doubling of rice production during the period to make the region largely self-sufficient in rice. More efficient production has brought down the price of this staple by about 50% in real terms over the last three decades.

How much of this increase was due to research? And how much was due to research carried on in international institutions? We cannot simply compare yields of modern varieties to those of traditional varieties; modern varieties are typically grown under more favourable conditions, with higher levels of inputs. Attributing all the yield gains to research thus conflates genetic improvement with other factors. Moreover, much of the research effort in rice has taken place in national institutions. It would be misleading to attribute it all to international institutions.

We made an attempt to estimate the net gains from the adoption of modern varieties by using costs and returns data for the traditional and modern varieties for selected countries, collected through sample household surveys by IRRI and NARS partners in the late 1990s. Aggregate costs and returns data for rice cultivation are also available for Indonesia in the *Statistical Year Book* (1998). The data are summarized in Table 5.8. It may be noted that the average yield for the traditional varieties reached 2.28 t ha^{-1} in the late 1990s, about 54% higher than the level of 1.48 t ha^{-1} that prevailed in 1966 when IR8 was released. This might have been the growth in rice yield had there been no rice research efforts by IRRI and NARS (the counterfactual). The yield of modern varieties for the late 1990s stood at 4.38 t ha^{-1}, using an average yield from both irrigated and rainfed ecosystems. The yield is over 5.0 t ha^{-1} when it is grown under irrigated conditions but would be at least 1 t lower when grown under favourable rainfed conditions. The numbers indicate a yield gain of 2.1 t ha^{-1} when the farmer shifts land from the traditional to modern varieties. The farmer however incurs an additional cost on account of higher input use – in particular, chemical fertilizers, irrigation charges, labour and pesticides. This additional cost is estimated at 1.16 t ha^{-1} (2.68–1.52) in rice equivalents. Thus, about 55% of the yield gains from the adoption of MVs are lost on account of higher use of inputs. The net yield gain is therefore estimated at 0.94 t ha^{-1}, equivalent to US$150 at the price prevailing in the domestic markets in the late 1990s.

In the late 1990s MV rices expanded to about 72 million ha of rice area (71%) in South and Southeast Asia. This number combined with the estimated yield gains per hectare from the adoption of MVs suggests that the annual gains from the adoption of modern varieties now stand at about US$10.8 billion. The amount is nearly 150 times the annual investment made in rice research by IRRI and NARS together.

Although these calculations are somewhat crude, they clearly indicate the enormous rate of return on investment in rice research in Asia. A separate assessment of research impact for Latin America suggests that benefits from international research are in the neighbourhood of $500 million per year.

The other benefit of genetic improvement research which we have not included in the above calculation is the reduction in the growing period for the new generations of modern varieties that enables farmers to increase cropping intensity. Although the yield potential of rice did not increase since the introduction of IR8, rice breeders have succeeded in reducing the crop maturity period substantially without reducing the yield potential. The Vietnamese research system now produces varieties with a duration of 90–100 days, which have contributed to a substantial increase in the rice cropping intensity in South Vietnam. The data from IRRI's experimental farm show that the crop maturity period has been reduced from 135 days, for IR8 and IR5, to 114 days for IR64, increasing the yield per day from 47 kg for IR8 to 60 kg for IR64, for the dry season. For the wet season, the yield per day increased from 35 kg for IR20 to 42 kg for IR72.

Table 5.8. Estimates of the net gains from the adoption of modern rice varieties.

	Rice yield (kg ha^{-1})		Cost in rice equivalent (kg ha^{-1})		Net gain from the adoption of MV	
Country	MV	TV	MV	TV	kg ha^{-1}	US$ ha^{-1}
Bangladesh	3980	1970	2614	1600	996	149
West Bengal, India	4174	1921	2631	1475	1097	155
Vietnam	4805	2297	4044	2419	883	120
Philippines	3780	2100	2363	1579	896	170
Indonesia	5176	3093	1759	521	845	156
Average	4383	2276	2683	1519	943	150

Note: For Indonesia, the figures for modern varieties (MV) are for Java, where adoption rate is almost complete, while the figures for traditional varieties (TV) are for Kalimantan, where most of the area is grown with traditional varieties. The traditional varieties fetch a higher price in the market because of better quality. The yields for traditional varieties are adjusted for the price premium over the modern varieties.
Source: *Statistical Yearbook of Indonesia*, 1998 and IRRI, costs and returns data from farm household surveys (unpublished).

Another major achievement of the genetic improvement research is the incorporation of host plant resistance against major insects and diseases. A study by Evenson (1998a) for India estimates that conventional breeding for insect resistance has produced a yield gain of 10–14% and for disease resistance 7–10%. These gains are reflected in higher and more stable yields, the benefits of which are subsumed in the calculations made in Table 5.8.

To conclude, as long as international research contributes in a measurable degree to an increase in rice productivity, the economic payoffs will be overwhelmingly large. Can we be certain that there is a measurable contribution? The best evidence comes from farmers' choices about which varieties to grow. The evidence on adoption and diffusion of MVs show that farmers continue to increase their use of varieties developed with the help of the international research system. So long as this is true, and so long as this observation is valid, and so long as new research products continue to flow into farmers' fields, the case for international rice research is outstanding.

Conclusions

Several observations seem warranted. First, there appears to be little evidence of a slowdown in the rate of varietal releases by national programmes. Second, there is little evidence of a declining role for international institutions in generating these released varieties. IRRI remains an important source of germplasm, both for direct use by farmers and as elite material for use in breeding programmes. IRRI-developed and IRRI-derived materials account for large fractions of area planted to rice in South and Southeast Asia, and their overall importance shows no sign of fading. The same appears to be true for CIAT in Latin America.

The Green Revolution, far from having finished in the 1970s, has continued well into the 1990s and beyond. A popular conception of the Green Revolution would hold that the major episode of diffusion took place in the period from the mid-1960s to the mid-1970s, with little gain thereafter. In fact the data suggest a far more protracted episode of technological change. If the diffusion of modern varieties had halted in 1980, the area currently planted to MVs would be about half of what it is today.

Moreover, the diffusion of MVs in the 1980–2000 period was not simply based on momentum from the previous 20 years. Instead, a substantial proportion of this expansion took place through the development of new varieties with new characteristics – primarily disease and pest resistance, improved grain quality, and shorter duration.

Overall, the message is clear: international rice research continues to have extremely large economic payoffs.

References

Cabanilla, V.L., Hossain, M. and Khush, G.S. (2000) Diffusion of breeding materials and genetic composition of improved rices in South and Southeast Asia: 1965–1998. *Social Science Division Working Paper*. International Rice Research Institute, Los Baños, Philippines.

Chaudhary, R.C., Seshu, D.V., Alluri, K., Cuevas-Perez, F., Lopez, V.C. and Khush, G.S. (1999) *INGER-derived Rice Varieties Directly Released in Various Countries.* IRRI, Los Baños, Philippines.

Dalrymple, D.G. (1986) *Development and Spread Of High-yielding Rice Varieties in Developing Countries.* Bureau for Science and Technology, Agency for International Development, Washington, DC.

Evenson, R.E. (1998a) Rice varietal improvement and international exchange of rice germplasm. In: Pingali, P.L. and Hossain, M. (eds) *Impact of Rice Research*. Thailand Development Research Institute and International Rice Research Institute, Los Baños, Philippines, pp. 51–82.

Evenson, R.E. (1998b) The economic value of genetic improvement in rice. In: Dowling, N.G. *et al.* (eds) *Sustainability of Rice in the Global Food System*. Pacific Basin Study Center and International Rice Research Institute, Los Baños, Philippines, pp. 303–320.

Evenson, R.E. and Gollin, D. (1997) Genetic resources, international organizations and improvement in rice varieties. *Economic Development and Cultural Change* 45, 471–500.

Hargrove, T.R. (1979) The diffusion and adoption of semi-dwarf rice cultivars as parents in Asian rice breeding programs. *Crop Science* 19, 571–574.

Hargrove, T.R. and Cabanilla, V.L. (1988) Twenty years of rice breeding. *Biological Science* 38, 675–681.

Hargrove, T.R., Coffman, W.R. and Cabanilla, V.L. (1980) Ancestry of improved cultivars of Asian Rice. *Crop Science* 20, 721–727.

IRRI (1980) *Annual Report for 1979.* International Rice Research Institute, Los Baños, Philippines.

IRRI (1983) *The International Directory of Rice Workers.* Library and Documentation Center, International Rice Research Institute, Los Baños, Philippines.

IRRI (1990) *Program Report for 1989.* International Rice Research Institute, Los Baños, Philippines.

IRRI (1993) *Rice Research in a Time of Change: IRRI's Medium Term Plan for 1994–98.* International Rice Research Institute, Los Baños, Philippines.

IRRI (1996) *IRRI Toward 2020: IRRI Strategy.* International Rice Research Institute, Los Baños, Philippines.

IRRI (2000) *Program Report for 1999.* International Rice Research Institute, Los Baños, Philippines.

IRRI (2001) *The Rewards of Rice Research: IRRI Annual Report 1999–2000.* IRRI, Los Baños, Philippines.

Khush, G.S. (1995) Modern varieties – their real contribution to food supplies and equity. *GeoJournal* 35, 275–284.

Khush, G.S. and Hossain, M. (2000) *Technologies for Increasing Food Production.* International Rice Research Institute, Los Baños, Philippines.

Paris, T., Singh, A., Luis, J., Hossain, M., Singh, H.N., Singh, S. and Singh, O.N. (2000) Incorporating gender concerns in participatory rice breeding and varietal selection. In: Lilja, N. *et al.* (eds) *Assessing the Impact of Participatory Research and Gender Analysis.* CIAT, Cali, Colombia, pp.109–121.

Philippine Rice Research Institute (PhilRice) (2000) *Annual Report 1999.* PhilRice, Munoz, Nueva Ecija, Philippines.

Raab, R.T., Denning, G.L. and Cruz, R.R.V. (1998) The impact of IRRI's training program: a different perspective. In: Pingali, P.L. and Hossain, M. (eds) *Impact of Rice Research.* Thailand Development Research Institute (TDRI), International Rice Research Institute (IRRI), Los Baños, Philippines.

Appendix 5.1

Rice area, production and yield

Selected rice-consuming and producing countries	Rough rice						Area planted to modern varieties (%)
	Production ('000 t)		Area ('000 ha)		Yield (t ha^{-1})		
	1990	1999	1990	1999	1990	1999	
World	**520,053**	**596,485**	**146,933**	**155,128**	**3.5**	**3.8**	–
Asia	**479,480**	**540,621**	**132,328**	**138,503**	**3.6**	**3.9**	**74**
Bangladesh	26,778	29,857	10,435	10,470	2.6	2.9	60
Cambodia	2,500	3,800	1,740	1,961	1.4	1.9	11
China (including Taiwan)	191,615	200,499	33,519	31,720	5.7	6.3	100
India	111,517	131,200	42,687	44,800	2.6	2.9	73
Indonesia	45,179	49,534	10,502	11,624	4.3	4.3	81
Japan	13,124	11,469	2,074	1,788	6.3	6.4	100
Korea, DPR	3,570	2,343	650	580	5.5	4.0	100
Korea, Rep. of	7,722	7,271	1,244	1,059	6.2	6.9	100
Laos	1,491	2,103	650	718	2.3	2.9	34
Malaysia	1,960	1,934	681	674	2.9	2.9	93
Myanmar	13,972	17,075	4,760	5,458	2.9	3.1	76
Nepal	3,502	3,710	1,455	1,514	2.4	2.4	36
Pakistan	4,891	6,900	2,113	2,400	2.3	2.9	100
Philippines	9,885	11,388	3,319	3,978	3.0	2.9	89
Sri Lanka	2,538	2,692	828	829	3.1	3.2	91
Thailand	17,193	23,272	8,792	10,000	2.0	2.3	30
Vietnam	19,225	31,394	6,028	7,648	3.2	4.1	82

Continued

Appendix 5.1 Rice area, production and yield (*continued*)

Selected rice-consuming and producing countries	Rough rice						Area planted to modern varieties (%)
	Production ('000 t)		Area ('000 ha)		Yield (t ha^{-1})		
	1990	1999	1990	1999	1990	1999	
Africa	**12,407**	**17,602**	**6,099**	**7,842**	**2.0**	**2.2**	–
Côte d'Ivoire	660	1,162	572	750	1.2	1.5	–
Egypt	3,167	5,816	436	655	7.3	8.9	–
Guinea	424	750	436	500	1.0	1.5	–
Liberia	185	210	200	163	0.9	1.3	–
Madagascar	2,420	2,637	1,165	1,227	2.1	2.1	48
Mali	338	589	231	330	1.5	1.8	–
Nigeria	2,500	3,397	1,208	2,050	2.1	1.7	–
Senegal	181	240	73	96	2.5	2.5	–
Sierra Leone	504	247	393	213	1.3	1.2	–
Tanzania	740	676	385	474	1.9	1.4	–
Latin America	**15,565**	**24,045**	**6,183**	**6,611**	**2.5**	**3.6**	**40**
Brazil	7,419	11,779	3,945	3,810	1.9	3.1	25
Colombia	2,117	2,059	521	431	4.1	4.8	87
Cuba	474	420	155	145	3.1	2.9	100
Dominican Rep.	428	563	89	125	4.8	4.5	81
Ecuador	840	1,290	269	366	3.1	3.5	59
Europe	**2,404**	**3,238**	**449**	**581**	**5.4**	**5.6**	–
Australia	**924**	**1,410**	**105**	**140**	**8.8**	**10.1**	–
USA	**7,080**	**9,546**	**1,142**	**1,442**	**6.2**	**6.6**	**100**

Appendix 5.2

Rice consumption data

Selected rice-consuming and producing countries	Total rice consumption (rough rice equivalent) ('000 t)		Milled rice consumption per capita (kg per head year^{-1})		Daily calorie supply per capita (no.)		Rice in total calorie supply (%)	
	1990	**1998**	**1990**	**1998**	**1990**	**1998**	**1990**	**1998**
World	**454,349**	**511,675**	**57.7**	**58.1**	**2704**	**2792**	**21**	**21**
Asia	**413,723**	**464,143**	**88.9**	**86.6**	**2540**	**2699**	**35**	**32**
Bangladesh	25,639	28,001	156.2	149.7	2074	2050	75	73
Cambodia	2,167	2,713	167	168.9	1960	2078	79	75
China (incl. Taiwan)	162,223	172,507	93.7	91.6	2711	2972	35	31
India	97,480	118,345	76.4	80.4	2275	2466	33	32
Indonesia	40,407	46,176	147.4	149.3	2604	2850	56	52
Japan	11,946	11,361	64.5	60	2895	2874	24	22
Korea, DPR	2,256	2,358	73.5	67.4	2468	1899	31	37
Korea, Rep. of	6,674	6,537	103.8	94.6	3100	3069	36	33
Laos	1,046	1,333	168.1	172.2	2121	2175	70	70
Malaysia	2,359	2,937	88.2	91.5	2778	2901	31	31
Myanmar	12,760	14,222	210	213.2	2626	2832	78	73
Nepal	2,987	3,087	106.1	90.1	2398	2170	41	38
Pakistan	3,433	3,629	19.2	16.3	2341	2447	8	7
Philippines	9,104	10,409	100.1	95.2	2396	2288	41	41
Sri Lanka	2,486	2,550	97.3	92.2	2200	2314	43	39
Thailand	8,937	9,855	107.2	109	2125	2462	50	44
Vietnam	15,382	19,201	153.8	165.1	2198	2422	71	67

Continued

Appendix 5.2. Rice consumption data (*continued*)

Selected rice-consuming and producing countries	Total rice consumption (rough rice equivalent) ('000 t)		Milled rice consumption per capita (kg per head year^{-1})		Daily calorie supply per capita (no.)		Rice in total calorie supply (%)	
	1990	**1998**	**1990**	**1998**	**1990**	**1998**	**1990**	**1998**
Latin America	**16,998**	**18,271**	**26**	**24.5**	**2677**	**2799**	**10**	**9**
Brazil	9,156	9,281	41.3	37.3	2743	2926	15	13
Colombia	1,641	1,733	31.3	28.3	2419	2559	13	11
Cuba	754	916	47.3	55	3092	2473	15	22
Dominican Rep.	515	508	48.3	41.2	2211	2277	21	18
Ecuador	648	851	42.1	46.6	2498	2725	17	16
Guyana	101	92	85.1	72.3	2272	2476	33	28
Peru	1,333	1,468	41.2	39.5	1946	2420	22	17
Surinam	54	24	89.8	39	2449	2633	34	14
Uruguay	48	52	10.3	10.6	2545	2866	4	4
Africa	**15,129**	**20,269**	**16.5**	**18.1**	**2342**	**2439**	**7**	**7**
Europe	**2,781**	**4,296**	**3.7**	**3.9**	**3379**	**3217**	**1**	**1**
Australia	**170**	**258**	**6.7**	**9.3**	**3216**	**3191**	**2**	**3**
USA	**2,595**	**3,679**	**6.8**	**9**	**3483**	**3757**	**2**	**3**

Appendix 5.3

Leading varieties in 12 Asian countries, 1998

Country	Variety	Year released	Rice area covered (%)	Source of information	Country of origin	Parentage	IRRI content
Bangladesh							
	BR11(Mukta)	1980	17		Bangladesh	IR20/IR5	IRRI parent
	BR14(Gazi)	1983	8		Bangladesh	IR5/BIPLAB	IRRI parent
	BR3(Biplab)	1973	7		Bangladesh	IR506/LATISAIL	IRRI parent
	IR8	1966	5		IRRI	PETA/DGWG	IRRI cross
	BR11	1980	3		Bangladesh	IR20/IR5	IRRI parent
Cambodia				2000 Impact questionnaire			
	IR66	1987	9		IRRI	IR 13240/IR 9129	IRRI cross
	KESAR	1993	2		IRRI	IR 24632/IR 31868	IRRI cross
	NEANG MINH		2		Cambodia	TRADITIONAL	No IRRI
	PHKA KHNEY		2		Cambodia	TRADITIONAL	No IRRI
	BANLA PHDAU		2		Cambodia	TRADITIONAL	No IRRI
India (eastern Madhya Pradesh)							
	Swarna	1982	30		India	VASISTA/MAHSURI	Other IRRI
	Safri 17		20		India	TRADITIONAL	No IRRI
	Mahomaya	1994	10		India	ASHA/KRANTI	Other IRRI
	RAMIKAJAR		10		India	TRADITIONAL	No IRRI
	IR36	1976	5		IRRI	IR 2042/CR 94-13	IRRI cross

Continued

Appendix 5.3. Leading varieties in 12 Asian countries, 1998 (*continued*)

Country	Variety	Year released	Rice area covered (%)	Source of information	Country of origin	Parentage	IRRI content
India (Kapurthala)							
	PR111	1993	27		India	IR54/PR106	IRRI parent
	PUSA44	1993	25		India	IARI5901/IR8	IRRI parent
	PR106	1978	20		IRRI	IR8/PETA//BELLE PATNA	IRRI cross
	PR113	1998	10		India	IR8//RP2151/IR8	IRRI parent
	IR8	1966	7		IRRI	PETA/DGWG	IRRI cross
India (Tamil Nadu)							
	CO37		14		India	TN1/CO29	No IRRI
	CO43	1982	14		India	DASAL/IR20	IRRI parent
	CO45	1991	5		India	R.HEENATI/IR3403	IRRI parent
	CO46	1997	2		India	T7/IR20	IRRI parent
	CO47	1999	1		India	IR50/CO43	IRRI parent
Indonesia							
	IR64	1985	30		IRRI	IR 5657/IR 2061	IRRI cross
	CISADANE	1980	2		Indonesia	PELITA I-1/B2388	Other IRRI
	MEMBERAMO	1995	3		Indonesia	B6555/BARUMUN	Other IRRI
	PB42	1980	1		IRRI	IR 2042/CR 94-13	IRRI cross
	IR36	1976	1		IRRI	IR 2042/CR 94-13	IRRI cross
Laos							
	TDK1	1993	34		IRRI	SPT 7149/IR 13423	IRRI cross
	RD6	1977	19		Thailand	Irradiated KDML105	No IRRI
	KDML105	–	7		Thailand	TRADITIONAL	No IRRI
	DOKMAY	–	–		Laos	TRADITIONAL	No IRRI
	MEUNG NGA	–	–		Laos	TRADITIONAL	No IRRI

Malaysia						
	MR84	1986	77	Malaysia	CR261/MR50	No IRRI
	MR167		6	Malaysia	Y978/PTB18//MR71	Other IRRI
	MR77		7	Malaysia	67009-5/ZENITH//IRON171	No IRRI
	IR42	1977	3	IRRI	IR 2042/CR 94-13	IRRI cross
	SEMERAK		2	Malaysia	TRADITIONAL	No IRRI
Myanmar						
	THEEDAT YIN(IR13240)	1990	20	IRRI	IR30/IR36	IRRI cross
	SWETHWEYIN (IR9224)	1985	20	IRRI	IR7531/IR36	IRRI cross
	MANAWTHUKHA	1977	19	India	Selection from MAHSURI	No IRRI
	SHWEWARTUM	1972	17	IRRI	IR5 MUTANT	IRRI cross
	INMAYEBAW	–	5	Myanmar	TRADITIONAL	No IRRI
Pakistan						
	SUPER BASMATI	1996	60	Pakistan	BASMATI320/IR661	IRRI parent
	BASMATI385	1985	25	Pakistan	BASMATI370*4/TN1	No IRRI
	KS282	1982	7	Pakistan	BASMATI370/IR95	IRRI parent
	IR6	1971	7	IRRI	SIAM 29/DGWG	IRRI cross
	BASMATI198	1972	1	Pakistan	BASMATI370/TN1	No IRRI
Philippines						
	IR64	1985	30	IRRI	IR 5657/IR 2061	IRRI cross
	PSBRC14 [a]	1992	12	Philippines	IR 18348/C1064	IRRI parent
	PSBRC28 [a]	1995	2	IRRI	IR 28239/IR 64	IRRI cross
	PSBRC18 [a]	1994	3	IRRI	IR 24594/IR 28222	IRRI cross
	PSBRC34	1995	-	Philippines	BURDAGOL/ farmer's local selection	No IRRI

Continued

Appendix 5.3. Leading varieties in 12 Asian countries, 1998 (*continued*)

Country	Variety	Year released	Rice area covered (%)	Source of information	Country of origin	Parentage	IRRI content
Sri Lanka							
	BG300	1987	22		Sri Lanka	BG367/IR841/BG276	Other IRRI
	BG352	1992	12		Sri Lanka	BG380/BG367	Other IRRI
	BG94	1978	12		Sri Lanka	IR262/LD66	IRRI parent
	BG350	1986	7		Sri Lanka	BG94///BG401/80-3717// BG94	Other IRRI
	BG450	1985	6		Sri Lanka	BG12/IR42	IRRI parent
Thailand							
	RD6	1977	28		Thailand	Irradiated KDML105	No IRRI
	KDML105	1959	23		Thailand	TRADITIONAL	No IRRI
	SPR60	1987	1		Thailand	LEUANG TAWNG/ C4-63//IR8	IRRI parent
	RD23	1981	1		Thailand	RD7/IR32//RD1	Other IRRI
	RD10	1981	–		Thailand	Irradiated RD1	No IRRI
Vietnam (South Vietnam)[b]							
	IR64	1985	20		IRRI	IR 5657/IR 2061	IRRI cross
	OM997	1994	9		Vietnam	COLOMBIA/IR64	IRRI parent
	IR50404	1992	3		IRRI	IR 33021/IR 32429	IRRI cross
	IR56279		6		IRRI	missing parents – CP135	IRRI cross
	DT10	1990	14		Vietnam	C4-63 Irradiated	No IRRI

[a] Source: Philippine survey in four villages on Technology, Income Distribution and Poverty study, 1997.
[b] Source: Impact of modern technology on rice production and its role in income distribution and poverty alleviation in Vietnam.

Ecological Diversity and Rice Varietal Improvement in West Africa 6

T.J. DALTON AND R.G. GUEI

Research on rice improvement in West Africa originated more than 65 years ago in national agricultural research programmes, hybridization was initiated in 1951, and collaboration with international agricultural research centres (IARCs) in the 1960s. Regional rice research is currently conducted by international and national agricultural research centres with support from global germplasm networks and several bilateral development organizations. Although several case studies have highlighted the impact of national rice research and development activities, the impact of genetic enhancement and varietal improvement has not been documented on a regional scale.

In order to estimate the impact of regional rice research activities in varietal improvement, this study focuses on seven of the most important producers of rice within the region: Côte d'Ivoire, Ghana, Guinea, Mali, Nigeria, Senegal and Sierra Leone. Combined, these countries produce approximately 91% of all the region's rice and represent about 92% of the total area under rice cultivation. Production ranges from the highly productive irrigated perimeters of the Sahel in Mali, Nigeria and Senegal to the mangrove swamps along the southwestern shores of Guinea, Senegal and Sierra Leone. The upland rainfed production system remains the most important in terms of area coverage, followed closely by the rainfed lowlands and distantly by the irrigated areas found in the humid and Sahel regions (Table 6.1). A significant area of deep-water floating ecology is still cultivated along West Africa's major rivers – the Niger in Mali, the Benoué in Nigeria, and to a lesser extent in Sierra Leone. Mangrove rice production is restricted to the southwestern coast of the region.

Table 6.1. Study countries and West Africa regional production and area summary.

	Area (>'000 ha)	Production (>'000 t)	Yield (t ha^{-1})	% area in ecology: Rainfed upland	Rainfed lowland	Irrigated lowland	Mangrove swamp	Deep-water floating
Nigeria	1784	3122	1.75	35	45	12	0	8
Guinea	445	668	1.50	69	19	1	11	0
Côte d'Ivoire	750	1223	1.63	74	19	7	0	0
Sierra Leone	289	392	1.35	67	22	2	7	2
Mali	302	463	1.53	3	25	32	0	40
Ghana	96	202	2.10	9	81	10	0	0
Senegal	70	160	2.29	5	43	45	7	0
Study total	3737	6229	1.66	45.6	34.2	11.1	2.0	7.2
Regional total	4084	7408	1.69	43.3	35.2	12.1	2.8	6.9

Source: FAO, 1999 (1996 data); WARDA Task Force Estimates; IAEG Germplasm Impact Survey Results, 1999.

The size of the rice economy in West Africa exceeds US$2.75 billion annually and is largely composed of US$1.85 billion in production plus US$0.9–1.0 billion in imports.[1] Production patterns differ dramatically across the region and we have disaggregated rice improvement by production ecology. The recent history of regional varietal improvement research is presented so as to provide an overview of genetic enhancement strategies. We then estimate the total regional resources invested in varietal improvement and hybridization strategies by national programmes in 1998. Finally, we describe the production and release of improved varieties, diffusion patterns, and derive an estimate of the aggregate productivity gain due to varietal improvement in 1998.

History of International Collaboration[2]

Several countries have long histories of varietal improvement facilitated by bilateral collaboration: Côte d'Ivoire, Mali and Senegal, and Guinea. The primary collaborative partner for the first three countries was French led, originally through the *Institut de recherches agronomiques tropicales* (IRAT), and later through the *Centre de coopération internationale en recherche agronomique pour la développement* (CIRAD). In 1963, IRAT recorded its first crosses in Senegal, where a collaborative programme existed in the Cassamance until 1973. Through this effort, eight crosses were registered for the rainfed lowland and upland ecolo-

[1] Using a regional import parity price of US$225 per tonne.

[2] Much of this discussion draws on *WARDA in Transition: Highlights 1988–1989*. WARDA, Bouaké, 1990.

gies of the Cassamance region (CIRAD-CA, 1993). By contrast, the programme in Côte d'Ivoire lasted for over 20 years, and 44 separate varieties were catalogued, many of which were released recently by the Ivorian Government. In Mali, collaboration lasted for a short period in the early 1970s and for an 8-year period in the 1980s and 1990s. From the first effort, four varieties were catalogued and one released in 1980 for the deep-water floating ecology, while over 250 fixed lines were produced in the second effort, many of which are currently under evaluation by the national programme in the rainfed lowlands.

In Guinea, large-scale bilateral assistance has been, and continues to be, received from the North Korean Government. The North Korean Government developed the Kilissi agricultural research station in the early 1980s to serve the region. Considerable improvement has occurred in breeding iron-toxicity-tolerant varieties for the rainfed lowlands.

Concurrent with these two examples of bilateral collaboration, many nations collaborated with international agricultural research institutes – the West Africa Rice Development Association (WARDA), the International Institute of Tropical Agriculture (IITA) and the International Rice Research Institute (IRRI) – to foster germplasm exchange and increase national capacity in rice improvement.

WARDA was constituted in 1970 by 11 countries with the assistance of the United Nations Development Programme (UNDP), the Food and Agriculture Organization of the United Nations (FAO) and the Economic Commission for Africa (ECA), and later expanded its membership to 17 countries. The original mandate was very broad and included efforts in research, training, development and policy formation. Early research emphasized adaptive field trials using direct introductions of exotic materials from outside the region. In the early 1980s, serious flaws in WARDA's varietal improvement strategy began to become apparent as exotic introductions failed to stem the imbalance between supply and demand, and most were found to be unsuitable for the farming conditions within the region (WARDA, 1990). A CGIAR external review in 1983 concluded that 'West Africa cannot rely upon the importation of rice technologies developed elsewhere ... [and] only a few of the technologies developed on experiment stations have been broadly adopted by farmers'.

In 1986, the Association joined the Consultative Group on International Agricultural Research (CGIAR) and committed itself to a reorientation of its research programme. In 1988, WARDA moved its headquarters from Liberia to Côte d'Ivoire, and initiated the development of a new strategic plan for the 1990–2000 period. This new strategic vision, focusing on an interdisciplinary problem-solving approach to technology development, represented a major shift away from the original mandate of the Association.

In addition to developing an integrated regional programme with national collaborators, WARDA also embarked in a collaborative agenda with IITA. During meetings in 1987 and 1988, it was agreed that all varietal improvement work would be shifted to WARDA by the end of 1990. A subsequent external review of WARDA in 1993 recommended that the International Network for Genetic Evaluation of Rice for Africa (INGER-Africa) be relocated to WARDA in order to reinforce national agricultural research systems' (NARS) linkages and to target the WARDA Task Force mechanism for broader germplasm dissemination. Although discussions on the transfer of INGER-Africa from IRRI/IITA were initiated in 1988, the transfer of INGER-Africa to WARDA was delayed until April 1997.

National and Regional Resources for Genetic Improvement

In many nations, the history of varietal improvement began before the advent of the CGIAR. In the 1960s and 1970s, these programmes were the first to incorporate early advances in germplasm improvement into their varietal development programmes. The most notable example of a long-term varietal improvement programme is in Nigeria, where the National Cereals Research Institute (NCRI), and its predecessor, has contributed to varietal development for nearly 50 years.

The national programme in Nigeria has the greatest number of postgraduate trained rice research scientists within the region. The first official rice varietal release (BG 79) occurred in 1954, for the rainfed lowland ecology. Since then, 50 more rice varieties have been released at fairly regular intervals and recent estimates indicate that more than 90% of the rice area is planted to improved varieties. The most sizable yield advances have occurred in the rainfed lowland and irrigated ecologies.

In comparison with the national focus of the Nigerian programme, Rokupr Rice Research in Sierra Leone has served as the locus for regional mangrove rice improvement. British and Sierra Leonean scientists have been working at the station since 1934 (Matlon *et al.*, 1998). Early efforts focused on varietal adaptation to saline and sulphate acid conditions found in the mangrove swamps and, in the 1960s, advanced into hybridization. This effort was reinforced by WARDA from 1976 until 1993. Rokupr has produced a number of extremely popular varieties for the mangrove swamp areas and these varieties have been diffused through the mangrove rice research network to other nations, some of which (Guinea Bissau and the Gambia) are not covered in this study, and into the inland valleys[3].

[3] Dr Shar Fomba, from the Rokupr Rice Research Station, explained that varieties that do well in the mangrove areas are generally adapted to rainfed lowland conditions, since they are very similar to the mangrove areas except that acidity and salinity problems are absent.

Despite isolated cases of a strong and concerted effort on varietal improvement, regional human capacity in rice improvement remains limited (Table 6.2). Of the 106 scientists actively participating in rice research in the national programmes (in 1998), 19% possess a BSc degree, 51% a Master's degree and less than 30% a PhD. National research programmes are supported by WARDA, where one-third of scientific effort is allocated to varietal improvement, a figure consistent with that of national programmes.

The majority of disciplinary strength lies in cultivar development activities which are largely composed of varietal screening by breeders for general phenological acceptability, with supporting efforts in assessing abiotic and biotic stress tolerance by entomologists, pathologists, weed scientists and soil specialists. A large amount of scientific time is allocated to the 'other' category by WARDA for strategic research in biotechnology (1.3 scientist-years), physiology (2.9 scientist-years), systems analysis and modelling (0.4 scientist-years), farmer preferences for plant traits (0.4 scientist-years), grain quality analysis (0.4 scientist-years) and germplasm conservation. On a regional scale, 70% of all plant improvement activities are allocated to cultivar development work, 9% to basic research, nearly 9% to pre-breeding work, and 3% to genetic resource management.

While most countries have a team of scientists allocated to varietal improvement, limited operating funds and weak capital infrastructure for genetic resource management hinders national efforts in plant breeding and genetic conservation. While specific breakdown of national programme investment in varietal improvement was not available for any country, a few were able to provide total annual expenditure on rice research activities. Nigeria, for example, spent approximately

Table 6.2. Human capacity allocated to varietal improvement in West Africa, 1998 (scientist-years).

Country	Breeding	Agronomy	Entomology	Pathology	Weed science	Soils	Other	Total
Nigeria	2.2	1.9	0.4	0.1	0.2	0.2	0.3	5.3
Guinea	6.2	2.4		0.8				9.4
Côte d'Ivoire	2			0.1			1.0	3.1
Sierra Leone	4	0.3	0.4	0.6		0.6	0.4	6.3
Mali	2.8	0.6	0.3	0.3	0.3		0.1	4.4
Ghana	1.7	1.6		0.6	0.3	1.1	0.2	5.5
Senegal	0.6	0.3		0.2	0.2	0.4		1.7
NARS Total	19.5	7.1	1.1	2.7	1	2.3	2	35.7
WARDA	2.3	1.0	0.3	0.3	0.3	0.2	5.5	9.9
Regional total	21.8	8.1	1.4	3	1.3	2.5	7.5	45.6

Source: IAEG Germplasm Impact Survey Results, 1999; WARDA 1999 Workplans and Budget.

US$144,000 for their entire rice research programme in 1996 and US$132,000 in 1997 in nominal terms[4] (Ojehomon *et al.*, 1999). Total expenditures at Rokupr Rice Research Station are about US$120,000 per year based upon historical recall (Sama S. Monde, Director General, personal communication). By contrast, WARDA allocates slightly more than US$2.2 million per year to varietal improvement. Based upon these figures, direct financial investment within the region for varietal improvement does not exceed US$3.2 million annually, omitting bilateral collaboration[5]. The ratio of investment in rice varietal improvement to the value of regional production is slightly less than one-fifth of 1% annually.

National breeding strategies

Currently, few national programmes pursue large-scale breeding programmes and fewer maintain *ex situ* germplasm collections (Table 6.3). Nigeria, for example, requests germplasm annually from advanced research institutes or draws upon *in situ* released varieties for crossing purposes. In Mali, as in most other countries, crossing blocks are regenerated biannually, thus limiting maximum size. The largest block size was found at the Malian irrigated rice research station in Niono, where 1500 varieties are retained. Most countries, however, maintain crossing blocks of 100–250 entries. Only 87 crosses were made by national programmes in 1998, most of which occurred in Guinea, but many scientists indicated that they frequently skipped years between crosses to allow evaluation of early generations.

Table 6.3 summarizes the usage of germplasm by country and by rice-producing ecology[6]. The first row presents the crossing block size and the number of crosses made in 1998. The subsequent rows present the approximate breakdown of the crossing blocks into genetic sources as a percentage of the total size in row 1, and the second figure is the percentage of material used in the hybridization activities.

[4] When denominated in national currency, investment increased between the two periods, but these figures reflect weakening terms of trade of the Naira to the US dollar.

[5] This figure represents the upper limit of research investment in varietal improvement, since the figures from NARS included all expenditure on rice research, and not just varietal improvement.

[6] The Ivorian national programme was restructured in 1998, and virtually no research was conducted. Ghana does not maintain a crossing block and does not engage in hybridization.

Table 6.3. *Ex-situ* germplasm collections and active crossing programmes in West African NARS, 1988.

	Nigeria[a]	Mali			Guinea		Senegal	
	Irrigated and rainfed lowland	Irrigated	Rainfed lowland	Upland	Rainfed lowland	Upland	Rainfed lowland	Upland
Block size/crosses made in 1998	0/2	1500/0	200/0	100/0	25/30	24/43	100/6	100/6
Source of genetic material in crossing block / material used in hybridization (% of collection)								
National advanced lines	20/0		80/0	80/0	64/49	13/17	20/10	20/10
National released cultivars	20/75				32/45	4/7	20/40	20/40
CGIAR lines	30/0	90/0			4/6	79/69	30/10	30/10
Advanced lines from other countries	20/25			10/0			20/40	20/40
Wild relatives			5/0	5/0				
Landraces[10]	10/0	10/0	5/0			4/7	10	10
Other sources[b]				5/0				

[a]Approximate size of previous collection.
[b]Primarily from CIRAD collections.

[10] Some consider the term 'landrace' to be correct only for livestock. The term has gained popular usage in the literature on crop genetic diversity and recurs with such well-known experts as Jack Harlan, T.T. Chang and Mike Jackson. In this study, we use the term synonymously with 'traditional variety' in order to describe cultivars that originated in West and Central Africa and have not been passed through generative germplasm development procedures.

Regional production and release of improved varieties

Despite limited regional resources invested annually in varietal improvement, 197 improved varieties have been released with more than 122 targeted for release in the next 5 years (2000–2004, Table 6.4). Varietal

Table 6.4. Varietal release history by country and ecology.

Country and ecology	Pre-80	1980–84	1985–89	1990–94	1995–99	2000–04	Total
Nigeria							
Irrigated			7		4	6	17
Rainfed lowland	7	4	3	1		7	22
Upland	1	1	6	5		9	22
Mangrove						1	1
Guinea							
Rainfed lowland			3		5	6	14
Upland				2	2	5	9
Mangrove	6		1	4	2		13
Côte d'Ivoire							
Irrigated	3	3	1		5	2	14
Rainfed lowland	4	3	1		1	2	11
Upland	6	7			14	4	31
Sierra Leone	3		4				7
Irrigated							
Rainfed lowland			6				6
Upland	1		7		1		9
Mangrove	5		1	2		6	14
Mali							
Irrigated	4	2	3	3	1	11	24
Rainfed lowland		2	1			11	14
Upland	1					10	11
Floating	3	2				2	7
Ghana							
Irrigated			1		1	11	13
Rainfed lowland		2	4		1	11	18
Upland						10	10
Senegal							
Irrigated	2			3	2	7	14
Rainfed lowland	2	1	2		4		9
Upland		3			1		4
Mangrove			1		3	1	5
Total							
Irrigated	12	5	16	6	13	37	89
Rainfed lowland	13	12	20	1	11	37	94
Upland	9	11	13	7	18	38	96
Floating	3	2				2	7
Mangrove	11		3	6	5	8	33
Grand total	48	30	52	20	47	122	319

Source: IAEG Germplasm Impact Survey Results, 1999.

release lapsed during the early 1980s and the early 1990s, but on average, about eight varieties per year have been released since 1980. Release data prior to 1980 were limited and often inconsistent between nations.

Regional collaboration has produced a considerable number of new varieties for two ecologies: the mangrove swamps and irrigated lands. These two ecologies have the highest number of releases per hectare of cultivation, 0.30 and 0.11, respectively, but relatively limited area. The other three ecologies – the rainfed lowlands, uplands and floating ecologies – have produced just 0.04, 0.03 and 0.02 releases per hectare despite covering, with the exception of the deep-water floating area, the majority of the rice-growing area.

CGIAR germplasm has played an extremely important role in the production of new varieties and in the improvement of rice productivity. The six mutually exclusive content indicators comprise:

1. 'Centre Cross–Centre Selection' – varieties that were crossed by a CGIAR centre and whose subsequent selection was conducted by a CGIAR breeder
2. 'Centre Cross–NARS Selection' – varieties that were crossed by a CGIAR centre and whose subsequent selection was conducted by a national-programme scientist
3. 'NARS Cross–Center Ancestor' – varieties produced by a national programme using a CGIAR parent or grandparent
4. 'NARS Cross–NARS Ancestor' – varieties produced by a national programme using a parent produced by a national programme
5. 'Landrace' – selections from landraces (traditional varieties), which were then formally released (not necessarily originating in country of release)
6. 'Unknown Improved Content' – improved varieties of unknown heritage.

The content of the released varieties was, for the most part, discernible with precision. The overall tabulation of the centre content indicates that 54 of the 197 released varieties in the study (27%) are a direct result of CGIAR germplasm enhancement, an additional 31 of these varieties (16%) have parents or ancestors developed by the CGIAR (categories 2 and 3). In fact, all but three of the varieties in these groups combined are in category 3, with NARS having made the crosses.

A large number of varieties, 80 of the 197, were also developed by national programmes without any direct or indirect involvement of the CGIAR (category 4); 38 of the 80 varieties were produced wholly by IRAT and 13 by the Rokupr Rice Research Station. The Rokupr station also produced several more mangrove varieties using materials developed by IRRI, in conjunction with WARDA. Most of the remainder originate from Asian national programmes.

The third most important source of released varieties is traditional varieties introduced from outside the region and also transferred from one nation to another within the region. These purified landraces (category 5) consisted of 27 out of 197, or 14% of accessions. It was impossible to recover the pedigree and origination of five varieties.

Deriving the mechanism by which released varieties were introduced into the national programmes, especially those with CGIAR content, is problematic. Centres, national programmes and bilateral assistance often promoted the same varieties through different mechanisms. For example, IRRI-developed varieties were tested through the International Rice Testing Programme (IRTP), the WARDA coordinated varietal trials, and also exchanged by IITA breeders with the national-programme breeders. It is clear, however, that the 'old' WARDA played an extremely important role in promoting the exchange of varieties developed by regional programmes and also in the transfer of landraces around the region.

Improvement and diffusion of new varieties

Approximately the same number of varieties has been released in the irrigated, rainfed lowland and upland ecologies, but adoption rates for these ecologies differ dramatically (Table 6.5). The irrigated ecology has the highest adoption rate of improved varieties and has benefited the most from the introduction of materials developed in Asia. This ecology is the most homogeneous in the region and, in the case of those irrigation schemes located in the Sahel, the most similar to controlled Asian production systems. In addition, this ecology was the first to take advantage of the semi-dwarf gene found in Dee-Geo-Woo-Gen. This is the case for the irrigated schemes in Senegal, where farmers widely adopted a descendant of Taichung Native 1, and more recently one of IR36. In Mali, the most popular variety cultivated is Kogoni 89–1, a cross of a local cultivar found in the *Office du Niger* with IR34. Despite the incorporation of the semi-dwarfing trait into modern varieties, 20% of the irrigated area in Nigeria is still planted to varieties developed prior to incorporation of this trait (sub-category 'Pre-Modern'), largely holdovers from screening and diffusion mechanisms in the 1950s and early 1960s. Pre-modern varieties occupy negligible irrigated areas in other countries.

Adoption and diffusion patterns for the rainfed lowland ecology are dramatically different. Pre-modern varieties cover a very large area in many nations as do traditional varieties,[7] indicating the difficulty of

[7] Coverage by traditional varieties is calculated as the remainder of 'Pre-Modern' and 'Modern'.

Table 6.5. Diffusion rates for introduced varieties (% of area cultivated).

	Total rice area (>'000 ha)	Ecology and variety type[a]							
		Upland		Rainfall lowland		Irrigated lowland		Mangrove	Deep-water
		Purified landrace	Hybridized	Pre-modern	Modern	Pre-modern	Modern	All	All
Nigeria	1784	31	67	60	37	20	80		
Guinea	445		20		55			39	
Côte d'Ivoire	750	43	7		30	2	93		
Sierra Leone	289	58	2	65	17		57	80	
Mali	302	10	0	5	15	3	96		70
Ghana	96			15	65	0	65		
Senegal	70	0	20		30		100		

[a]Varieties are distinguished in two ways. With regard to the uplands, 'purified landrace' refers to traditional varieties from Africa and outside the region that were purified and diffused by formal mechanisms, while 'hybridized' refers to those developed through deliberate hybridization activities and diffused. With regards to the rainfed and irrigated lowlands, 'pre-modern' refers to varieties developed before the successful hybridization of Dee-Geo-Woo-Gen and the introduction of the semi-dwarfing gene, while 'modern' refers to those varieties developed after hybridization with the semi-dwarfing gene. Most introduced varieties in the mangrove ecology were improved through hybridization. 49% of the impact in the deep-water floating ecology is due to a purified landrace from Asia and the remainder to hybridization.

Source: IAEG Germplasm Impact Survey Results, 1999.

developing modern semi-dwarfing varieties for this ecology. In only Guinea and Ghana do modern lowland varieties cover more area than traditional and pre-modern varieties. Until recently, rice was a minor crop in Ghana and when the national agricultural policy began to emphasize rice, the national agricultural research service turned to the CGIAR for its germplasm. Of the 80% of improved varieties planted in the rainfed ecology in Ghana, more than half of that area is planted to CGIAR materials and falls into the modern sub-category. Guinea's success can be tied to large-scale screening of CGIAR germplasm combined with hybridization of Chinese and local cultivars.

Adoption rates in the uplands are the lowest for all ecologies and extremely low for hybridized modern varieties. In Côte d'Ivoire, approximately 50% of the upland area is planted with introduced varieties; however, the two most popular varieties, Iguape Cateto and Moroberekan, are pureline landraces introduced as early as 1960. When these traditional varieties are removed, the adoption of improved varieties is approximately 7%. Mali only has a very small area of upland rice under cultivation and few improved varieties have been adopted. Less than 2% of upland area in Sierra Leone is planted with varieties developed through hybridization, but 58% is covered with purified landraces; the remainder is covered with unknown local varieties.

Low adoption rates in the uplands have commonly been attributed to weak extension and varietal diffusion mechanisms. The widespread adoption of purified landraces, however, indicates that varieties have been introduced successfully and have diffused. The lack of adoption of modern, hybridized varieties may therefore be attributed to varietal development programmes that have not produced varieties that outperform local cultivars in terms of yield, stability, duration, stress resistance or consumption characteristics. Upland rice has the oldest cultivation history in the region, with *Oryza sativa* probably introduced by Portuguese explorers over 500 years ago on their return from India, while *Oryza glaberrima* has been domesticated for at least 2500 years (Carpenter, 1978). 'Traditional' varieties, well adapted to local growing conditions, have been developed from subsequent introductions of *O. sativa* and centuries of farmer selection.

Extensive evidence exists for the widespread adoption of improved varieties in the mangrove ecology. In the early 1990s, adoption studies indicated a 19% adoption rate for improved varieties in the mangrove swamp areas. Current estimates, derived in this study, indicate that the rate has more than doubled to 39%. In Sierra Leone, much of this increase is attributable to varieties released after the 1992 study, and high adoption rates are due also in part to farmer proximity to the Rokupr Station. In the deep-water ecology of Mali, the high adoption statistics are linked to the widespread usage of the traditional deep-water Khao Gaew cultivar from Thailand. This variety was promoted early on by WARDA's deep-water station located in Mopti, Mali.

At the national level, the overall adoption rates of improved varieties is mixed. Appendix 6.2 presents a country by ecology breakdown of varieties currently cultivated on a large scale, excluding local cultivars, but including traditional varieties transferred from other countries and released. Nigeria has the highest rates of varietal adoption.[8] Most of the varieties that are widely cultivated in Nigeria were introduced more than 15 years ago. In the rainfed lowlands, 81% of the area planted to improved varieties are by those released prior to 1974. In the upland and irrigated ecologies, 53% and 75% of all area is planted to varieties released during the early 1980s. In contrast, only 25%, 3% and 31% of irrigated, rainfed lowland and upland area is planted to varieties released in the 1990s, offering weak causal evidence of the time-lag from release to adoption and the pace of varietal turnover.

Financial impact of varietal improvement under uncertainty

Considerable uncertainty exists in the calculation of the financial impact of varietal improvement in West Africa. Uncertainty exists in three major forms: prices, adoption rates and national rice areas. In order to calculate the regional benefits of varietal improvement under uncertainty, Monte Carlo techniques were used to fully exploit variability in the uncertain estimates and to derive a realistic distribution of probable impact. This technique samples all combinations of possible prices, adoption rates and FAO estimates, and subsequently calculates the gains from varietal improvement. This process continues until no additional information is gained from re-sampling the uncertain input parameters. The output contains the range between high and low values plus the probability of each plausible outcome, and hence confidence in a particular value. Further description of the approach is presented in Appendix 6.1.

[8] Unfortunately, few adoption studies are available to corroborate high adoption rates and, where available, these studies are usually restricted to limited geographical coverage (Ojehomon *et al.*, 1999). None the less, researchers from IITA have indicated that most farmers claim to know the names of their rice varieties as either FARO or ITA cultivars (Victor Manyong, personal communication). In addition, Nigeria has the longest history of varietal improvement plus an agricultural policy environment which long subsidized crop inputs. Regardless of this anecdotal and historical evidence, considerable uncertainty exists on Nigerian adoption rates.

Overall, the cumulative impact of genetic enhancement and transfer contributed US$374 million to the regional rice economy in 1998 (Fig. 6.1). In the most conservative case, varietal improvement contributed just US$156 million to the regional economy and at the most optimistic it could have contributed up to US$848 million. There is less than a 10% probability that the actual value is below US$243 million and less than 30% that it is below US$290 million. Figure 6.1 indicates that there is a high probability that the impact lies between US$330 million and US$386 million and much less certainty that it extends upwards, given the high degree of uncertainty in prices, adoption rates and FAO area estimates.

The breakdown and distributional impact upon nations and ecologies, as measured at the median values, is presented in Table 6.6. Varietal improvement has increased farm revenues, on average, by US$100 per hectare, but much more so in irrigated and rainfed lowland areas. The greatest financial impact has occurred in the irrigated

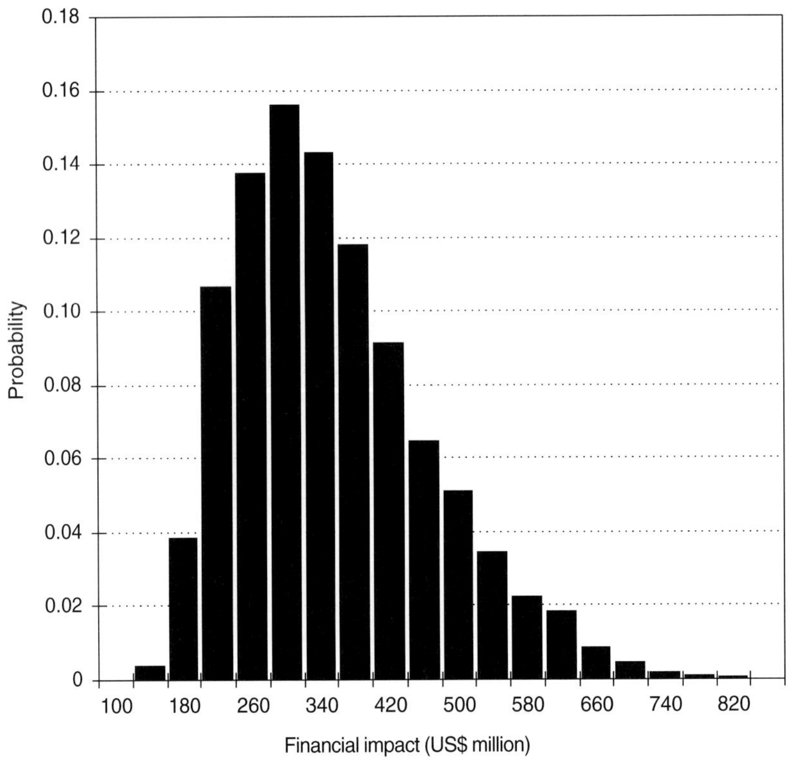

Fig. 6.1. Probability distribution of the regional financial impact of varietal improvement and transfer in 1998.

Table 6.6. Distributional impact of varietal improvement (1998 US$ thousands).

Total gains	Rainfed upland	Rainfed lowland	Irrigated lowland	Mangrove swamp	Deep-water floating	Total	Gain ha^{-1} (US$)
Nigeria	24,041	145,168	39,251			208,461	117
Guinea	7,600	19,001		3,605		30,206	68
Côte d'Ivoire	13,658	7,014	22,474			43,147	58
Sierra Leone	9,738	7,482	818	1,578		19,616	68
Mali	24	299	27,861		$8,677	36,862	122
Ghana		29,659	1,498			31,157	325
Senegal	0	304	4,516			4,819	69
Total	55,062	208,927	96,419	5,132	$8,677	374,267	100
Gain ha^{-1} (US$)	32	163	232	69	$32	100	

Source: IAEG Germplasm Impact Survey Results, 1999.

lowlands. On average, gross revenues per hectare in the irrigated areas have increased by US$232 as a result of varietal improvement. This is followed distantly by the rainfed lowland areas at US$163 ha^{-1}, the mangrove swamps (US$69), deep-water floating ecology (US$32) and the uplands (US$32).

On a national scale, productivity impacts are highly variable but the benefits are much greater for those countries with larger irrigated or lowland areas. These results have profound distributional implications for farm households throughout the region, in that small-scale farmers in the uplands, mangrove swamps and those in highly variable deep-water floating areas have not benefited to the same degree as those in more favourable rice-growing ecologies. In Côte d'Ivoire, for example, the gain from varietal improvement in the uplands is approximately 16–18% of short-run net revenues per hectare, while in the lowlands it represents about 48% of short-run net revenues (figures calculated from Dalton, 1999a).

Productivity gains may be attributed to CGIAR institutes, national programmes and traditional varieties from within and outside the region (Table 6.7). The most important financial source of genetic enhancement comes from varieties developed by national programmes in both Africa and Asia (category 4, NARS Cross–NARS Parent). Of the 103 varieties included in the financial analysis, 38% fall into this class and account for nearly 39% of the financial impact. Many of the varieties found in this category are successful varieties developed in Asia and cultivated in the irrigated and, to a lesser extent, lowland ecologies. In addition, successful varieties developed in Asia and Africa have also drawn upon CGIAR-developed varieties as parents in their crosses.

Over 25% of all varieties currently in use were directly developed by CGIAR institutes, including WARDA, IITA and IRRI, and these varieties contributed about 29% of the total financial gain. Only one country, Mali, does not use varieties directly developed by the CGIAR, but its most popular irrigated variety uses IR36 as a parent. The total number of CGIAR-related varieties climbs to nearly 40%, and contributes over 46% to the gain, when combined with the second and third categories. These categories include varieties developed by national programmes using CGIAR parents. Impact from CGIAR sources is distributed widely across all nations in the study. In addition to the direct and indirect role of CGIAR breeding programmes in improving national rice production, WARDA is particularly responsible for the spread of traditional varieties that have increased yields in the most difficult ecologies found within the region.

The study highlights two cases of disproportionate impact. National programmes have been able to take strong advantage of varieties developed by international institutes and incorporate these vari-

Table 6.7. Financial contribution by centre content indicators.

Centre content indicator	Minimum	Maximum	Mean	Share of financial gain (%)	Share of adopted varieties (%)
	(1998 US$ thousands)				
Centre Cross–Centre Selection	31,539	277,860	107,655	28.8	25.2
Centre Cross–NARS Selection	1,510	7,119	3,583	0.0	0.9
NARS Cross–Centre Ancestor	26,024	116,956	59,050	16.7	13.6
NARS Cross–NARS Parent	62,345	330,506	146,014	39.1	37.9
Landrace	19,903	94,591	43,361	12.1	20.4
Unknown	4,787	28,172	12,099	3.2	1.9

eties into their breeding programmes (Category 3, NARS Cross–Centre Ancestor). The best example of this type of relationship occurred in Mali, where the national programme crosses local varieties with IRRI-bred irrigated lines. At the other extreme, pureline landraces still account for about 23% of all currently cultivated released varieties, but they proportionately contribute far less of a productivity gain, just 13.4% (Category 5, Landrace).

Conclusions

The history of varietal improvement in West Africa may be characterized as disjunct, with many actors and overlapping activities often simultaneously pursuing similar objectives until the late 1980s. In the 1990s, varietal improvement activities were centralized at WARDA, and new collaborative mechanisms developed in order to facilitate varietal improvement and to target the more difficult rice production ecologies of the rainfed lowlands and uplands. The resulting impact of the centralization and the renewed collaborative varietal development mechanism may be responsible, in part, for the large numbers of varieties in the pipeline for release over the next 5 years. None the less, case studies conducted in several countries over the past 8 years indicate that the returns on investment in varietal development have always exceeded 20% annually and, in select cases, upwards of 100% per year.

This study found limited financial and human investment in regional varietal improvement. Approximately 36 scientist-years in the national programmes and 10 at WARDA are allocated to rice varietal improvement, but financial investment does not exceed US$3.2 million annually. Despite limited investment in varietal improvement, 197 varieties have been released over the past 20 years, and over half of these varieties have generated sizable gains in rice productivity at the farm and national level.

Using conservative adoption estimates, and relying upon historical data on national costs of rice imports as the opportunity cost of forgone production, the study determined that varietal improvement contributes, on average, US$374 million to the regional economy and this may be as much as $848 million per year. Over 43% of this gain is attributable to CGIAR germplasm improvement programmes, either as direct varietal products or as parents used by national breeding programmes in the creation of released varieties. In addition to the direct role of CGIAR programmes in varietal development, this study has identified a second important role of the CGIAR in coordinating germplasm exchange, including regional and exotic landraces, as well as varieties developed by national programmes.

Varietal gains have largely occurred in the more favourable rice producing ecologies: the irrigated and rainfed lowlands. According to national research and extension authorities, adoption rates in the irrigated ecologies are close to 100% and about 62% in the rainfed lowlands. Per hectare gains in output are approximately US$232 and US$163, respectively. While the production of improved materials for the mangrove ecology has been impressive, per hectare gains have been about US$69, largely due to the high productivity of the ecology even when cultivated to traditional varieties. Almost all impact in the floating ecology is due to the transfer of Asian deep-water materials.

By contrast, gains in the uplands have been much more modest, especially outside Nigeria. While adoption rates in Nigeria are high, the yield gains per hectare have been slight. In other upland rice producing countries, adoption rates of improved materials do not exceed 30% when introduced traditional varieties are excluded. In the upland rice growing belt of West Africa – Côte d'Ivoire, Liberia, Sierra Leone and Guinea – almost all popular released varieties are pureline landraces: LAC 23, Moroberekan, Iguape Cateto, Ngovie, Ngiema Yakei and OS6, and productivity has increased by only \$32 ha^{-1}.

The widespread coverage of introduced varieties (including landraces) indicates that transfer of varieties into a highly heterogeneous production ecology is possible. In the next 5 years, 37 new varieties for the uplands are expected to be released including low-management *O. sativa* cultivars and several which have begun to exploit *O. glaberrima* gene pools. Many of these varieties have shown productivity gains of 24% over local varieties in widespread farmer-controlled evaluations (Dalton, 1999b). The financial value of such a gain in just Guinea, Côte d'Ivoire and Sierra Leone, assuming a 10% adoption rate would amount to nearly US$8 million per year, and nearly US$20 million at 25% adoption rates.

In addition to the upland ecology, 31 new varieties are available for release in the rainfed lowlands. Many of these varieties target stress-prone areas where introduced Asian germplasm did not generate productivity gains. These two ecologies represent strategically important areas for future research and extension activities.

This chapter has highlighted the substantial impact of regional varietal improvement programmes by estimating the incremental gain to regional rice production, holding constant crop and resource management interventions. Substantial additional benefits to regional rice productivity may be attributed to crop husbandry, fertilizer application and timing recommendations for irrigated and lowland rice, but these are beyond the scope of this chapter. This chapter has not attempted to derive any indirect benefits from varietal improvement at the farm or community level. Nevertheless, without regional efforts in varietal improvement, the regional balance-of-payment deficit for rice imports would have been 40% higher or an additional 658,000 ha of land would have been needed to be under rice cultivation to maintain consumption at current levels.

References

Byerlee, D. and Moya, P. (1993) *Impacts of International Wheat Breeding Research in the Developing World, 1966–1990.* CIMMYT, Mexico.

CIRAD-CA (1993) *Variety's Descriptive Form: Rice.* Seed Laboratory, Montpellier, France.

Carpenter, A.J. (1978) The history of rice in Africa. In: Buddenhagen, I. W. and Persley, G.J. (eds) *Rice in Africa.* Academic Press, New York.

Dalton, T.J. (1999a) Farmer's evaluation of rice varieties distributed in the Danané PVS program, 1998. Unpublished Technical Report. WARDA, Bouaké.

Dalton, T.J. (1999b) Crop budgets for rice-based cropping systems in four areas of Côte d'Ivoire. Unpublished Technical Report. WARDA, Bouaké.

Matlon, P., Randolph, T. and Guei, R. (1998) Impact of rice research in West Africa. In: Pingali, P. and Hossain, M. (eds) *Impact of Rice Research. Proceedings of the International Conference on the Impact of Rice Research, 3–5 June 1996, Bangkok.* Thailand Development Research Institute, Bangkok, Thailand, and International Rice Research Institute, Manila, Philippines, pp. 383–401.

Ojehomon, V.E.T., Wayagiri, J., Tiamiyu, S.A., Maji, A.T. and Misari, S.M. (1999) *Impact of upland and lowland rice varieties in Nigeria.* Paper submitted to the National Agricultural Research Project (NARP), Abuja.

WARDA (1990) *WARDA in Transition.* WARDA, Bouaké.

WARDA (1999) *WARDA Workplan and Budget.* WARDA, Bouaké.

Appendix 6.1

We used Monte Carlo techniques to derive a probabilistic estimate of the financial impact of genetic enhancement. Several elements required to estimate the financial impact of genetic enhancement are either determined stochastically or could not be determined with reasonable precision: rice prices, adoption rates, national rice areas. In most cases, adoption rates were estimated by national research and extension authorities rather than through statistical sampling procedures. Focus group discussions with researchers, extension and national authorities were held in all nations. Using a Delphic approach, adoption rates were elicited variety by variety. Rather than arbitrarily deflating the figures provided by national authorities, their estimates were considered the upper adoption limit.

Since the analysis is largely determined by the results in Nigeria, extreme caution was taken to conservatively estimate this source of financial impact and actual adoption values were limited to be above 30% of the reported values and lower than 100% of the reported values, with the mean value of 60% and the most likely value of 50%. These four values are used to construct a probability density function that may be interpreted as adoption rates ranging somewhere between 30% and 100% of the reported values, but most likely to be about 55%, with a 20% chance that they are above 80%. This assumption does not rule out the chance that the adoption rates are equal to those reported.

Less conservative adoption estimates were assumed for the other countries. In the case of the remaining countries, adoption ceilings were estimated at 100% of the reported values and the floor at 50%. Mean values were estimated at 80% of the reported and the most likely values at 85%. In addition to price and adoption rates, the FAO estimates of area under rice cultivation are also uncertain. In this instance, the FAO could be seen as both underestimating areas or overestimating actual rice areas. Therefore, it was assumed that actual rice areas vary by a standard deviation of 5% about the reported value.

Distributions and distributional moments used for uncertain parameters

These ten sources of uncertainty were simultaneously modelled using Monte Carlo techniques in order to derive a distribution of probable outcomes on the financial gain from varietal improvement. The distributions and their moments are presented in the table overleaf. The

Distributions and distributional moments used for uncertain parameters

	Distribution	Minimum	Mean	Maximum	Standard deviation	Variance	Skewness	Kurtosis
Import parity prices ($ t^{-1})								
Nigeria	Beta	109	382	708	158	25095	0.147	1.946
Guinea	Beta	150	258	421	62	3850	0.337	2.249
Côte d'Ivoire	Beta	123	308	518	60	3594	0.191	2.809
Sierra Leone	Beta	193	336	530	49	2439	0.328	3.072
Mali	Beta	155	330	536	60	3615	0.142	2.761
Ghana	Beta	114	310	562	101	10238	0.204	2.186
Senegal	Beta	101	210	348	45	2016	0.143	2.485
Adoption rates (%)								
Nigeria	Beta	30%	60%	100%	0.17%	0.03%	0.231	2.067
General	Beta	51%	80%	100%	0.11%	0.01%	–0.304	2.262
FAO area estimates (%)	Normal	82%	100%	117%	0.05%	0.002%	0.002	2.979

Source: FAO Agrostat, 1970–1996 for rice prices.

moments of these distributions were determined through the iteration of the Delphic approach. Individual responses were recorded and the spread of responses provided the minimum and maximum levels and central moments. The results of the Monte Carlo exercise indicate the range and confidence for values on the financial impact of rice varietal improvement in West Africa. It values the yield gain attributable to varietal improvement and assumes that rice yields would not have increased beyond the level of locally available cultivars[9].

[9] In the case of the Sahelian irrigated schemes, this is the value of the gain over the earliest introduced varieties, in other words, a Type II technological change as described by Byerlee and Moya (1993). In humid areas this gain is in relation to local traditional varieties.

Appendix 6.2

Economically important varieties grown within the region[a] (synonyms given in parentheses)

	Upland	Rainfed lowland	Irrigated lowland	Mangrove	Deep-water
Nigeria	Ex-China	FARO 1 (BG79)	FARO 37 (ITA 306)		
	FARO 11 (OS 6)	FARO 8 (MAS 2401)	FARO 35 (ITA 212)		
	FARO 43 (ITA 128)	FARO 9 (SIAM 29)	FARO 44		
	FARO 46 (ITA 150)	FARO 12	FARO 50 (ITA 230)		
	FARO 48 (ITA 301)	FARO 15			
		FARO 18 (Tjina)			
		FARO 35 (ITA 212)			
		FARO 37 (ITA 306)			
		FARO 44			
		FARO 51 (Cisadane)			
Guinea	CK 12	CK 21		WAR 1	
	CK 5	CK 211		WAR 73	
	CK 7	CK 30		WAR 77	
	CK 8	CK 4		RD 15	
		CK 43		B 38-D2	
		CK 73		ROK 5	
		CK 92			
Côte d'Ivoire	IAC 164	Gambiaka	BG 90-2		
	IAC 165	Fossa	IR 5		
	IRAT 144	Bouake 189	IR 8		
	IDSA 6		Bouake 189		
	Moroberekan				
	Iguape Cateto				

Sierra Leone	ROK 3 (Ngiema Yakei) ROK 16 (Ngovie) ROK 17 (LAC 23) ROK 20 (IRAT 161)	ROK 24 (Suakoko 8) ROK 27 (Warkaiyo) ROK 29 ROK 30	ROK 11 (ADNY 2) ROK 14 (MANG 2)	CP 4 ROHYB 4 ROK 10 ROK 22 ROK 5 WAR 77-3-2-2 WAR 81-2-1-2	not available
Mali	Dourado Precoce	Gambiaka BG 90-2	Seberang BG 90-2 Kogoni 89-1		D 52-37 BH 2 Gambiaka DM 16 Khao Gaew
Ghana		GR 17 (IET 2885) GR 19 (C168) GR 20 GR 21 GR 18 GR 22 GRUG 7	GR 22 GRUG 7		
Senegal	DJ 11-509 DJ 8-341	BG 90-2 DJ 684D	SAHEL 201 SAHEL 202 SAHEL 108 (ITA306) I Kong Pao Jaya		

[a]Excludes traditional varieties that have not passed through a derivative varietal development process.

Impacts of CIMMYT Maize Breeding Research 7

M. MORRIS, M. MEKURIA AND R. GERPACIO

Maize genetic improvement work is carried out at two of the 16 international agricultural research centres (IARCs) that are members of the Consultative Group on International Agricultural Research (CGIAR). The International Maize and Wheat Improvement Centre (CIMMYT), headquartered in Mexico, holds a global mandate for maize improvement research and targets lowland tropical, subtropical, mid-altitude and tropical highland environments throughout the developing world. The International Institute of Tropical Agriculture (IITA), headquartered in Nigeria, holds a regional mandate for maize improvement research and targets mainly humid tropical and moist savannah zones of western and central Africa. This chapter discusses impacts of CIMMYT's maize breeding programme; impacts of IITA's maize breeding programme are discussed in Chapter 8.

This chapter draws on three regional studies – one each for Latin America, eastern and southern Africa, and Asia (see Morris and López-Pereira, 1999; Gerpacio, 2001; Hassan *et al.*, 2001). Information was collected through a survey of maize breeding organizations in 37 developing countries. Questionnaires were completed for 104 public breeding institutes and seed production agencies, as well as for 267 private seed companies. In terms of geographical coverage, the survey concentrated on countries targeted by the CIMMYT maize programme. All of the important maize-producing regions in the developing world were included, except for West and Central Africa (where the CGIAR mandate for maize genetic improvement is held by IITA), northern China (where farmers grow mainly temperate materials that are not targeted by

CIMMYT), and West Asia and North Africa (omitted for logistic reasons). Collectively, the countries included in the survey account for about 95% of the area planted to maize in non-temperate production environments of Latin America, eastern and southern Africa, and Asia.[1]

Why Maize is Different from Other Crops

Distinctive characteristics of maize

Maize (*Zea mays* L.) differs from other crop species in a number of respects that affect the way genetic improvement activities are organized and carried out, as well as the process by which modern varieties (MVs) are taken up by farmers and diffused across the countryside.[2] To understand the impact of international breeding efforts, it is important to understand the characteristics of maize that distinguish it from other crop species.

Open pollination

Maize is an open-pollinating species, unlike other leading cereals such as wheat and rice, which are self-pollinating. When self-pollinating species reproduce, the pollen that fertilizes a given ovary to produce a viable seed almost always comes from the same plant, so each generation of plants retains the essential genetic identity of the preceding generation. By contrast, when maize reproduces, genetic material is exchanged between neighbouring plants. Unless pollination is carefully controlled, a field of maize that is harvested and replanted will result in a field in which all of the resulting maize plants will differ from the preceding generation and from each other.

Importance of hybrid vigour

When maize reproduces, much depends on whether the pollen grain used to fertilize a given kernel comes from the same plant or from a different

[1] In China, the survey covered only the five southern provinces in which maize is grown in non-temperate production environments (Guangxi, Guizhou, Hunan, Sichuan, Yunnan). Although varietal releases and seed sales data were not collected in northern China, the aggregate MV adoption rate was estimated based on information provided by the Chinese national maize programme.

[2] Throughout this report, the term *varieties* is used in a generic sense to refer to both open-pollinated varieties of maize as well as hybrids. The term *modern varieties* (MVs) is used to refer to open-pollinated varieties and hybrids that have been improved by a formal breeding programme.

plant. When maize plants self-fertilize, the resulting progeny are often characterized by undesirable traits, such as reduced plant size and low yields. But when maize plants cross-fertilize, some of the resulting progeny have desirable traits, such as increased plant size and high yields. Commonly referred to as 'hybrid vigour', this phenomenon is attributable to the complementary action of favourable alleles and is exploited by plant breeders in their efforts to develop commercial varieties.

Multiple end uses

No other cereal can be used in as many ways as maize. Virtually every part of the maize plant has economic value, including the grain, leaves, stalk, tassel and sometimes even the roots. In view of the multiple end uses, it is not surprising that farmers grow thousands of varieties featuring unique combinations of desirable traits. Maize is not the only cereal that is genetically diverse, but what distinguishes maize from most other cereals is the extent to which genetic diversity is actively managed at the household level. In developing countries, it is not uncommon to find the same farmer growing three, four, and sometimes even more, distinct maize varieties, each carefully selected to satisfy a specific food, feed, or industrial use.

Variability of maize production environments

Maize is the world's most widely grown food crop. Maize is cultivated at latitudes ranging from the equator to approximately 50°N and 50°S, at altitudes ranging from sea level to over 3000 m elevation, under temperatures ranging from extremely cool to very hot, under moisture regimes ranging from extremely wet to semi-arid, on terrain ranging from completely flat to precipitously steep, in many different types of soil, and using a wide range of production technologies. The extreme variability between many of these production environments gives rise to strong genotype × environment (G×E) interactions.

Implications for breeding research

The distinctive characteristics of maize have important implications for genetic improvement efforts.

Farmer breeding

Because maize is an open-pollinated species, new genetic combinations are continuously being formed in farmers' fields through natural outcrossing. In many parts of the world, farmers understand that the genetic composition of their cultivars changes with every cropping cycle. When the time comes to select seed for replanting in the following season, they are careful to choose materials that exhibit desirable traits (Morris *et al.*, 1999). Some farmers go a step further and deliberately generate new genetic combinations by planting seed of different varieties within the same plot or in adjacent plots to encourage cross-pollination. Alternatively, through a process known as *rustification* or *creolization*, farmers may acquire seed of an MV and apply selection pressure to alter its characteristics to better meet local production and/or consumption requirements. Although maize is not the only crop subjected to farmer selection pressure, no other major grain can be manipulated as rapidly as maize.

Emphasis on hybrids

The distinctive biological characteristics of maize have not only encouraged farm-level breeding activity, but they have also had an important influence on institutional breeding efforts. Because the physical separation of the male and female flowers in maize makes controlled cross-pollination relatively easy, and because hybrid vigour in maize is so pronounced, many maize improvement programmes have concentrated almost exclusively on the development of hybrids. The focus on hybrids as a way of achieving genetic gains makes sense from a scientific point of view, but it also makes sense from an economic point of view, since hybrids are an attractive business proposition.

Location specificity of improved germplasm

In industrialized countries, maize is grown mainly in temperate environments; in developing countries, it is grown mainly in non-temperate environments. This difference has important implications for the flow of improved technology. Maize germplasm that performs well in temperate regions generally cannot be introduced into non-temperate regions without undergoing extensive local adaptation. This means that with maize, MVs developed for use in the USA, Western Europe, and northern China are of little direct use in developing countries.

Implications for germplasm diffusion

The distinctive characteristics of maize not only influence breeding efforts, but they also have important implications for the dissemination of improved germplasm.

Critical importance of seed

More so than with any other crop, with maize the dissemination of improved germplasm is critically dependent on the availability and affordability of high-quality seed. Because the genetic composition of maize plants grown from farm-saved seed can change considerably from generation to generation, if farmers want to be certain of maintaining a high level of genetic purity, they must purchase fresh seed for each cropping cycle. This is true even for open-pollinated varieties.

Need for an effective seed industry

Since genetically pure maize seed is too costly and technically difficult for farmers to produce, the fact that fresh seed must be acquired for each cropping cycle means that MVs can disseminate only with the support of a viable seed industry (Morris, 1998). This can present a bottleneck, particularly in developing countries where subsistence farming is still common, because seed companies rarely find it profitable to sell to subsistence farmers. Such farmers are often neglected by the seed industry and thus do not have reliable access to sufficient quantities of high-quality seed.

Investment in Maize Breeding Research

The International Maize and Wheat Improvement Centre (CIMMYT)

As the holder of a global mandate for maize genetic improvement, CIMMYT is a major player in international breeding efforts. The organization of maize breeding activities at CIMMYT is consistent with the Centre's mission to strengthen local breeding capacity in developing countries. The CIMMYT maize breeding programme does not produce finished varieties that can be delivered directly to farmers. Rather, the CIMMYT maize breeding programme seeks to develop intermediate products for use by national breeding programmes – improved germplasm showing high yield potential, good agronomic characteristics, resistance to important biotic and abiotic stresses, and/or enhanced nutritional quality. CIMMYT scientists accomplish this goal by collecting, evaluating and preserving a wide range of maize germplasm; by improving materials in their own breeding plots; and by managing an international testing network through which sets of experimental materials are distributed to key sites around the world for evaluation by local collaborators. In return for growing these experimental materials under specified levels of management and recording key performance data, the collaborators are free to request additional seed of the most promising materials for use in their own breeding programmes. The CIMMYT-managed international testing networks thus provide national breeding programmes with ready access to germplasm and information that they would not be able to generate on their own.

Judged strictly in terms of numbers of researchers, CIMMYT is a minor actor in the global maize breeding industry. The CIMMYT maize programme currently includes about 30 scientists, of whom approximately 25 engage in breeding or breeding support (including genetic resources conservation and management). Quite a few national agricultural research organizations in the South, many public universities in the North, and practically all of the leading private seed companies thus have larger maize breeding programmes than CIMMYT.

How much does CIMMYT invest in maize genetic improvement? The question is not as straightforward as it seems, because 'maize genetic improvement' can be defined broadly or narrowly. Since CIMMYT is first and foremost a plant-breeding institute, it could be argued that CIMMYT's entire budget is devoted in one way or another to the improvement of its two mandate crops. Yet certain activities carried out by CIMMYT staff have little direct connection to plant breeding (e.g. farming systems research, crop and resource management research, policy research, networking and training activities), so it could also be argued that something less than the Centre's entire budget is spent on crop improvement research.

Figure 7.1 shows the evolution of CIMMYT's expenditures on maize genetic improvement under two sets of assumptions. In Scenario 1, it is assumed that CIMMYT's entire budget is dedicated to crop improvement research and that it can be allocated between maize and wheat in proportion to the relative sizes of the maize and wheat programme budgets. In Scenario 2, it is assumed that the proportion of CIMMYT's entire budget that can be allocated to maize improvement research is proportional to the number of senior maize programme staff among all senior staff. The budget data have been adjusted for inflation by converting to 1996 constant dollars. CIMMYT currently invests between US$7.5 million and US$18.5 million per year on maize genetic improvement (including genetic resources conservation and management). Over time, numbers of maize programme staff have declined as a proportion of total staff with the diversification of CIMMYT's research portfolio, so Scenario 2 probably represents a better measure of CIMMYT's current investment in maize breeding research.

Public national breeding programmes

Public national breeding programmes play an important role in the global maize breeding industry, supporting nearly 1000 senior breeders

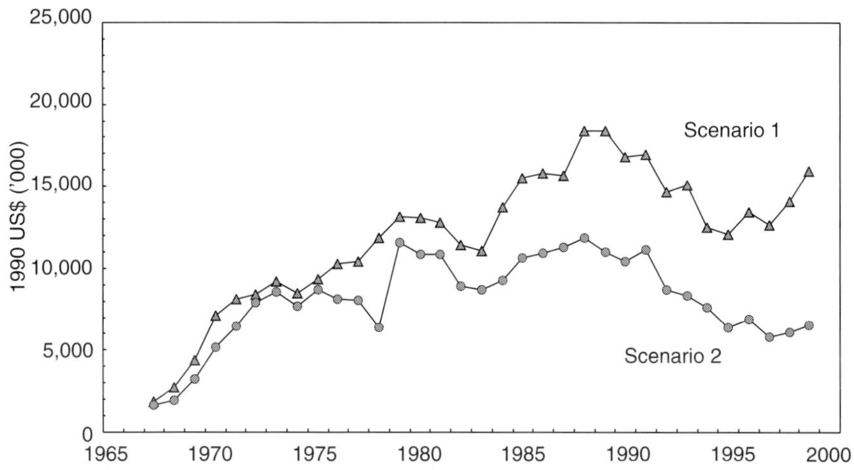

Fig. 7.1. CIMMYT maize research expenditures, 1967–1999.

worldwide (Table 7.1). These breeders are fairly evenly distributed across all developing regions, with the exception of China, which claims a disproportionately large share.[3] The organization of public breeding programmes varies considerably by region, however. Public breeding activities in Latin America and Asia are generally more decentralized, with larger numbers of relatively small breeding programmes, whereas in eastern and southern Africa they are generally more centralized, with smaller numbers of larger breeding programmes.

Private seed companies

Following rapid expansion in recent years, the private sector has become a major player in the maize breeding industries of many developing countries. Private seed companies today employ over 400 senior maize breeders worldwide (Table 7.2). Nearly 60% of these breeders are employed by multinational companies, a marked increase from earlier years, when most maize breeding work was still being carried out in national companies. In contrast with the public sector, however, private-sector breeding capacity is not distributed evenly throughout the

[3] Since the China data in Table 7.2 refer only to the five southern provinces of China in which maize is grown in non-temperate production zones, they do not include an additional 1500 Chinese breeders working in central and northern China. When these additional breeders are included, two out of every three maize breeders in the developing world are Chinese.

Table 7.1. Public-sector maize research investment indicators, developing countries, late 1990s.

	Number of countries surveyed	Public maize breeding programmes	Maize scientists (FTEs)[a]	Maize scientists per programme	Maize scientists per million ha maize area	Maize scientists per million t maize production
Latin America	18	49	290	5.9	10.2	4.2
Eastern and southern Africa	12	4	109	27.3	7.6	4.1
East, South and Southeast Asia	7	116	505	4.4	26.3	11.0
All regions	37	169	904	5.3	14.6	6.4

[a]FTEs = full-time equivalents.
Source: CIMMYT maize research impacts survey.

Table 7.2. Private-sector maize research investment indicators, developing countries, late 1990s.

	Number of countries surveyed	Private seed companies with breeding programmes		Private-sector maize researchers		Maize scientists per million ha maize area	Maize scientists per million t maize production
		National	Multinational	National	Multinational		
Latin America	18	65	27	101	109	7.4	3.1
Eastern and southern Africa	12	10	2	10	35	3.1	1.7
East, South and Southeast Asia	7	23	21	70	96	8.7	3.6
All regions	37	99	51	174	240	6.7	3.0

Source: CIMMYT maize research impacts survey.

developing world. Latin America and Asia (with the exception of China) support a large number of private seed companies, reflecting not only the presence in those regions of important commercial maize sectors, but also a friendlier business climate. Private seed companies are much less common in eastern and southern Africa, reflecting the relative scarcity in this region of commercial maize sectors, as well as generally more challenging business environments.

Products of Maize Breeding Research

The principal output of maize breeding programmes is improved cultivars, so varietal releases represent one obvious productivity measure. CIMMYT maintains two varietal releases databases – one for varieties developed by public breeding programmes, and one for varieties developed by private seed companies. The temporal coverage of these two databases differs. The public-sector varietal releases database contains information about approximately 1250 varieties and hybrids released since the mid-1950s by public breeding programmes in 37 developing countries.[4] The private-sector varietal releases database contains information about approximately 1025 varieties being sold by private seed companies during the late 1990s in the same 37 countries. Since many private companies do not maintain records going back into the distant past, with the private sector it was not possible to compile a complete list of all varieties developed since 1966, the year in which CIMMYT was established. Private seed companies therefore were asked to provide information only about varieties they were currently selling. In most instances, these consisted of relatively recent hybrids developed during the 1990s.

Public-sector releases

Public maize breeding programmes have been very productive, developing and releasing a steady stream of varieties (Fig. 7.2). On aggregate, the rate at which varieties are released has grown steadily over time and shows no sign of slowing down. Taking varietal releases as a crude measure of research output, this suggests that public maize breeding programmes have not suffered any significant decline in productivity.

[4] Here the discussion relates only to varieties released since 1966, the year in which CIMMYT was officially established.

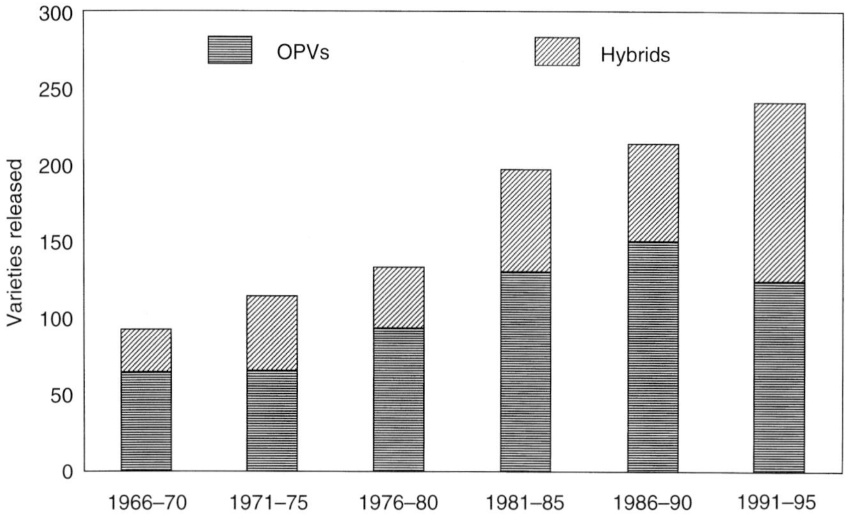

Fig. 7.2. Public-sector maize varietal releases, 1966–1995. (Source: CIMMYT global maize impacts survey.)

Since 1966, public maize breeding programmes in developing countries have developed and released nearly twice as many open-pollinated varieties (OPVs) as hybrids, reflecting the traditional emphasis in the public sector on breeding open-pollinating materials (Fig. 7.2). However, the ratio of OPVs to hybrids has changed over time in response to changes in the prevailing philosophy about the suitability of hybrids for small-scale farmers. The proportion of hybrids among public-sector releases increased sharply during the 1990s, and during the most recent period (1996–1998) hybrids actually outnumbered OPVs by a slight margin.

To what extent have public maize breeding programmes in developing countries made use of CIMMYT germplasm? This question is not easy to answer. Use of CIMMYT germplasm is challenging to track for at least three reasons:

1. Defining 'CIMMYT germplasm' is often very difficult. Modern maize breeding is truly international, and today most breeders routinely work with source materials obtained from all over the world. Screening and evaluation require a great deal of teamwork, since materials must be evaluated in multiple locations. In this context, it is not always clear how credit for the breeding effort should be attributed, so the definition of 'CIMMYT germplasm' becomes very blurred.
2. Breeders who use CIMMYT source materials themselves may not know exactly how much CIMMYT germplasm is actually present in a

finished cultivar. Modern maize improvement methods typically involve repeated cycles of selfing, crossing and backcrossing. Selection strategies vary widely and frequently change. Because of the complex and frequently *ad hoc* nature of the breeding process, the precise genetic composition of finished varieties cannot be known with certainty. Even if the source materials can be identified, their relative contribution may be unknown.

3. Even when breeders know how much CIMMYT germplasm is present in a finished variety, they may not be willing to reveal this information. Most commercial maize varieties now have closed pedigrees, meaning that information about their genetic background is not publicly available. Breeding programmes, especially commercial programmes that respond to economic incentives, have an interest in keeping pedigrees closed, because once the genetic background of a variety becomes public knowledge, other breeders will be able to copy the variety. In the past, public breeding programmes were rarely concerned with earning profits from sales of their germplasm products, so they were usually willing to provide pedigree information. More recently, the situation has changed. With the strengthening of intellectual property rights, many public breeding programmes have adopted closed-pedigree policies.

Despite these complicating factors, an effort was made to document the use of CIMMYT germplasm. Survey respondents were asked whether the varieties developed by their breeding programmes had been developed using CIMMYT source materials, defined as materials that had been improved by the CIMMYT maize programme. Non-CIMMYT materials that might have been obtained from the CIMMYT germplasm bank but that had not been improved by CIMMYT breeders (e.g. landraces and improved materials developed by other breeding programmes) were not considered.

Use of CIMMYT germplasm by public breeding programmes has been extensive (Fig. 7.3). Of all publicly bred maize varieties released from 1966 to 1998, over one-half (52%) contained CIMMYT germplasm. Excluding varieties adapted for temperate environments (which are not targeted by the CIMMYT maize programme), the proportion containing CIMMYT germplasm was even higher (54%). The use of CIMMYT germplasm by public breeding programmes has increased over time. During the most recent period, 64% of all public-sector varietal releases contained CIMMYT germplasm (73% when temperate materials are excluded). Belying predictions that CIMMYT's role would decline as national programmes gained in strength, the CIMMYT maize programme continues to represent an important source of breeding materials for public breeding programmes.

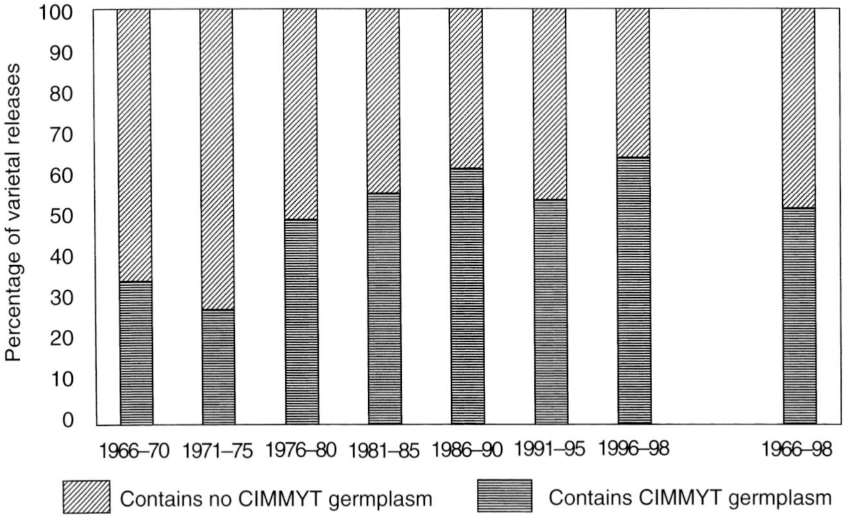

Fig. 7.3. Use of CIMMYT germplasm by public breeding programmes. (Source: CIMMYT global maize impacts study.)

Private-sector releases

Since the private-sector varietal releases database contains only information about varieties being sold during the late 1990s, it cannot be used to draw conclusions about the past productivity of private breeding programmes. But while the historical coverage may be incomplete, the use of CIMMYT germplasm by private breeding programmes has clearly been substantial. Of all private-sector maize varieties being sold during the late 1990s, 60% contained CIMMYT germplasm. The proportion varied greatly by region, however. In Latin America, 73% of all private-sector varieties contained CIMMYT germplasm (and fully 89% of all varieties adapted to non-temperate production environments). In other regions, the use of CIMMYT germplasm by private companies was much more modest. In eastern and southern Africa, 9% of the varieties developed by private breeding programmes contained CIMMYT germplasm, and in Asia, 19% of the varieties developed by private breeding programmes contained CIMMYT germplasm.

As expected, private breeding programmes have focused almost exclusively on developing hybrids. Fully 98% of all proprietary materials sold during the late 1990s were hybrids.

Diffusion of Improved Germplasm

The data on varietal releases attest to the productivity of maize breeding programmes in developing countries and show that breeders both in the public and private sectors have made extensive use of CIMMYT germplasm. What these data do not reveal, however, is the extent to which farmers have taken up MVs. For that, it is necessary to examine varietal adoption and diffusion patterns. Because of the difficulties inherent in estimating the adoption of improved germplasm, we present two types of data that relate to the uptake of MVs. First, we present information about commercial seed sales. Although seed sales do not provide a direct measure of the area planted to MVs, seed sales data nevertheless provide important information about the strength of the demand for MVs. Next we turn to direct estimates of the area planted to MVs.

Sales of Commercial Maize Seed

Table 7.3 shows sales of commercial maize seed for 1996–1997 reported by the public seed agencies and private companies that participated in the CIMMYT survey.[5] The seed sales data are noteworthy in four respects. First, maize seed is big business in the developing world; sales for the industry as a whole exceeded half a million tonnes in 1996–1997. Second, the size of the commercial maize seed industry varies tremendously between regions. Latin America represents by far the largest regional market, followed by East, South and Southeast Asia, with eastern and southern Africa trailing behind. Third, with the significant exception of China, the maize seed industry has effectively been privatized; at the global level, private seed companies outsell public seed agencies by more than 2:1 (this ratio increases to nearly 10:1 when China is excluded). Fourth, the market for maize seed is dominated by hybrids; in all three regions, sales of OPV seed account for less than 10% of the total market share.

Of all maize seed sold in 1996–1997, about one-fifth (21%) was seed of varieties developed and released by public breeding programmes, and about four-fifths (79%) was seed of varieties developed and released by private breeding programmes. Privately bred varieties were highly favoured in Latin America (accounting for nearly 97% of all seed sales within the region) and in eastern and southern Africa (accounting for nearly 93% of all seed sales within the region). Use of public- and private-sector varieties was more evenly balanced in Asia, although variability within the region was great; most of the seed sold in China (also parts of India) was seed of public varieties, while most of the seed sold in other countries was seed of private varieties.

[5] Consistent with the rest of this report, the data for China include only the five southern provinces in which maize is grown in non-temperate production environments.

Table 7.3. Commercial maize seed sales, by type of seed and seed organization, 1996/97.

	Public sector			Private sector			Total		
	OPVs	Hybrids	Total	OPVs	Hybrids	Total	OPVs	Hybrids	Total
Latin America	4,700	4,500	9,200	14,400	280,700	295,100	19,100	285,200	304,300
Eastern and southern Africa	1,300	1,800	3,100	1,800	37,400	39,200	3,100	39,200	42,300
East, South and Southeast Asia[a]	1,700	94,000	96,200	3,200	67,800	71,000	4,900	162,300	167,200
All regions	7,700	100,300	108,500	19,400	385,900	405,300	27,100	486,700	513,800

[a]Southern China only.
Source: CIMMYT maize research impacts survey.

The seed sales data provide direct evidence that CIMMYT germplasm is being used extensively. Of all the commercial maize seed sold during 1996–1997 in developing countries (not including northern China), approximately 60% consisted of varieties developed using CIMMYT germplasm. Focusing more directly on environments targeted by the CIMMYT maize programme, of all commercial maize seed sold during 1996–1997 in non-temperate areas (i.e. excluding northern China, Argentina and South Africa), 71% was seed of varieties developed using CIMMYT germplasm. The global totals conceal considerable variability at the regional level, however. Sales of CIMMYT-derived varieties were extensive in the enormous Latin American market (where 76% of all seed sold during 1996–1997 consisted of varieties developed using CIMMYT germplasm), but they were more modest in the smaller markets of eastern and southern Africa (21% of all seed containing CIMMYT germplasm) and in Asia (16% of all seed containing CIMMYT germplasm).

Adoption of MVs

How extensive is the area planted in the developing world to maize MVs? Respondents to the CIMMYT survey were asked to provide estimates of the area under three categories of materials: (i) cultivars grown from farm-saved seed (including landraces, farmers' traditional varieties, and older OPVs and hybrids grown from advanced-generation seed that has been recycled more than three times); (ii) newer OPVs grown from commercial seed that has been recycled up to a maximum of three times; and (iii) hybrids grown from newly purchased commercial seed.

Table 7.4 presents estimates of the area under each of the three germplasm categories during the late 1990s.[6] Overall, of the 94.2 million ha planted to maize in the countries covered by the CIMMYT and IITA surveys, approximately 58.8 million ha (62.4%) were planted to MVs. Excluding Argentina and South Africa, where maize is grown mainly in temperate environments, of the 65.7 million ha planted to maize in non-temperate environments, approximately 31 million ha (47.2%) were planted to MVs.

These MV adoption estimates are conservative, since they refer only to area planted with recently purchased commercial seed. In many parts of the developing world, much of the area under farm-saved seed is planted with advanced-generation seed of MVs. Because of natural outcrossing combined with farmer selection pressure, plants grown from advanced-generation seed of MVs may not bear much resemblance to

[6] Table 7.4 includes summary data from the IITA impacts survey.

Table 7.4. Maize area planted to improved OPVs and hybrids, developing countries, late 1990s.[a]

	Total maize area[b] (million ha)	Area planted using farm-saved seed[c] (%)	Area planted using commercial seed		
			OPVs[d] (%)	Hybrids (%)	All MVs (%)
Latin America	27.1	55.1	5.0	39.9	44.9
w/o Argentina	24.5	59.6	5.3	35.1	40.4
Eastern and southern Africa	14.9	47.5	6.9	45.7	52.6
w/o South Africa	10.9	64.1	8.3	27.6	35.9
Western and Central Africa	8.2	63.2[e]	33.1[f]	3.7[f]	36.8[e]
East, South and Southeast Asia[g]	42.3	17.6	12.3	69.6	82.4
w/o China	20.5	35.3	22.1	42.6	64.7
All regions	94.2	37.6	11.5	51.0	62.4
All non-temperate regions	65.7	52.8	14.8	32.4	47.2

[a]Data refer to different years: Latin America 1996; eastern and southern Africa 1997; East, South and Southeast Asia 1997; western and Central Africa 1998.
[b]Includes only countries covered by the CIMMYT and IITA surveys, plus northern China.
[c]Includes landraces, farmers' traditional varieties, and older OPVs and hybrids grown from advanced-generation seed recycled more than three times.
[d]Includes area grown from commercial OPV seed that has been recycled up to a maximum of three times.
[e]Based on results of IITA impacts study.
[f]Estimated based on results of 1992 CIMMYT impacts study.
Source: CIMMYT global maize impacts survey.

plants grown from newly purchased commercial seed, but they still contain some improved germplasm. In particular, they may retain certain highly desirable traits. The MV adoption estimates in Table 7.4 fail to account for the improved germplasm found in farm-saved seed, so they understate the true impact of international breeding efforts.[7]

Adoption of MVs containing CIMMYT germplasm

The seed sales data can be combined with the MV adoption data to derive estimates of the area planted to MVs developed using CIMMYT germplasm (Table 7.5). In 1996/97, of the 58.8 million ha planted to

[7] When MV adoption data for maize are compared with MV adoption data for other crops, it should be remembered that the data for other crops typically *include* the area planted using advanced-generation farm-saved seed. Thus when it comes to defining what constitutes an MV, maize is held to a stricter standard.

Table 7.5. Maize area planted to MVs developed using CIMMYT germplasm, developing countries, late 1990s.[a]

	Maize area[b] (million ha)	Maize area under MVs (%)	Maize area under MVs ('000 ha)	Seed with CIMMYT germplasm (%)	Maize area under MVs with CIMMYT germplasm ('000 ha)
Latin America	27.1	44.9	12,171	80.9	9,842
w/o Argentina	24.5	40.4	9,899	92.8	9,183
Eastern and southern Africa	14.9	52.6	7,834	20.8	1,630
w/o South Africa	10.9	35.9	3,910	36.7	1,433
Western and central Africa	8.2	36.8[c]	3,013	67.0[d]	2,019
East, South and Southeast Asia	42.3	82.4	34,851	20.7	7,222
w/o China	20.5	64.7	13,244	38.2	5,062
All regions	94.2	62.4	58,805	36.1	21,210
All non-temperate regions	65.7	47.2	31,001	58.7	18,195

[a]Data refer to different years: Latin America 1996; eastern and southern Africa 1997; East, South and Southeast Asia 1997; western and Central Africa 1998.
[b]Includes only countries covered by the CIMMYT and IITA surveys.
[c]Based on results of IITA Impacts Study.
[d]Estimated based on results of 1992 CIMMYT Impacts Study
Source: CIMMYT global maize impacts survey.

MVs in the countries covered by the CIMMYT and IITA surveys, about 21.2 million ha (36.1%) were planted to varieties that were developed using CIMMYT germplasm. Restricting the focus to non-temperate environments targeted by the CIMMYT maize programme, of the 31.0 million ha planted to MVs in these environments, about 18.2 million ha (58.7%) were planted to varieties that were developed using CIMMYT germplasm.

Economic Benefits of CIMMYT's Maize Breeding Programme

What economic benefits have been generated by CIMMYT's maize breeding programme? The benefits attributable to a crop-breeding programme are typically estimated as the additional production value that results from adoption of MVs developed by the breeding programme.[8] Three key parameters are needed to calculate this value: (i) the area planted to MVs developed by the breeding programme; (ii) the productivity gains associated with adoption of these MVs; and (iii) the price of the crop. For simplicity, productivity gains are often expressed in terms of yield per unit land area, even though yield is not always the most appropriate measure.[9]

Estimates of the area planted to CIMMYT-derived MVs have been presented earlier in this chapter. Representative maize prices for each country can be calculated based on international reference prices and transport cost data. Thus what remains is to estimate the yield gains associated with the adoption of CIMMYT-derived MVs. In theory, yield gains associated with adoption of an MV can be expressed as the difference between the yield obtained with the farmer's current variety (which may be a landrace, a local variety or an older MV) and the yield obtained with the MV, holding all other inputs constant. In practice, this difference is often difficult to measure, for at least three reasons (see Chapter 3 for further discussion of this issue).

[8] Additional benefits accrue to food and feed processors (who experience increased demand for their services), agricultural labourers (who face increased employment opportunities), and other groups that are affected via price- or income-transmitted multiplier effects. These types of 'indirect' benefits are difficult to measure and therefore have not been not considered here.

[9] The benefits of improved germplasm may also be reflected in earlier maturity, enhanced grain quality, improved quantity or quality of fodder, or better tolerance of biotic or abiotic stresses (which may allow the crop to be grown in places or at times where it could not be grown before, or with fewer inputs even if the grain yield remains unchanged).

1. Yield gains vary depending on the type of MV adoption that is taking place. For example, yield gains will differ greatly depending on whether the farmer is replacing a landrace with an improved OPV, a landrace with a hybrid, an older improved OPV with a newer improved OPV, an improved OPV with a hybrid, or an older hybrid with a newer hybrid.
2. Yield gains vary depending on environmental factors, including agroclimatic conditions and farmer management practices. The same MV may deliver significant productivity gains in favourable production conditions and modest productivity gains in unfavourable production conditions.
3. Yield gains achieved in farmers' fields come not only from adoption of MVs; yield gains come also from adoption of improved crop management practices, which frequently interact with MVs. In estimating the economic benefits attributable to plant breeding research, it is therefore necessary to distinguish between the 'germplasm effect' on yields and the 'crop management effect'. Relatively little empirical research has been done on this topic. One reasonable approach is to assume that improved germplasm and improved management practices have each contributed about 50% to observed yield gains in cereal crops (Bell *et al.*, 1995; Thirtle, 1995; Fuglie *et al.*, 1996).

Estimating yield gains becomes even more complicated when MV adoption is considered in a dynamic context. Thus far we have been discussing discrete, one-off yield gains realized when one variety is replaced by another variety. Over time, this process is normally repeated as farmers continually replace older varieties with newer varieties, resulting in a series of discrete yield gains that cumulatively can be expressed as an average rate of yield gains through time. Measuring yield gains in a dynamic context is further confounded by three additional factors:

1. The yield advantage associated with MVs often erodes over time as a result of seed recycling, especially with cross-pollinating crops such as maize.
2. Where regular replacement of MVs is taking place, yield gains over time will depend on the rate of varietal replacement.
3. In the absence of the breeding programme being evaluated, technical change would still take place. For example, even if the CIMMYT maize breeding programme had not existed, farmers in developing countries would still have had access to MVs developed by NARS and/or private seed companies. Therefore the benefits attributable to CIMMYT's maize breeding programme should not be estimated relative to the situation that existed when CIMMYT came into existence; rather, they should be estimated relative to what would have happened in the absence of CIMMYT. Thus it is necessary to develop a counterfactual scenario.

Numerous case studies have confirmed that the size of the yield gains associated with initial adoption of maize MVs is highly variable, as is the average rate of yield gains achieved over time as farmers subsequently replace older MVs with newer MVs (for a summary of the evidence, see Morris *et al.*, 1999). Because maize-based cropping systems found in developing countries are so diverse, this variability is considerably greater than for most other major cereals. In subsistence-orientated cropping systems, where farmers are adopting MVs along with improved crop management practices for the first time, yield increases of 100% or more are common. In modern commercial production systems, in which farmers are regularly replacing high-yielding single-cross hybrids that are grown with high levels of management, yield increases of 5–15% are more typical.

In the absence of detailed, location-specific information about the many parameters needed to estimate yield gains, it is difficult to calculate precisely the economic benefits attributable to CIMMYT's maize breeding programme. For illustrative purposes, however, it may be useful to establish lower and upper bounds on these benefits and to demonstrate how they are likely to be affected by changes in key parameters.

Table 7.6 presents a range of benefits attributable to CIMMYT's maize breeding programme under different assumptions. Based on the results of the global survey, the area planted to CIMMYT-derived MVs is assumed to be 21.2 million ha.[10] The average farm-level price of maize is conservatively estimated at US$120 t^{-1} (import parity basis). Depending on the yield gain parameter (columns 1 and 2), gross benefits associated with adoption of CIMMYT-derived MVs (germplasm effect plus crop management effect) range from US$1.3 billion to US$4.0 billion per year (column 3). Assuming that 50% of the yield gain is attributable to the germplasm effect and 50% to the crop management effect, then net benefits associated with adoption of CIMMYT-derived MVs (germplasm effect only) range from US$668 million to US$2.0 billion per year (column 4).

The net benefits estimates shown in Table 7.6, column 4, overstate the impacts of CIMMYT's maize breeding programme because they include the breeding contribution made by other research organizations. To isolate the benefits generated by CIMMYT's breeding programme, it is additionally necessary to estimate the proportion of the germplasm effect associated with adoption of CIMMYT-derived MVs that can be credited directly to their CIMMYT germplasm content.

[10] Including an estimated 2 million ha planted to CIMMYT-derived MVs in western and central Africa.

Table 7.6. Added production value attributable to CIMMYT's maize breeding programme (US$ million year^{-1}).

Yield gain attributable to adoption of CIMMYT-derived maize MVs (germplasm + crop management effects) (%)	Yield gain (t ha^{-1})	Gross benefits from adoption of CIMMYT-derived maize MVs (US$ million year^{-1})	Net benefits attributable to germplasm effect of MV adoption (US$ million year^{-1})	Contribution of CIMMYT germplasm 25% – Net benefits generated by CIMMYT maize breeding programme	50%	75%
15	0.525	1336	668	167	334	501
25	0.875	2227	1114	278	557	835
35	1.225	3118	1559	390	770	1169
45	1.575	4009	2004	501	1002	1503

Assumptions:
Area planted to maize CIMMYT-derived MVs in developing countries: 15.556 million ha.
Average yield of MVs 3.5 t ha^{-1} (implies average yield of non-MVs: 1.2 t ha^{-1}).
Proportion of yield gain attributable to 'germplasm effect': 50%.
Average price of maize grain: US$120 t^{-1}.
Source: Calculated by the authors.

In the absence of detailed information about the breeding history of maize MVs, it is not possible to formulate pedigree-based rules for assigning credit among different research organizations. Therefore, benefits were calculated using a range of plausible values for the parameter that denotes the contribution of CIMMYT materials (these values are shown at the top of Table 7.6, columns 5–7). Under the most conservative assumption (25% of the germplasm effect attributable to CIMMYT), and depending on the average yield gain, CIMMYT's maize breeding programme generates from US$167 million to US$501 million per year in benefits in developing countries. Under the most generous assumption (75% of the germplasm effect attributable to CIMMYT), and once again depending on the average yield gain, CIMMYT's maize breeding programme generates from US$501 million to US$1.5 billion per year in benefits. Averaging out different adoption scenarios around the world, the value of benefits generated by the CIMMYT maize programme probably falls somewhere around the middle of this range, i.e. from US$557 million to US$770 million per year.

These estimated benefits figures are based on the following assumptions:

- the genetic composition of maize cultivars changes rapidly in farmers' fields through natural outcrossing, so crops grown using advanced-generation farm-saved seed that has been recycled many times can no longer be characterized as MVs
- not all maize MVs contain CIMMYT germplasm, so only part of the total area planted to maize MVs can be considered planted to CIMMYT-derived MVs
- farm-level yield gains are attributable partly to the effect of improved germplasm and partly to the effect of changes in farmers' management practices (often including increased use of fertilizer)
- many CIMMYT-derived MVs also contain improved germplasm developed by NARS and/or private seed companies, so the yield gain associated with adoption of these MVs cannot be attributed entirely to the CIMMYT germplasm.

Although the figures in Table 7.6 are admittedly somewhat speculative, they point toward an important conclusion: even under conservative assumptions, the CIMMYT maize breeding programme pays for itself many times over. One factor contributing to this result is simply the global importance of maize. Considering the vast area that is planted to maize, CIMMYT-derived varieties do not have to achieve complete dominance in order to generate attractive returns to the CIMMYT breeding effort; current adoption rates already translate into enormous economic benefits.

Future of International Maize Breeding

International maize breeding efforts have generated enormous benefits. Modern varieties currently cover nearly two-thirds of the area planted to maize in developing countries, bringing increased incomes to millions of maize-producing households and lower food prices for even greater numbers of maize consumers. The widespread diffusion of maize MVs is particularly impressive given the distinctive characteristics of maize compared with other leading cereals. Because maize is an open-pollinated crop, farm-saved seed quickly loses its genetic purity, so farmers who wish to grow MVs must replace their seed regularly. For this reason, MVs of maize can disseminate only in the presence of an efficient seed industry – something that is still lacking in most developing countries.

Against a backdrop of declining public support for maize research, international agricultural research centres continue to play a vital facilitating role in support of international breeding efforts. By serving as the hub of a global network of germplasm improvement and exchange, CIMMYT has been particularly effective in promoting international flows of improved germplasm, as evidenced by the widespread use of CIMMYT materials in both public and private breeding programmes. Thanks to strong links with local breeding programmes in both the public and private sectors, the CIMMYT maize programme has achieved enormous payoffs from a very modest investment. The economic benefits generated by CIMMYT's maize breeding effort (additional production value) are estimated to range between US$167 million and US$1.5 billion per year.

In recent years, policy reforms have paved the way for greater participation by the private sector in the maize seed industries of many developing countries. Increased privatization has brought generally positive results, but at the same time there are grounds for concern. The accelerating cost of crop genetic improvement research, coupled with the growing importance of intellectual property rights, is rapidly changing the rules of the plant breeding game. Fearful of conceding advantage to potential competitors, many of the large corporations that currently dominate the global maize seed industry are becoming less enthusiastic about sharing information, technology and germplasm. As a result, maize breeding is rapidly being transformed from a collaborative activity undertaken for the common good into a competitive activity undertaken for shareholder profit. Since most public breeding programmes depend heavily on the free exchange of germplasm and information, this trend raises questions about the future role of the IARCs and of the international breeding system in general.

The privatization of national maize seed industries also raises questions about the distributional impacts of technical change. In spite of all the progress that has been made in disseminating the products of modern plant breeding programmes, considerable challenges remain to be overcome if maize MVs are to reach the poorest of the poor. Over one-third of the developing world's maize area (nearly half of the maize area in non-temperate environments) is still planted to farm-saved seed of uncertain genetic background and variable quality. In many instances, farmers continue to use farm-saved seed not because improved germplasm is unavailable; rather, the problem is that small-scale farmers located in isolated rural areas are not attractive customers for profit-orientated commercial seed producers. As the IARCs and their national programme partners reposition themselves in the rapidly evolving global seed industry, they will be challenged to come up with creative approaches to reaching the millions of small-scale farmers who still do not enjoy full access to the fruits of international breeding efforts.

References

Bell, M.A., Fischer, R.A., Byerlee, D. and Sayre, K. (1995) Genetic and agronomic contributions to yield gains: a case study for wheat. *Field Crops Research* 44, 55–65.

Fuglie, K., Ballenger, N., Day, K., Klotz, C., Ollinger, M., Reilly, J., Vasavada, U. and Yee, J. (1996) *Agricultural Research and Development: Public and Private Investments Under Alternative Markets and Institutions.* Agricultural Economic Report No. 735. Economic Research Service, US Department of Agriculture, Washington, DC.

Gerpacio, R. (ed.) (2001) *Impacts of International Maize Breeding Research in Asia, 1966–1998.* CIMMYT, Mexico.

Hassan, R.M., Mekuria, M. and Mwangi, W.M. (2001) *Maize Breeding Research in Eastern and Southern Africa: Current Status and Impacts of Past Investments by the Public and Private Sectors, 1966–1997.* CIMMYT, Mexico.

Morris, M.L. (ed.) (1998) *Maize Seed Industries in Developing Countries.* Lynne Rienner, Boulder, CO, and CIMMYT, Mexico.

Morris, M.L. and López-Pereira, M.A. (1999) *Impacts of Maize Breeding Research in Latin America, 1966–1997.* CIMMYT, Mexico.

Morris, M.L., Risopoulos, J. and Beck, D. (1999) *Genetic Change in Farmer-recycled Maize Seed: A Review of the Evidence.* CIMMYT Economics Program Working Paper 99-07. CIMMYT, Mexico.

Thirtle, C.G. (1995) Technological change and the productivity slowdown in field crops: United States, 1939–78. *Southern Journal of Agricultural Economics* 17 (December), 33–42.

Impact of IITA Germplasm Improvement on Maize Production in West and Central Africa

8

V.M. Manyong, J.G. Kling, K.O. Makinde, S.O. Ajala and A. Menkir

Maize is one of the major cereals in West and Central Africa (WCA). Maize accounts for a little over 20% of domestic food production in Africa, a proportion that has increased over time as maize has replaced other food staples, particularly sorghum and millet, in West Africa (Smith *et al.*, 1994). Maize has also increasingly become a major source of cash for smallholders (Smith *et al.*, 1997). Trends in maize production indicate a steady yearly growth. For example, the annual growth rate of production of maize was 4.07% in West and 2.35% in Central Africa. This was achieved to a large extent through expansion of the area planted to maize (calculated from FAOSTAT, 2000). However, maize yields also increased. In the years 1989–1991 the average maize yield of 1.2 t ha^{-1} in Africa was twice the maize yield estimated for the 1950s, before improved technology became more generally available (Byerlee and Heisey, 1997). Widespread adoption of improved maize varieties contributed to remarkable changes in the farming systems of the Guinea savannahs and semi-arid zone of WCA within less than 20 years. In the savannahs, maize is no longer a backyard crop but a major cereal crop grown for both cash and food (Eckebil, 1994; Fajemisin, 1994; Smith *et al.*, 1997).

The International Institute of Tropical Agriculture (IITA) has a regional mandate for maize research in WCA. IITA works in partnership with national agricultural research systems (NARS) to develop and disseminate improved maize technologies that meet the requirements of the major clients, the small-scale farmers.

Maize Improvement at IITA: Past and Present

Early efforts were directed to the development of high-yielding, open-pollinated maize varieties with resistance to the prevailing major diseases in the humid forest and moist savannahs of WCA, such as maize streak virus (MSV), blight, rust and leaf spot.

Significant efforts were also made in the breeding of early- and extra-early-maturing improved varieties identified from regional trials. These varieties enabled maize production to expand into the Sudan savannahs, where the short rainy season had hitherto precluded maize cultivation, and allowed double-cropping systems in areas with a long rainy season.

A hybrid maize programme launched in 1979 led to the development and release of first generation inbred lines and hybrids for WCA in 1983. The spillover effect of this release was the formation of seed companies to market hybrid maize. In 1993, each of the three seed companies operating in Nigeria (Premier, UTC and UAC) officially included IITA's open-pollinated and hybrid maize varieties in their seed catalogues.

Downy mildew (DM) on maize, caused by the fungal pathogen *Peronosclerospora sorghi* (Weston and Uppal, C.G. Shaw) was widely reported in Africa in the 1970s and has become a serious threat to maize production in parts of Nigeria, Zaire (now Democratic Republic of Congo), Mozambique and Uganda (Kling *et al.*, 1994). Development of DM-resistant varieties has been a major priority in the breeding programme at IITA since the early 1980s. Working with NARS, a number of widely adopted DM-resistant varieties have been developed and released by IITA.

Striga infests almost 21 million ha of land in Africa. Combating *Striga* has been one of the focal issues for the IITA maize research team since the mid-1980s. Major achievements have been made in breeding both for tolerance to *Striga* and for reduced emergence of the parasite. Efforts are now under way to identify the mechanisms of resistance in these new varieties and inbred lines. The Rockefeller Foundation has provided funds for a collaborative project between IITA, International Maize and Wheat Improvement Centre (CIMMYT) and the Kenyan national programme to map the genes for resistance to *Striga* in IITA germplasm. The research is ongoing and scientists expect to have a map of one population ready soon. This should enable them to initiate marker-assisted backcrossing of the resistance genes into diverse, adapted populations.

On-farm trials of *Striga*-resistant maize varieties and other methods for integrated control are currently being conducted by NARS scientists in many countries of WCA with support from the West and Central

African Maize Network (WECAMAN), the African Maize Stress project (funded by the United Nations Development Programme (UNDP) and the International Fund for Agricultural Development (IFAD)), and the Korean-funded OAU/SAFGRAD project. The main objective is to promote the adoption of *Striga*-resistant varieties in rotation with selected legume cultivars that are efficient in causing suicidal germination of *Striga hermonthica*.

High-yielding, improved varieties are sometimes not adopted by farmers because they may be susceptible to storage weevils or may lack the quality desired by end-users for local food preparations. IITA has worked closely with scientists in Benin Republic to understand the mechanisms of weevil resistance and to develop a variety acceptable to farmers and consumers. Recently, breeding efforts have been initiated to enhance the micronutrient content of maize varieties, in an effort to combat the widespread diseases of iron deficiency anaemia and corneal blindness caused by vitamin A deficiency. Another major human health concern is the problem of aflatoxin contamination in stored maize. IITA is presently engaged in a project funded by Rotary Club International and GTZ (German technical Cooperation) to use integrated approaches for reducing the deleterious effects of aflatoxin. A collaborative project has also been initiated with the USDA to use molecular techniques for the transfer of aflatoxin resistance into adapted maize varieties.

IITA works with WECAMAN to coordinate international trials of improved maize germplasm. In addition, several tonnes of breeders' seed are distributed each year in bulk samples of 1–5 kg for multiplication by NARS. IITA scientists are actively involved in regional networks and in a number of training activities intended to enhance the capacity of NARS scientists to develop their own technologies and to promote adoption of those technologies by farmers. Another important activity supported by WECAMAN is the promotion of community seed production schemes to meet farmers' demands for high quality seed.

In addition to its breeding programme, IITA maintains in its gene bank about 500 local accessions of maize collected from different countries in WCA. These accessions have been extensively used to develop drought-tolerant and *Striga*-resistant maize populations.

Methodology

The study had a target of 12 countries: nine in West Africa and three in Central Africa. The participating countries were selected because they represented over 95% of the area cultivated to maize in WCA. However, only 11 countries were effectively covered (see Appendix 8.1).

A survey questionnaire was mailed to the leaders of national breeding programmes on maize. Information requested included a list of maize varieties released in the country during the period 1970–1998, characteristics of the released varieties, parent materials, area planted with different varieties in 1998 or the preceding year, organizations involved in the development and distribution of varieties, the contribution of IITA to human capital development, NARS investments in maize research, input and output prices, and planting material regulations in the study countries. Descriptive statistics such as means and frequency of events were used to analyse the data. Two models were used to estimate the benefits.

Gross economic benefit (GEB) was calculated to broadly quantify the economic impact of the international breeding programme on maize. The formula for the calculation of GEB is as follows:

$$\mathrm{GEB} = \sum_{j=1}^{m} \sum_{i=1}^{n} \left(A_j * I_{ij} * y_{ij} * p_{ij}\right) \tag{1}$$

with

$$p_{ij} = P_{ij} / E$$
$$y_{ij} = Y_{ij} - L_{ij}$$

where GEB = gross economic benefit, i = maize variety, j = country, A = total area planted to maize (ha), I = percentage of area cultivated to i, y = yield advantage (kg ha^{-1}), p = farm gate price in US dollars in a major maize-growing area of the country, Y = average farm yield for an improved variety (kg ha^{-1}), L = average farm yield for a local variety (kg ha^{-1}), P = farm gate price in a local currency in the major maize-growing area of the country, and E = exchange rate of local currency to US$1.

The second model was further developed from the GEB formula to measure the impact on food security as follows:

$$N = \sum_{j=1}^{m} \sum_{i=1}^{n} \left(A_j * I_{ij} * y_{ij} * w_i * e_i\right) / d \tag{2}$$

with

$$d = r * 365$$

where N = number of persons with adequate food supply from the total incremental production due to the use of improved varieties, w = percentage of waste for the maize product i, e = energy content for the maize product i, d = required yearly energy per capita, and r = required daily energy per capita.

The leaders of the national programmes mentioned some difficulties in completing the questionnaire, since detailed statistics were not available (for example, on the area planted to each improved maize variety) and there was a lack of time series data on maize. In such circumstances, the study relied on secondary sources from the literature or 'expert opinions'. In other cases, the information was supplied by only a few countries (or private organizations) and so was incomplete, making the analysis difficult for aspects such as investments in maize research. As a result, average values from a few countries were applied to all the countries. Nevertheless, this research provides valuable information for measuring the impact of IITA's improved maize germplasm on production in West and Central Africa.

Results

Patterns of varieties released

Number of released varieties

From 1965 to 1997, public maize research programmes released a total of 186 maize varieties while the private sector released 81 maize varieties in those countries included in this study (Table 8.1). The public sector in Nigeria released the most varieties, followed by that in Burkina Faso, Senegal and Benin. Mali and Togo released the lowest number of maize varieties (see Appendix 8.1 for details). The number of improved varieties released in West Africa was 75% of the total, while that released in Central Africa was 25%. None of the countries in this second sub-region mentioned the involvement of the private sector in maize research and seed production.

From both the private and public sectors, the bulk of the varieties were released after the 1980s (Table 8.1). This is correlated with the establishment of Consultative Group on International Agricultural Research (CGIAR) centres. For example, agricultural research at IITA started in the late 1970s. The impact of the IITA research became apparent in the 1980s, when many improved varieties derived from IITA materials were released by NARS. Although only four out of 11 coun-

Table 8.1. Number of maize varieties released per decade in West and Central Africa from 1965 to 1997.

Decade	Public maize ($n = 11$)	Private maize ($n = 5$)
1965–79	17	5
1980–89	54	36
1990–97	115	40

Note:
n = number of countries.

tries mentioned the private sector, its role in maize seed production has been increasing since 1980. A large proportion of maize varieties released by the private sector contained materials from the public sector (Fig. 8.1). Therefore, the public sector is playing an essential role in the development of the private seed companies for maize in WCA.

Sources of germplasm for the released varieties

Another major impact of IITA is in the proportion of its germplasm that was incorporated in the released varieties. Three major sources of germplasm have been used in the development of the released varieties: IITA, CIMMYT and landraces. The average proportion of germplasm incorporated in the released varieties appears in Table 8.2. Materials from IITA and CIMMYT have been the major sources of germplasm used in the development of the released maize varieties since the creation of CGIAR. In the 1970s, IITA and CIMMYT supplied nearly 95% of the germplasm. In the 1990s, the percentage of germplasm from these centres is still as high as 60%. While materials from CIMMYT were the main source of germplasm in the maize varieties released in the 1970s,

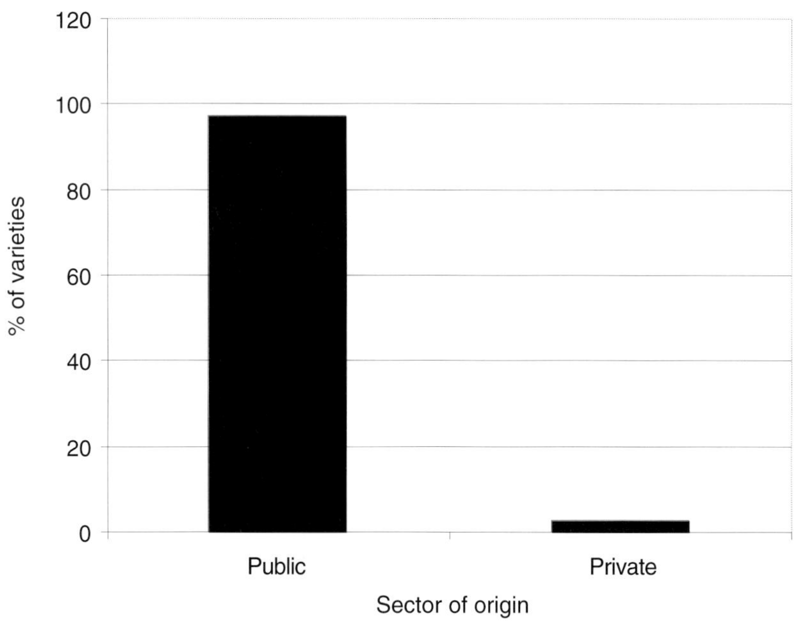

Fig. 8.1. Source of materials used by the private sector in the released maize varieties in West and Central Africa, 1965–1997 ($n = 81$).

Table 8.2. Average percentage of germplasm incorporated per source in the maize varieties in West and Central Africa from 1965 to 1997.

Source	1965–69 (n = 2)	1970–79 (n = 14)	1980–89 (n = 49)	1990–97 (n = 115)
IITA	0.00	15.00	10.94	49.17
CIMMYT	0.00	72.86	48.16	10.92
Landrace	50.00	3.58	12.74	15.74
Others	50.00	8.56	28.16	24.17
Total	100.00	100.00	100.00	100.00

Note:
n = number of varieties released.

IITA became the major source of germplasm for NARS in WCA in the 1990s.[1] The national breeding programmes also made extensive use of local materials in the development of maize varieties. For example, the proportion of germplasm from landraces in the released varieties was about 16% in the 1990s.

Analysis of the trend of the released maize varieties by source germplasm indicated some striking differences (Fig. 8.2). The trend for the release of the IITA-based maize varieties remained positive over the period of this study. The use of landraces and CIMMYT-based varieties shows a negative trend. Therefore, IITA has been the major driver of the steady growth in the number of varieties released over the last 30 years. This trend was expected because IITA has had the mandate for maize research in WCA since 1980. Before that, NARS in the surveyed countries released an average of 0.83 varieties each year. Since IITA has had the leadership of maize research in WCA, the number of releases by NARS has risen to 5.19 varieties per year. Increases in the average number of varieties released every year could be associated with the impact of IITA in West and Central Africa. IITA also had an impact on the efficiency with which NARS developed new varieties from the landraces. The number of landrace-based varieties released every year before 1980 averaged 0.43. This increased to an average of 1.48 varieties (or an increase of 352%) after IITA took the mandate for WCA.

[1] It can be difficult, particularly in maize, to identify or categorize the sources of germplasm in released varieties. The categories used here and the assignment of varieties into different categories reflects the best judgement of NARS maize scientists. The respective contributions of IITA and CIMMYT, in particular, are hard to identify because of long-standing patterns of cooperation and germplasm exchange.

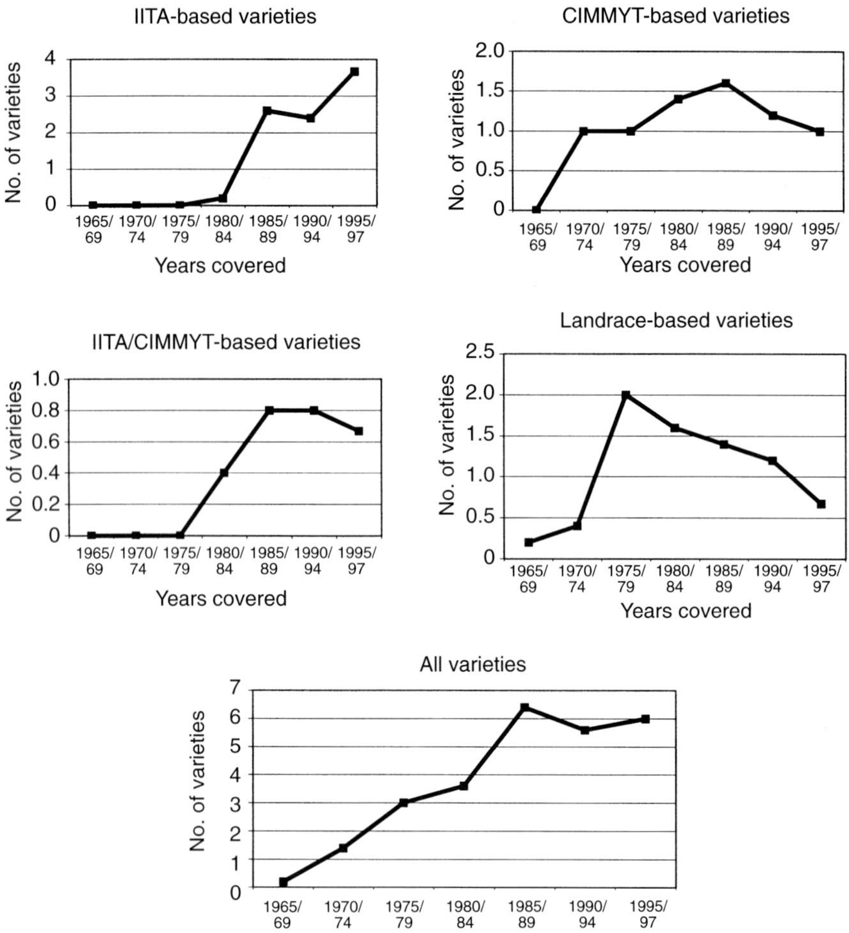

Fig. 8.2. Average number per year of maize varieties released every 5 years by source germplasm in West and Central Africa, 1965–1997.

Patterns of use of germplasm from the CGIAR future harvest centres

There was a significant difference in the use of maize germplasm from the CGIAR centres. While the bulk of IITA germplasm was used directly or with some improvement for local adaptation, the CIMMYT germplasm was used as a source of key traits that required adaptation to local conditions (Table 8.3a, b). The strategy used by IITA to support the weak national breeding programmes by supplying nearly finished products could have contributed to such a high level of direct use of the germplasm emanating from a CGIAR centre. Also, the dominant mate-

Table 8.3. Patterns of use of maize germplasm from IITA and CIMMYT in the varieties released by NARS in West and Central Africa (% of varieties).
(a) IITA

Category	1970–79 (*n* = 2)	1980–89 (*n* = 5)	1990–98 (*n* = 57)	All (*n* = 64)
1	0	20	23	22
2	50	40	25	27
3	50	40	52	51
Total	100	100	100	100

(b) CIMMYT

Category	1970–79 (*n* = 11)	1980–89 (*n* = 24)	1990–98 (*n* = 13)	All (*n* = 48)
1	46	50	70	54
2	27	37	15	29
3	27	13	15	17
Total	100	100	100	100

Notes:
n = number of varieties released.
Categories:
1, Basic germplasm (substantial improvement done after being received from IITA or CIMMYT); 2, selection from IITA or CIMMYT varietal trials, with some improvement for local adaptation; 3, direct use of IITA or CIMMYT material, no additional improvement done except seed multiplication.

rials used in the development of new varieties were from open-pollinated lines (Fig. 8.3) since many poor farmers in WCA cannot afford to buy new seed every cropping season as is the case for hybrid varieties.

Maize varieties released in the sub-region were based on 54 lines from IITA and 23 lines from CIMMYT. The IITA-based germplasm possesses one or more desirable traits, such as resistance to biotic constraints (streak, borer, DM and *Striga*); different maturity groups (Fig. 8.4); grain colour (yellow and white); grain texture (dent and flint), as well as high and stable yields (Table 8.4, see also Appendix 8.2). Much of the past effort focused on the development of resistance to streak, a devastating maize disease in the humid zones, along with white colour to meet end-user requirements for human consumption, and late maturity to exploit the long growing season of the moist savannahs. More efforts in the future should focus on breeding for resistance to stem borers, low soil nitrogen, *Striga* and drought. The last two traits are critical to move maize to a new frontier, namely to the drier savannahs, where *Striga* and drought are important constraints to maize production.

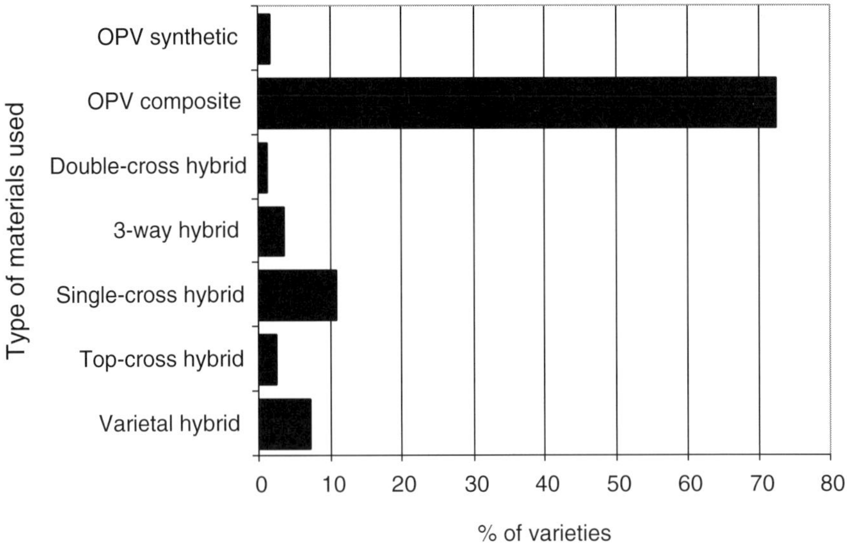

Fig. 8.3. Type of materials used in the maize varieties released by NARS in West and Central Africa 1965–1997 ($n = 164$).

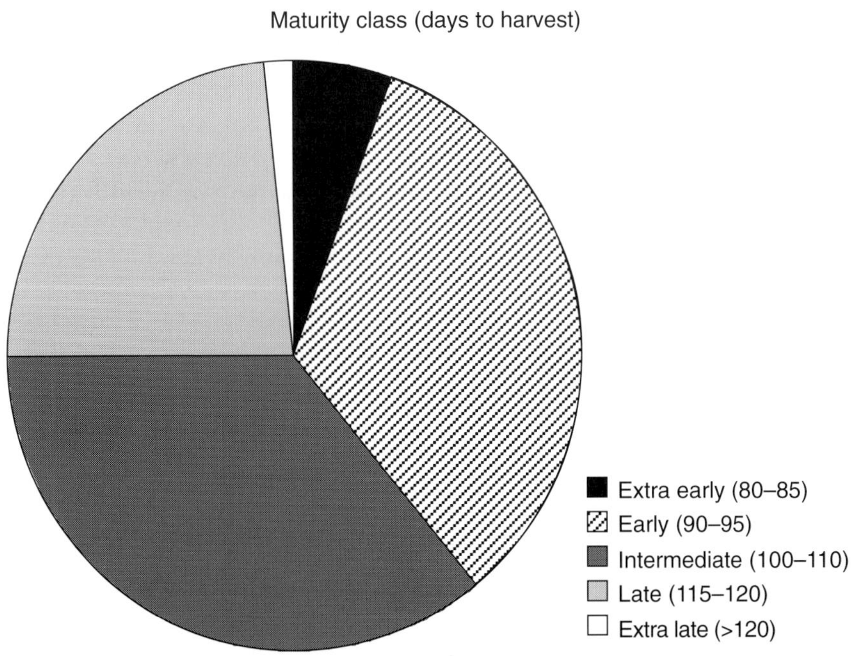

Fig. 8.4. Maturity class for the maize varieties released by the public sector in West and Central Africa, 1965–1997 (% of varieties, $n = 67$).

Table 8.4. Positive attributes of IITA maize germplasm used by national breeding programmes in West and Central Africa, 1965–1997.

	Source germplasm (n = 54)	
Attribute	*N*	%
Increased yield	54	100
Streak resistance	45	83.3
Downy mildew resistance	8	14.81
Borer resistance	2	3.70
Striga resistance	3	5.56
Yellow colour	15	27.78
White colour	35	64.8
Extra-early maturity	2	3.70
Early maturity	14	25.93
Intermediate maturity	2	3.70
Late maturity	38	70.37

Notes:
n, total number of source germplasm from IITA; *N*, number of source germplasm from IITA with the specified positive attribute.

More recently, effort has focused on early- and extra-early-maturing varieties. This reflects the ecological shift in the relative importance of maize research and production in West Africa from the forest and moist savannahs to the dry savannahs (Manyong *et al.*, 1996). Earliness is an escape mechanism against drought in the dry savannahs and also allows for at least two crops to be grown in a relay mode in the short monomodal rainfall pattern of that ecology. This ecological shift is also evident in the large proportion of varieties that are adapted to the northern Guinea and Sudan savannahs (42% for both public and private maize) compared with those for the lowland forests (25% for public and 10% for private).

Regulatory measures for the movement of maize germplasm

There are no consistencies in regulatory measures for the movement of maize germplasm for the development of new varieties across the countries surveyed. All the countries recognized that local testing is required before a variety is released. Usually a public agency performs the test and three countries out of 11 mentioned a period of 3 years for the process of registration/release to be completed. Two-thirds of the countries declared that they must certify maize seed in order for it to be sold commercially and a public agency is in charge of this task in 89% of the countries. Maize seed can be sold commercially before it has been officially released/registered or without 'truthful labelling' in 33% of the countries. Currently there are laws/legislation in progress for plant variety protection in about half of the countries (56%). Those laws apply to maize in 86% of those countries. There are no laws for plant patent in 67% of the countries. Farmers can legally multiply seed for their own use (all countries). They can also multiply seed for sale to others in 90% of the countries. Breeders from public research organizations do not receive any royalty from private companies for the use of public open-pollinated varieties and hybrids. Though it was shown earlier in this chapter that 97% of varieties sold by the private sector are from the public sector. Currently, only half of countries do authorize import of commercial maize seed. There is an import duty (tax) to pay; however, most of the respondents were not aware of the percentage of value that is charged. There is no restriction at all in the import of maize seed for research purposes in all the countries. All imported maize germplasm is subject to quarantine. From the above findings, it is clear that a common legislation is needed in West and Central Africa in order to facilitate a free and safe movement of germplasm, which could result in a quick release of a high number of improved maize varieties to farmers.

Resources

Human resources

Public organizations working on maize were found in all the countries included in this study. However, private or parastatal seed organizations were identified only in Nigeria (three organizations), Senegal (two), Chad (one), Burkina Faso (five) and Ghana (eight). Most of these private maize seed companies in WCA are community-based seed production systems managed by farmers' groups and non-governmental organizations (Badu-Apraku *et al.*, 1999).

The total number of employees working in maize improvement was about 54 staff-year equivalents (SYE) per country in the public sector and 136 SYE in the private sector (Table 8.5, see also Appendix 8.3). Senior and intermediate researchers and administrators represented about 12% of the staff for private maize and 10% for public maize. Of the total human resources involved in maize research in the private sector, 16% were working in research, 65% in seed production and 19% in distribution and administration (Appendix 8.3). The same analysis for public maize resulted in 60.7% for research, 30.1% for seed production and 9.2% for distribution and administration. It is obvious that the public sector had more human resources in research than the private sector.

Table 8.5. Personnel working on maize research systems for the study countries in West and Central Africa, 1997–1998 (SYE).

Category	Maize public (n = 10)	Maize private (n = 3)
Senior researchers and administrators	52.0	25.5
Intermediate-level researchers and administrators	60.75	26
Technicians and other support personnel	116.80	87.45
Farm and casual labourers	272.50	270.68
Total	502.05	409.63

Notes:
SYE, Staff year equivalent; *n*, number of countries (Nigeria is not included for the public maize, while Ghana and Burkina Faso are not included for the private maize).

This fact could explain why the private sector used a large share of maize materials that were first developed by the public institutions (Fig. 8.1). The involvement of the public sector in seed distribution is probably due to the weakness of the private sector. This needs to be addressed if a strong, efficient and sustainable seed industry is to develop in WCA.

Financial resources

Research programmes, both public and private, were requested to provide data on the total investments in maize research in 1997–1998. Total investments included average annual salary (including all benefits) and other support costs (including fixed, operational and administrative costs) for four categories of staff (senior researchers and administrators, intermediate level researchers and administrators, technicians and other support personnel, and farm and casual labourers). Only three out of 19 private organizations supplied data on annual salaries, and no organization gave information on other costs, despite the confidentiality promised in the use of such data. Therefore, the analysis on financial resources for NARS was done only for the public sector. Even in this case, not all the countries supplied all the required data, as shown in Table 8.6. In particular, respondents did not include the fixed costs, such as those for laboratories. Therefore the results below should be considered as being very rough estimates only.

The nine countries that completed the forms invested about US$530,000 for maize (Table 8.6). The annual salaries represent 80% of the cost while the share cost for senior and intermediate level staff was about 76%.

Table 8.6. Total investments in maize research for the public sector in selected African countries, 1997–1998.

Category of staff	Average annual salary (US$) maize (*n* = 9)	Other costs (US$) maize (*n* = 9)	Total maize (*n* = 9)
Senior staff	166,429.44	1,341.70	250,771.14
Intermediate staff level	143,917.11	10,083.51	154,000.62
Technicians and support staff	81,134.28	7,216.92	88,351.20
Farm and casual labourers	32,261.40	3,521.88	35,783.28
Total	42,374.23	105,164.00	528,906.24

Note:
n = number of countries.

Measurement of gains

Human capital development

Over the last 30 years, IITA has contributed immensely to capacity building in the national systems for maize research in WCA (Table 8.7). The average number of scientists trained per year was 16 (PhD, MSc, research associate, visiting student), representing about 12% of the trainees.

Table 8.7. Number of scientists trained at IITA on maize research in West and Central Africa, 1970–1998.

Category of training	1970–1979	1980–1989	1990–1998	Total
PhD level				20
Men	3	4	10	17
Women	–	–	3	3
MSc level	–	–	–	10
Men	2	4	3	9
Women	–	1	–	1
Research training associate				17
Men	7	8	1	16
Women	–	1	–	1
Visiting student research scholars				14
Men	7	6	1	1
Women	–	–	–	–
Group trainees				429
Men	25	273	107	405
Women	–	9	15	24
Total				490
Men	44	295	122	461
Women	–	11	18	29

Source: calculated from IITA, training programme.

Although women benefited from training as well, there was a gender bias in the manpower development since women represented only 6% of trainees. A meagre total of five female scientists, as opposed to 42 male scientists, benefited from individual training. The number of trainees increased from the 1970s to the 1980s, and decreased from the 1980s to the 1990s. There has been a specialization in the type of training offered by IITA over time. While the number of PhD and MSc graduates increased from the 1970s to the 1990s, visiting student research scholarships virtually disappeared in the 1990s so as to allocate the scarce resources for training to students who are undertaking a research project for a post-graduate degree. Also, the level of group training is declining, as it becomes more demand-driven in response to the specific needs of the NARS partners.

Merging data from Table 8.5 (details are in Appendix 8.3) and those from Table 8.7 allows an assessment of the impact of IITA on the capacity building within the NARS for maize research in WCA. The IITA trainees represented about 36% of senior researchers and 14% of intermediate level researchers working in maize research and seed production. Despite the substantial effort from IITA, the number of national scientists remains very low compared with the area planted to maize in the study area. The ratio of senior researchers to areas planted to maize was as low as one scientist per 106,780 ha in 1998. These results highlight the need to pursue individual, long-term training (PhD and MSc levels) for NARS to strengthen local and regional capacities for maize research in WCA.

Gross returns and impact on food security

A gross measure of economic benefits was calculated for 1998 by taking into consideration the area planted to improved varieties in each country and the yield advantage. The area planted to improved varieties was calculated from the results of farm surveys. If this information was not available, 'expert opinion' was sought, to provide the required estimates on areas planted to improved varieties. The yield advantage was calculated as the difference in farmers' fields between the average yield of improved and local varieties (or improved varieties older than 5 years). This is a conservative estimate of benefits since it uses the real yield from farmers' fields; it does not consider the potential yield of improved varieties from either on-station or on-farm research. However, the measurement of yield advantages did not take into consideration the number of years or cycles, that were required to develop the improved varieties, nor does it distinguish between genetic sources of yield gain and other inputs or differences in management. The average local price of maize at harvest in a major maize-growing area in each study country was converted into US dollars using an average of the exchange rate for 1998 for the computation of the gross returns. In order to assess the impact on food security, the incremental production was converted into energy by taking away the percentage of waste (15% for dry grain) and applying the energy content from the FAO food composition tables. A daily calorie intake of 2200 kcal per capita was considered to convert the total energy from the net incremental production into the number of people per year who could be fed, thanks to the improved varieties.

On average, about 8.2 million ha (37%) of the total area for the countries included in this study were planted to improved maize varieties (Table 8.8). Adoption of improved maize varieties resulted in a yield advantage of 45.3% over local varieties. As indicated above, this yield advantage is a gross figure. It does not account for more intensive use of fertilizers or other inputs. Also, it might be expected that modern varieties may be planted on particularly favourable plots. The increased production of maize grain reached 2,585,500 t year^{-1}. Such an additional output would meet the caloric needs of about 9.3 million people in 1996. This population size represents nearly 4% of the total population in the countries included in this study. Such a percentage is slightly above the 3% yearly population growth rate for Africa. The results from the economic analysis gave gross returns of about US$162 ha^{-1} of land planted to the improved maize varieties. Countries with a small yield advantage, such as Chad and Cameroon, also had the smallest returns on land and the smallest percentage of people benefiting from the improved varieties.

The range of benefits brought about by the use of germplasm from the CGIAR centres varied from one country to another. For example, the area planted to improved varieties of maize varied from 89% in Senegal to only 1.34% in Togo. In addition, the yield advantage varied from 17% in Cameroon to 100% in Togo. The gross economic benefit ranged from US$36 ha^{-1} in Chad to US$200 ha^{-1} in Nigeria for the areas planted to improved maize, and the proportion of the population to gain food security varied from 0.46% in Togo to 5.59% in Nigeria.

Conclusions

An average of 17 maize varieties were released per country by the public sector from 1965 to 1998. The CGIAR centres (IITA and CIMMYT) significantly contributed to the increasing number of varieties released by NARS since the late 1970s. Both IITA and CIMMYT were major sources of germplasm used by NARS for the development of the released varieties. The centres also contributed to increased production of varieties developed from local materials by NARS. Although public research institutions are supporting an emerging private sector, the private seed industry remains weak and hampers the development of the maize economy in WCA. Our survey indicated that considerable

Table 8.8. Estimated benefits from the use of improved maize varieties per country in West and Central Africa, 1998.

Country	Area planted in 1998		Average yield		Incremental production ('000 t)	GEB/ha^{-1} of improved varieties (US$)	Number of beneficiaries	
	Total ('000 ha)	Improved variety (%)	Local variety (t ha^{-1})	Advantage (%)			Total	Percentage of 1996 population
Benin	513.1	25.3	1.2	50.0	230.1	101.7	293,851	5.28
Burkina Faso	112.8	45.5	1.5	53.33	41.04	160.0	275,108	2.55
Ghana	550.0	53.0	1.5	33.33	145.75	108.7	550,783	3.09
Guinea	85.7	22.6	1.1	50.00	10.65	85.0	40,342	0.54
Mali	275.4	22.9	1.5	66.67	63.2	120.0	238,830	2.15
Nigeria	4,255.0	40.0	1.2	83.33	1,702.0	209.3	6,431,780	5.59
Senegal	138.8	89.2	1.07	24.3	32.19	46.0	121,637	1.43
Togo	381.6	1.3	1.0	100.0	5.13	154.5	19,386	0.46
Cameroon	374.2	28.0	1.8	16.67	31.44	48.0	118,810	0.88
DR of Congo	1,436.7	31.3	0.82	82.93	312.15	151.1	1,179,590	2.52
Chad	65.1	70.0	1.07	24.3	11.85	36.4	44,774	0.69
Total	8,188.4	36.8	1.27	45.3	2,585.50	162.3	9,314,890	3.77

Note: GEB = gross economic benefit.

progress was made through the release of varieties that combine one or more desirable traits. The study also highlighted the need for several new traits: resistance to *Striga* and borers, tolerance of low soil fertility, and drought resistance. The new traits are required in the second generation of varieties to expand the production of improved varieties.

IITA has contributed immensely to capacity building in sub-Saharan Africa (SSA). However, the ratio of senior researchers to area planted remains low. This highlights the need to pursue individual training schemes in order to increase the research capacity in SSA.

An estimate of gross benefits indicated, on average, that the area planted to the improved varieties was about 37% and the range in the percentage of yield advantage varied from 17% to 100%. The increased production in 1998 from the use of improved varieties was about 2.6 million t. This production could supply enough food during a year to about 9.3 million people, representing nearly 4% of the population in those countries included in this study. CGIAR centres such as IITA can still contribute to an increased rate of adoption of improved varieties in WCA.

Acknowledgements

IITA and the CGIAR/IAEG sponsored this study. Data collection was completed by the leaders of maize programmes in the study countries: Benin Republic (C.G. Yallou), Burkina Faso (J. Sanou), Cameroon (C.G. Zonkeng), Chad (M. Dabi), Congo Democratic Republic (M. Mbeya), Ghana (K. Obeng-Antwi), Guinea (S. Camara), Mali (N'tji Coulibaly), Nigeria (B.A. Ogunbodele), Senegal (A. Ndiaye) and Togo (E.Y. Mawule). Drs J.M. Fajemisin and B. Badu-Apraku coordinated data collection in some countries of West Africa. Professor R. Evenson, the CGIAR/IAEG principal investigator for the 'CGIAR germplasm study' provided timely advice that was useful in the writing-up of this report. Mrs R. Umelo edited the initial draft of this report. The authors appreciate the contribution of all to the completion of this study.

References

Badu-Apraku B., Hema, I., Thé, C., Coulibaly, N. and Mellon, G. (1999) Making improved seed available to farmers in West and Central Africa. In: Badu-Apraku, B., Fakorede, M.A.B., Ouedraogo, M. and Quin, F.M. (eds) *Strategy of Sustainable Maize Production in West and Central Africa. Proceedings of a Regional Maize Workshop, IITA-Cotonou, 21–25 April 1997.* Delar Tertiary Publishers, Ibadan, Nigeria, pp. 138–149.

Byerlee, D. and Heisey, P.W. (1997) Evolution of the African maize economy. In: Byerlee, D. and Eicher, C.K. (eds) *Africa's Emerging Maize Revolution.* Lynne Rienner Publishers, London, pp. 9–22.

Eckebil, J.P. (1994) New frontiers for food grain research for the 1990s. In: Menyonga, J.M., Taye Bezuneh, Yayock, J.Y. and Idrissa Soumana (eds) *Progress in Food Grain Research and Production in Semi-arid Africa.* OAU/STRC-SAF-GRAD, Ouagadougou, Burkina Faso, pp. 3–19.

Fajemisin, J.M. (1994) Regional approach to maize research for the semi-arid zone of West and Central Africa. In: Menyonga, J.M., Taye Bezuneh, Yayock, J.Y. and Idrissa Soumana (eds) *Progress in Food Grain Research and Production in Semi-arid Africa.* OAU/STRC-SAF-GRAD, Ouagadougou, Burkina Faso, pp. 157–168.

FAO (2000) FAOSTAT website: http://apps.fao.org/page/collections?subset=agriculture

Kling, J.G., Cardwell, K.F. and Kim, S.K. (1994) Advances in screening methods and breeding for downy mildew (*Peronosclerospora sorghi*) resistance of maize. In: Jewell, D.C., Waddington, S.R., Ransom, J.K. and Pixley, K.V. (eds) *Maize Research for Stress Environments. Proceedings of the Fourth Eastern and Southern Africa Regional Maize Conference.* CIMMYT Maize Programme, Harare, Zimbabwe, and Nairobi, Kenya, pp. 164–168.

Manyong, V.M., Smith, J., Weber, G.K., Jagtap, S.S. and Oyewole, B. (1996) *Macrocharacterization of Agricultural Systems in West Africa.* Resource and Crop Management Research Monograph No. 21. IITA, Ibadan, Nigeria.

Smith, J., Barau, A.D., Goldman, A. and Mareck, J.H. (1994) The role of technology in agricultural intensification: the evolution of maize production in the northern Guinea savanna of Nigeria. *Economic Development and Cultural Change* 42, 537–554.

Smith, J., Weber, G., Manyong, V.M. and Fakorede, M.A.B. (1997) Fostering sustainable increases in maize productivity in Nigeria. In: Byerlee, D. and Eicher, C.K. (eds) *Africa's Emerging Maize Revolution.* Lynne Rienner Publishers, London, pp. 107–124.

Appendix 8.1

Details on the number of maize varieties released by the participating countries in the IITA impact study on germplasm in West and Central Africa, 1965–1997

Country	Maize	
	Public	Private
West Africa	139	81
Benin	16	0
Burkina Faso	25	20
Ghana	17	14
Guinea	12	0
Mali	9	0
Nigeria	29	23
Senegal	22	16
Togo	9	8
Central Africa	47	0
Cameroon	13	0
Chad	14	0
D.R. Congo	20	0
Total	186	81

Appendix 8.2

Details of the positive traits of maize germplasm from IITA used by NARS breeding programmes in West and Central Africa, 1970–1998

	Positive attributes										
Germplasm	+Y	ST	SR	DMR	BR	Y	W	EE	E	I	L
TZESR-W	x		x				x		x		
TZECOMP4	x		x				x		x		
TZESR × GWA	x	x				x		x			
TZB-SR	x		x				x				x
DMR-ESRY	x		x	x		x			x		
TZPB-SR	x		x				x				x
DMR-ESRW	x		x	x			x		x		
TZEFY	x					x			x		
EV 8422 SR	x		x				x				x
EV 8430 SR	x		x				x		x		
EV 8431 SR	x		x			x			x	x	
EV 8444 SR	x		x				x				x
EV 8421 SR	x		x				x				x
POP 43 SR	x		x				x				x
POP 49 SR	x		x				x			x	
POOL 16 SR	x		x				x		x		
POP 63 SR	x		x				x				x
IKENNE 83 TZSRY	x	x			x					x	
TZMSR-W	x		x				x				x
SAMARU 83 TZSRY-1	x	x			x					x	
TZB	x						x				x
TZPB	x						x				x
TZSR-W	x		x				x				x
Ama TZBR-W	x			x		x				x	
TZBR-Eld 3 C2	x				x		x				x
TZL Comp 1 C4	x	x					x				x
Acr 92 TZE Comp 5-W	x	x				x		x			
Acr 91 Suwan 1-SR	x		x		x					x	
IWD STR C0	x	x					x				x
TZEE-Y-SR	x		x			x		x			
TZEE-W-SR	x		x				x	x			x
Acr 88 Pool 16-SR	x	x				x		x			
9033-26	x		x				x				x
9333-9B	x		x				x				x
TZSR-Y	x		x			x					x
TZESR-Y	x		x			x			x		
DMR-LSRW	x		x	x			x				x
DMR-LSRY	x		x	x		x					x

Continued

Appendix 8.2 (*continued*)

Germplasm	Positive attributes										
	+Y	ST	SR	DMR	BR	Y	W	EE	E	I	L
TZ 8843 DMRSR	x		x	x			x				x
SUWAN 2-SR	x		x	x		x					x
TZB-SR SGY	x		x			x					x
TZE COMP 3X4 C1	x	x				x		x			
TZE COMP 3 C1	x		x				x		x		
TZ 9043 DMRSR	x		x	x			x				x
8321-18	x		x				x				x
8321-21	x		x				x				x
8322-3	x		x				x				x
8644-27	x		x			x					x
8644-31	x		x			x					x
8644-32	x		x			x					x
8535-23	x		x				x				x
8425-8	x		x				x				x
8388-1	x		x				x				x
8505-6	x		x				x				x
Total	54	8	41	8	4	17	34	6	10	6	34

Note: +Y, more yield; STR, *Striga* resistance; SR, streak resistance; DMR, downy mildew resistance; BR, borer resistance; Y, yellow; W, white; EE, extra-early maturity; E, early maturity; I, intermediate maturity; L, late maturity; x indicates presence of the trait in the germplasm.

Appendix 8.3

Details of personnel working on maize research systems in the study countries in West and Central Africa, 1997–1998 (SYE)

Category of personnel	Maize public (n = 10)	Maize private (n = 3)
Senior researchers and administrators	52	25.50
Working in research	35.5	7.25
Working in planting material production	10.5	7.85
Working in distribution or administration	6	10.40
Intermediate level researchers and administrators	60.75	26
Working in research	33.75	6
Working in planting materials production	21	11
Working in distribution or administration	6	9
Technicians and other support personnel	116.80	87.45
Working in research	59.30	11.30
Working in planting material production	48.50	41.65
Working in distribution or administration	9	34.50
Farm and casual labourers	272.50	270.68
Working in research	176.50	42.0
Working in planting material production	71	205.18
Working in distribution or administration	25	23.50
Total		
Working in research	305.05	66.55
Working in planting material/production	151	265.68
Working in distribution or administration	46	77.40
Totals	502.05	409.63

Note: SYE, staff year equivalent; *n*, number of countries.

9 Impacts of Genetic Improvement in Sorghum

U.K. Deb and M.C.S. Bantilan

This chapter quantifies the impacts of sorghum genetic enhancement research, featuring the catalytic role of the International Crops Research Institute for the Semi-Arid Tropics (ICRISAT) in a research environment characterized by strong national agricultural research systems (NARS) and private sector institutions in Asia and weaker NARS in Africa. Impacts of sorghum improvement research are measured in terms of yield gain, reduction in unit production cost, technology spillover and improvement in yield stability. The results indicate substantial yield and stability gains accompanied by reductions in the unit costs of production from the adoption of improved sorghum cultivars. Countries with weak NARS, especially in Africa, benefited primarily from ICRISAT-developed varieties and through technology spillover. On the other hand, countries with strong NARS in Asia benefited largely from elite breeding materials developed by ICRISAT. An important policy implication arising from the study is the need for differential strategies for different regions to take into account the different research environments.

Introduction

Sorghum is a major world cereal crop, though not on the scale of rice or wheat. Some 85 countries cultivate sorghum in measurable quantities. The top five sorghum-growing countries, in terms of harvested area during 1994–1996, were India (1.23 million ha), Nigeria (0.60 million ha), Sudan (0.58 million ha), USA (0.39 million ha) and Niger (0.19 million ha). In terms of total sorghum production, the top five sorghum-producing countries are USA (1.62 million t), India (1 million t), Nigeria (0.66 million t), China (0.54 million t) and Mexico (0.42 million t). However, the highest sorghum yields were obtained in smaller producers: Italy (5898 kg ha^{-1}), France (5724 kg ha^{-1}) and Egypt (4620 kg ha^{-1}) (FAO, 1998).

Realizing the importance of the sorghum crop, donors and governments of different countries have invested substantial amounts of money to establish national and international research centres such as ICRISAT, which was founded in 1972. International research institutes in partnership with national research systems (both public and private) have made concerted efforts to develop improved sorghum cultivars and practices to increase yield and the social well-being of the producers and consumers of sorghum.

Human Resources Involved in Sorghum Improvement

Tables 9.1 and 9.2 report the level of scientific staffing involved in sorghum improvement. At ICRISAT, five sorghum breeders located in Asia and Africa are involved in breeding. Fifteen other scientists including agronomists, crop physiologists, genetic resources specialists, entomologists, pathologists and social scientists generate information for effective use by the breeders. In India, about 150 sorghum scientists in the public and private sector are working on this crop. In China, 200 scientists are working for sorghum improvement. The number of scientists working on sorghum in other Asian countries is also notable. However, in African countries the number of scientists working on this crop, with the exception of Ethiopia, Sudan and Kenya, is very low, generally between one and five scientists in each country. Usually African

Table 9.1. Number of sorghum scientists working in different countries (1999 or latest year).

Country	Latest year	Breeders	Agronomists	Seed technologists	Others*	Total
Asia						
China	1997	120	40	20	20	200
India	1998					150
Iran	1997	2	2	1	0	5
Pakistan	1997	5	9	0	0	14
Thailand	1998	11	12	6	7	36
Eastern and Central Africa[a]						
Burundi	1998	1	0	1	0	3
Eritrea	1998	1	4	0	0	5

Continued

Table 9.1. *Continued*

Country	Latest year	Breeders	Agronomists	Seed technologists	Others*	Total
Eastern and Central Africa[a] *Continued*						
Ethiopia	1998	15	15	0	20	50
Kenya	1998	3	8	0	7	18
Rwanda	1998					3
Sudan	1998	3	7	0	11	21
Uganda	1998	2	1	0	2	5
Southern Africa[b]						
Angola	1999	1	1			2
Botswana	1999	1	1	1 (all crops)	1	4
Lesotho	1999		1			1
Malawi	1999	1		1 (all crops)	1 (pests)	3
Mozambique	1999	1		2 (all crops)		3
Namibia	1999	1	1			2
Swaziland	1999		1			1
Tanzania	1999	2	2	1 (all crops)	1 (all crops)	6
Zambia	1999	2	1	1 (all crops)		4
Zimbabwe	1999	1	1	1 (all crops)	2 (pathology)	5
Western Africa[c]						
Burkina Faso	1991–92	2	3		3	8
Cameroon[d]	1991–92		1		3	4
Ghana[d]	1991–92		1		3	4
Mali	1991–92		3		4	7
Niger	1991–92		2		4	6
Nigeria	1991–92		1	5		6
Northern Africa						
Egypt	1998	13	8	1	3	25
West and Central Africa (Sorghum Research Network)[c]	1990–91					83
East Africa (Sorghum and Millet Network)[a]	1990–91					87

*Other disciplines that support varietal improvement, such as pathologists, entomologists, social scientists.

Notes:

[a]In East Africa, 70% of the researchers work on sorghum and millet as full-time researchers while 30% of them work on these two crops on a part-time basis and about 35% of the qualified scientists are based in two countries.

[b]For southern African countries, number of scientists indicates working both on sorghum and millet.

[c]In West and Central Africa, 38% are full-time researchers for sorghum, 62% are part-time researchers and about 25% of the qualified researchers are based at lead NARS.

[d]In both Cameroon and Ghana, one entomologist was working on a part-time basis for sorghum.

Sources: For Asia, ICRISAT Impact Monitoring Survey, 1997.
For Africa, ICRISAT Impact Monitoring Survey, 1998–2000.
For West Africa, Sanders *et al.* (1994) p. 53.

Table 9.2. Educational levels of sorghum scientists working in different countries (1999 or latest year).

Country	Latest year	BSc	MSc	PhD	Other	Total
Asia						
China	1997	108	44	18	30	200
India	1998					150
Egypt	1998	3	3	14	5	25
Iran	1997	2	3	0	0	5
Pakistan	1997	0	13	1	0	14
Thailand	1998	11	15	10	0	36
Eastern and Central Africa[a]						
Burundi	1998	1	2	0	0	3
Eritrea	1998	4	1	0	0	5
Ethiopia	1998	15	26	9	0	50
Kenya	1998	4	14	0	0	18
Rwanda	1998	3	0	0	0	3
Sudan	1998	1	4	16	0	21
Uganda	1998	0	3	2	0	5
Southern Africa[b]						
Angola	1999					2
Botswana	1999	1	1	2		4
Lesotho	1999	1				1
Malawi	1999		1	2		3
Mozambique	1999	2	1			3
Namibia	1999	1	1			2
Swaziland	1999	1				1
Tanzania	1999	1	3	2		6
Zambia	1999		3	1		4
Zimbabwe	1999	1	2	2		5
West Africa[c]						
Burkina Faso	1991–92					8
Cameroon	1991–92					4
Ghana	1991–92					4
Mali	1991–92					7
Niger	1991–92					6
Nigeria	1991–92					6
Northern Africa						
Egypt	1998	3	3	14	5	25
West and Central Africa (Sorghum Research Network, 18 Countries)	1991–92	33	27	23		83
East Africa (Sorghum and Millet Network, 8 Countries)	1990–91	29	31	27		87

Notes:
[a]In East Africa, 70% of the researchers work on sorghum and millet as full-time researchers while 30% of them work on these two crops on a part-time basis and about 35% of the qualified scientists are based in two countries.
[b]For southern African countries, number of scientists indicates working both on sorghum and millet.
[c]In West and Central Africa, 38% are full-time researchers on sorghum, 62% are part-time researchers and about 25% of the qualified researchers are based at lead NARS.
Sources: For Asia, ICRISAT Impact Monitoring Survey, 1997.
For Africa, ICRISAT Impact Monitoring Survey, 1998–2000.
For West Africa, Sanders *et al.* (1994) p. 53.

scientists are devoted to more than one crop, often sorghum and millet together. In other words, Asian NARS have devoted more resources to sorghum improvement than their African counterparts, not only in terms of quantity but also in terms of education levels.

Sorghum Genetic Enhancement Research: Objectives and Targets

Pre-breeding research

Collection, characterization and maintenance of landraces are essential for crop improvement, and ICRISAT has given high priority to this. As of December 1999, a total of 36,719 sorghum germplasm accessions from 90 countries have been conserved at ICRISAT. After collection and assembly, ICRISAT along with its NARS partners conducted evaluation trials to identify the useful traits available in the assembled germplasm. Scientists working on sorghum improvement request germplasm materials from ICRISAT. Based upon requests from different users, ICRISAT has distributed 239,742 items of sorghum germplasm to 99 countries (Kameswara Rao, Curator of Genetic Resources Unit, ICRISAT, personal communication). During evaluation trials, some landraces collected from different countries have been identified as superior to existing cultivars. A total of 23 varieties have been directly released from the distributed sorghum germplasm in 12 countries of Asia, Africa and Latin America (Table 9.3)

Sorghum breeding strategy at ICRISAT

Sorghum breeding research domains

During the preparation of ICRISAT's medium-term plan for 1994–1998, six sorghum research domains were explicitly defined for the first time. Table 9.4 summarizes the location and characteristics of each sorghum research domain.

Sorghum breeding research at ICRISAT

ICRISAT has been engaged in sorghum improvement since the early 1970s. Multidisciplinary teams of scientists are located in Asia at ICRISAT Centre (India); at regional centres in West Africa at Bamako (Mali) and Kano (Nigeria), eastern Africa at Nairobi (Kenya), southern Africa at Bulawayo (Zimbabwe), and in Latin America at El Batan (Mexico).

Table 9.3. Sorghum germplasm accessions or selections released as superior varieties in different countries.

Accession number	Country of origin	Country of release	Released name	Year of release	Remarks
IS 6928	Sudan	India	Moti	1978	Induced mutant
IS 2940	USA	Myanmar	Shwe-ni 2	1979	
IS 8965	Kenya	Myanmar	Shwe-ni 1	1979	
IS 302	China	Myanmar	Shwe-ni 10	1980	
IS 5424	India	Myanmar	Shwe-ni 8	1980	
IS 30468	Ethiopia	India	NTJ2	1980	
IS 18758	Ethiopia	Burkina Faso	E-35-1	1981	
IS 4776	India	India	U P Chari-1	1983	Forage sorghum
IS 9302	South Africa	Ethiopia	ESIP 11	1984	
IS 9323	South Africa	Ethiopia	ESIP 12	1984	
IS 2391	South Africa	Swaziland	MRS 13	1989	
IS 3693	USA	Swaziland	MRS 94	1989	
IS 8511	Uganda	Mozambique	Mamonhe	1989	
IS 23520	Ethiopia	Zambia	Sima	1989	
IS 9321	South Africa	Mexico		1990	
IS 9447	South Africa	Mexico		1990	
IS 9468	South Africa	Mexico		1990	
IS 9830	Sudan	Sudan	Mugawim Buda-2	1991	
IS 2923	USA	Botswana	Mahube	1994	
IS 23496	Ethiopia	Tanzania	Pato	1995	
IS 3541	Sudan	India	CS 3541		Converted Zerazera
IS 3924	Nigeria	India	Swarna		
IS 18484	India (AICSIP)[1]	Honduras	Tortillerio		

Source: N. Kameswara Rao *et al.* (1998).

The breeding concepts, objectives and the research approach involving partners have undergone several changes since ICRISAT was established. External environment, donors' perceptions, the national agricultural research systems (NARS) capacity and the ICRISAT research administration structures are some of the most important factors that caused these changes. At ICRISAT Patancheru, six different periods can be identified (Reddy *et al.*, 1998):

Phase 1: Wide adaptability and high grain yield (1972–75)
Phase 2: Wide adaptability and screening techniques (1976–1979)
Phase 3: Regional adaptations and resistance breeding (1980–1984)
Phase 4: Specific adaptation and resistance breeding (1985–1989)
Phase 5: Trait-based breeding and sustainable productivity (1990–1994)
Phase 6: Intermediate products and upstream research (1995–present).

Table 9.4 Sorghum research domains.

Domain	Production system characteristic	Major constraints	Locations
SG I (wide applicability)	Rainy season, multipurpose grain, stalk, fodder (fodder emphasis). Wide adaptability (June–August sowing)	Grain mould, shoot fly, headbug	West Africa (southern tier), India (Tamil Nadu, S. Karnataka, Andhra Pradesh)
SG 2 (dual purpose, specific adaptability)	Rainy season, dual purpose (grain and fodder). Specific adaptation (June sowing). Medium to late maturing types	Stem borer, grain mould, midge, shoot fly, drought	E. and southern Africa, India (Andhra Pradesh, N. Karnataka, Maharashtra, Madhya Pradesh, Gujarat), Latin America (some areas)
SG 3 (dual purpose, fodder emphasis)	Rainy season, dual purpose (fodder emphasis). Early maturing	Shoot fly, stem borer	W. Africa (northern tier), E. Africa (Yemen, Somalia), India (E. Rajasthan), Latin America (some areas), China and Iran
SG 4 (forage sorghum)	Rainy season, forage types (thin stalk, tillering), late-maturing	Stem borer, leaf diseases	India (N. Gangetic plain), Pakistan
SG 5 (early sowing *rabi*)	Post-rainy season (early sown before October). Bold grain types, dual purpose	Shoot fly, stalk rot, head bugs	India (S. Andhra Pradesh, S. Karnataka)
SG 6 (late sowing *rabi*)	Post-rainy season (late sown–mid/late October). Bold grain, photoperiod sensitivity require temperature insensitive		India (Gujarat, S. Maharashtra, N. Karnataka)
SG 7 (irrigated)	Irrigated sorghum		Iran, Egypt, Wadmadani (Sudan)
SG 8 (extreme altitude)	Others		(i) High altitude: China (ii) Low altitude: Indonesia, Brazil, Ecuador, Venezuela

There were some variations in research activities in Africa. Obilana (1998) documents the experience of sorghum breeding within the Southern Africa Development Community (SADC)/ICRISAT Sorghum and Millet Improvement Programme (SMIP). The strategy of the SADC/ICRISAT SMIP involved, first, the development and testing of improved technology of better varieties and hybrids (Phases I and II, 1983/84 to 1992/93) and second, technology transfer and exchange (1992/93 to1997/98).

Research Products

Intermediate products

The focus of sorghum research and development activities at ICRISAT has involved a massive screening programme, the development of suitable materials, and the understanding of genetics and mechanisms for resistance. In particular, ICRISAT breeders have worked towards the development of improved varieties and hybrids. Hybrids are generally known to have significantly higher productivity than their parental pure lines. The discovery of cytoplasmic-genetic male sterility (CMS) facilitated large-scale production of male-sterile lines (A-lines) when pollinated by the respective maintainer line (B-lines). Commercial hybrids are produced from these seed parents (A-lines) upon pollination by a restorer line (R-line). In recent years, considerable effort has gone into producing A-, B- and R-lines with desirable characteristics.

The team effort of scientists in sorghum breeding, entomology and pathology in screening materials for resistance to various biotic and abiotic stress factors has resulted in the development of various male-sterile lines.

ICRISAT organized more than 150 trials/nurseries and supplied 146,000 seed samples during 1986–1997. Of the seed samples distributed, Asia received 66%, followed by Africa (23%) and the Americas (10%). Partially converted lines were the most common category (13.5% of total supply) followed by restorers (9.8%), hybrids (9.4%), varieties (9.1%), maintainers (8.3%) and male-sterile lines (8.0%) (Reddy *et al.*, 1998). Both public research institutes and private seed producers in India have received seed samples from ICRISAT. The total number of seed samples supplied to Indian NARS increased to 14,310 in 1997 from 2131 in 1986. In addition to the seed samples and germplasm lines supplied from ICRISAT Patancheru, southern African countries obtained sorghum genetic materials (breeding lines, varieties, hybrid parents, hybrids) from the ICRISAT/SADC Sorghum and Millet Improvement project (SMIP). Obilana (1998) reported that SMIP conducted 608 collaborative trials and supplied 18,524 genetic material samples to 11 SADC countries during 1983/84 to 1997/98.

ICRISAT-bred varieties/hybrids and those derived from ICRISAT materials by the national breeders have been tested in the All India Coordinated Sorghum Improvement Project (AICSIP) trials since 1979/80. The average numbers of ICRISAT-derived varieties and hybrids entered into the advanced trials have increased over time. In collaboration with NARS, hybrids/varieties developed by ICRISAT have been tested in network trials for selecting for local conditions in Africa, Asia and Latin America. As a result, improved varieties and hybrids are released. The private sector is also marketing several hybrids developed from ICRISAT materials in India.

Release of improved cultivars

Table 9.5 shows the total number of improved sorghum cultivars (varieties and hybrids) released in different countries. A total of 405 improved sorghum cultivars are available in 43 countries of Asia, Africa and America. We have grouped the released sorghum cultivars as ICRISAT-bred (cultivars bred by ICRISAT breeders); ICRISAT-parent (cultivars developed by NARS based on ICRISAT parent materials and germplasm); ICRISAT-network (cultivars tested through ICRISAT network but not bred by ICRISAT scientists); and non-ICRISAT cultivars (cultivars released from other sources). A total of 146 sorghum cultivars were released from materials classified as ICRISAT-bred (64), ICRISAT-parent (29) or ICRISAT-network (53). Over time, the number of releases using ICRISAT-parent materials has been increasing. Up to 1975, 71 sorghum cultivars were released in 13 countries. These cultivars were mainly selections of landraces, except in India and China where hybrids were also released. The number of releases increased over time, and the number of countries having improved sorghum cultivars has also increased.

The first sorghum hybrid in India, CSH 1, was released in 1964. In 1999, 182 improved cultivars of sorghum were available in India for cultivation for grain, forage or dual purpose. Out of these 182 improved cultivars, 122 are 'notified' either by the national seed committee or by the state seed release committees. The remainder are mostly the research products of private seed companies. Out of these 122 notified cultivars, 23 were derived from ICRISAT materials. At least nine hybrids released by private seed companies are based on ICRISAT-parent materials. It is difficult to know the parentage of private hybrids due to confidentiality, but all private seed companies that have released hybrids in India have collaboration with ICRISAT. Private companies operating in India acknowledge that their research hybrids contain some input from ICRISAT material (ICRISAT/Rutgers University Study of Private Seed Sector, 1997).

Table 9.5. Number of released sorghum cultivars with ICRISAT content by period.

		1971–75 (1976–80)		1981–85 (1986–90)				1991–95 (1996–98)					
Country	Up to 1970	ICRISAT network	All sources	ICRISAT -bred	ICRISAT -parent	ICRISAT -network	All sources	ICRISAT -bred	ICRISAT -parent	ICRISAT -network	All sources	Total ICRISAT derived releases	TOTAL releases (up to 1998)
Asia													
China			4 (3)		0 (3)		5 (4)		2 (2)		6 (2)	7	24
India	9	0 (2)	16 (25)	2 (2)	2 (1)	1 (0)	31 (30)	3 (0)	9 (1)		43 (28)	23	182
Indonesia	5		4 (0)				2 (0)				2 (0)	0	13
Iran							0 (1)				2 (3)	0	6
Myanmar		0 (4)	0 (10)	4 (0)			8 (0)	3 (0)			3 (0)	11	21
Pakistan	3		1 (2)				0 (1)	0 (1)	0 (1)	0 (2)	0 (4)	4	11
Philippines								1 (0)		1 (0)	2 (0)	2	2
Thailand	2						1 (1)		1 (0)		3 (0)	1	7
Africa													
Botswana	4		3 (0)					1 (0)	2 (0)	1 (0)	4 (0)	4	11
Burkina Faso		1 (1)	1 (1)		0 (2)		0 (2)			1 (0)	1 (0)	5	5
Burundi						0 (2)	0 (2)					2	2
Cameroon					0 (1)		0 (1)					1	1
Chad					0 (1)		0 (1)					1	1
Côte d'Ivoire				0 (1)			0 (1)					1	1
Egypt	1		1 (1)								2 (3)	0	8
Ethiopia		0 (1)	0 (1)			3 (1)	3 (1)	0 (1)		1 (0)	1 (1)	7	7
Ghana				0 (1)			0 (1)					1	1
Kenya	1		1 (2)			4 (0)	4 (0)			2 (0)	2 (0)	6	10
Malawi							1 (0)	2 (0)			2 (0)	2	3
Mali								4 (0)			4 (0)	4	4
Mozambique	3			0 (2)		0 (3)	0 (5)	1 (0)			1 (0)	6	9
Namibia								0 (1)			0 (1)	1	1
Niger								2 (0)			2 (0)	2	2

Nigeria								4 (0)			4 (0)	4	4
Rwanda						2 (0)	0 (2)					2	2
Senegal		0 (1)	0 (1)									1	1
Sudan				1 (0)			1 (0)	5 (0)			5 (0)	6	6
Swaziland						0 (1)	0 (2)	2 (0)			1 (0)	3	3
Tanzania	1		2 (1)			0 (1)	0 (2)			1 (0)		2	6
Togo						0 (1)	0 (1)					1	1
Uganda	5		3 (0)								2 (0)	0	10
Zambia	1			1 (0)		0 (4)	1 (5)	1 (0)			2 (0)	6	9
Zimbabwe				0 (2)			1 (1)	0 (3)			1 (3)	5	6
Others													
Colombia									1 (0)	2 (0)	3 (0)	3	3
Costa Rica										1 (0)	1 (0)	1	1
Dominican Republic								1 (0)			0 (1)	1	1
El Salvador		0 (2)	0 (2)	1 (0)		0 (1)	1 (1)					4	4
Guatemala				1 (0)			1 (0)					1	1
Honduras				2 (0)		1 (0)	3 (0)					3	3
Mexico		0 (1)	0 (1)	0 (5)			0 (5)	2 (0)		1 (0)	3 (0)	9	9
Nicaragua				0 (1)		1 (0)	1 (1)					2	2
Panama										1 (0)	1 (0)	1	1
Paraguay										1 (0)	1 (0)	1	1
Total	35	1 (12)	36 (50)	12 (14)	2 (8)	12 (14)	64 (71)	32 (6)	15 (4)	13 (1)	104 (45)	146	405

Seven cultivars in China having ICRISAT parentage were released after 1987. In Pakistan, out of 11 improved cultivars, two (PARC SS1 and PARC SV1) have ICRISAT parentage, and two (PARC SH1 or CSH1, PARC SS2 or IRAT 204) were obtained through the ICRISAT network. One cultivar (Suphan Buri 1) was released in Thailand in 1993 from ICRISAT materials. Myanmar has released 21 varieties and 11 are directly from ICRISAT crosses (seven varieties) or direct introduction from the ICRISAT germplasm collection (four varieties). After 1982, all varieties released in Myanmar (seven) were direct introductions of ICRISAT crosses. Indonesia has released 13 improved sorghum varieties. Two were selections from local varieties and the other 11 were direct introductions. Pedigrees of these cultivars could not be traced. However, ICRISAT has had substantial collaboration with Indonesia.

Nigeria has four improved cultivars (ICSH 89002NG, ICSH 89009 NG, ICSV 111 and ICSV 400), and all are direct introductions from ICRISAT crosses after adaptive research trials. Egypt has seven recommended cultivars (four varieties and three hybrids). All cultivars released after 1980 in Botswana, Burkina Faso, Burundi, Cameroon, Chad, Ethiopia, Ghana, Mali, Mozambique, Namibia, Niger, Rwanda, Sudan, Tanzania, Togo and Zambia are either ICRISAT-bred or from the ICRISAT network. With the exception of one variety in Malawi and one hybrid in Zimbabwe, all released cultivars in Malawi and Zimbabwe in the 1980s and 1990s were from ICRISAT.

Varietal Diffusion and Adoption

Table 9.6 shows the rate of adoption of improved sorghum cultivars in different countries of the world. We consider four categories of modern variety adoption: percentage of area planted to ICRISAT crosses; percentage of area planted to varieties derived from ICRISAT parents; percentage of area planted to cultivars developed through ICRISAT networks; and percentage of area planted to non-ICRISAT cultivars. The rate of adoption is high in Asian countries, while it is comparatively low in African countries. Inter-country comparisons of adoption show that ICRISAT crosses are popular in African countries.

Extent of adoption

In India, the rate of adoption of improved sorghum cultivars in different districts is shown in Fig. 9.1. The rapid rate of adoption in Tamil Nadu and Maharashtra state is evident, while a very slow rate of adoption was observed in Rajasthan and Gujarat states. The rate of adoption

Table 9.6. Rate of adoption of different improved sorghum cultivars.

Country	Region	Year	Percentage of area planted to				
			ICRISAT-cross	ICRISAT-parent	ICRISAT-network	Others	All improved
Asia							
China		1998		9		89	98
India		1999	1	10	3	55	69
Iran		1995–96					87
Pakistan		1995–96					21
Thailand		1995–96		10			NA
Africa							
Angola	National	1997					17
Botswana	National	1997–98	33				33
Cameroon	Mayo Sava	1995	49				
	Diamare	1995	14				
	Mayo Danay	1995	12				
Chad	Guera, Mayo Kebbi, Chari Baguermi	1995	27				
	Guera	1995	38				
	Mayo Kebbi	1995	27				
	Chari Baguermi	1995	24				
Egypt		1995–96		5			35
Lesotho		1997	4				4
Malawi			10				10
Mali		1995	29				29
Mozambique			5				5
Myanmar		1995–96	10				10

Continued

Table 9.6. (*Continued*)

Country	Region	Year	Percentage of area planted to				All improved
			ICRISAT-cross	ICRISAT-parent	ICRISAT-network	Others	
Africa *Continued*							
Nigeria	Kano	1996–97	28				28
	Katsina		10				10
	Kaduna		29				29
	Jigawa		3				3
South Africa		1997					77
Sudan		1995–96	3			19	22
Swaziland		1997					50
Tanzania		1997					2
Zambia			35				35
Zimbabwe			36			36	

Source: ICRISAT Impact Monitoring Survey, 1997–2000; Ogungbile *et al.* (1999) for Nigeria; Rohrbach and Makhwaje (1999) for Botswana; SMIP (1999) for Angola, Lesotho, South Africa, Swaziland and Tanzania; Yapi *et al.* (1998) for Mali; Yapi *et al.* (1999) for Cameroon and Chad.

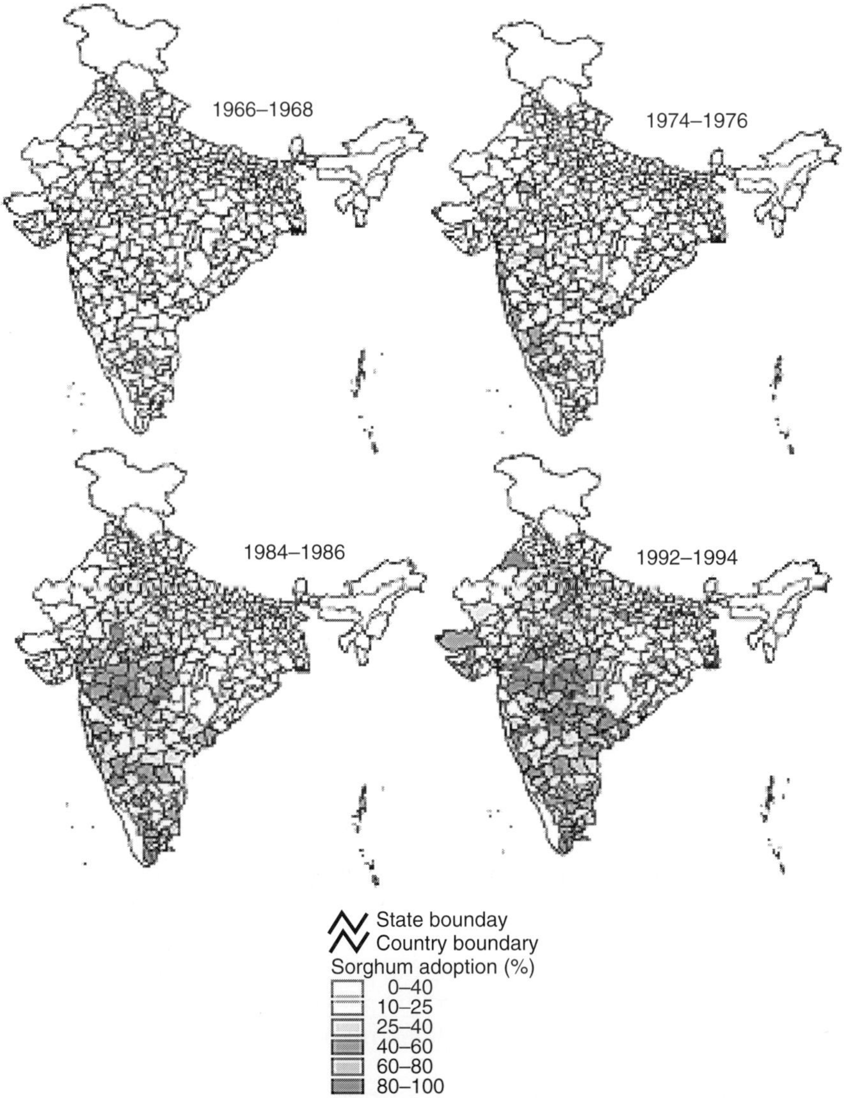

Fig. 9.1. Rate of adoption (%) of improved sorghum cultivars in India.

is higher (more than 80%) in central Maharashtra and in some districts of Andhra Pradesh. Later on we shall show that these districts have also gained in terms of yield increase. The trends in adoption of specific improved sorghum cultivars in India are shown in Fig. 9.2. The initial rapid adoption of CSH 1 is evident, as is the subsequent adoption of CSH 5, CSH 6 and CSH 9. MSH 51, popularly known as Mahyco 51, a cultivar from the private sector, has also been adopted by the

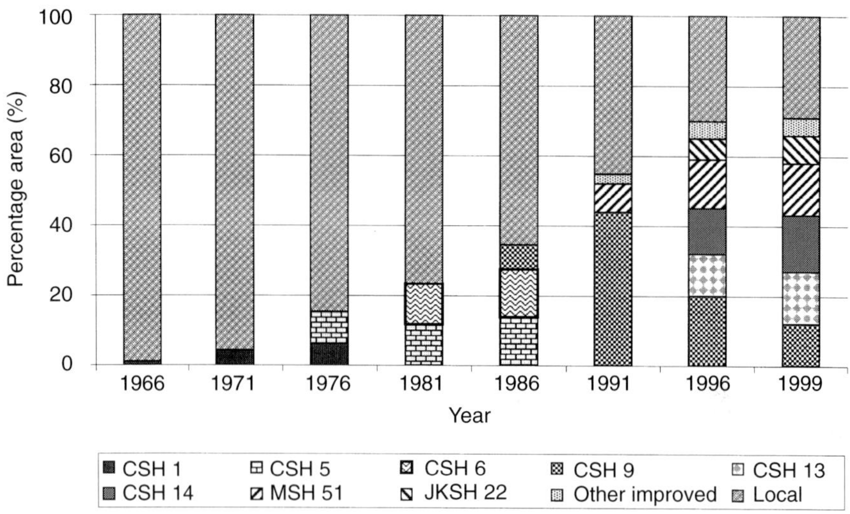

Fig. 9.2. Trends in adoption rate of different improved sorghum cultivars in India, 1966–1999.

farmers to a large extent. JKSH 22, another cultivar from the private sector, is also gaining ground. Improved open-pollinated varieties were less popular than the hybrids from the beginning (Rana *et al.*, 1997). Since the hybrids provide higher yield and are now readily available from a large number of private and public seed companies, the adoption of hybrids has taken off. Three phases in the spread of improved sorghum cultivars are observed in India. Until 1975, only CSH 1 was dominant, and it replaced traditional local cultivars. Between 1976 and 1986, the dominant improved sorghum cultivars were CSH 5 and CSH 6. This phase was characterized by the replacement of traditional and initial improved cultivars (CSH 1, CSH 2, CSH 4) by new cultivars (CSH 5, CSH 6). After 1986, the initial cultivars were replaced by new cultivars (CSH 9, MSH 51, JKSH 22) at a faster rate. During this period, Indian farmers made use of the large number of private-sector hybrids in the market.

In China, the rate of adoption of improved cultivars is the highest among all Asian countries. In 1995/96, essentially all of the sorghum area was under improved cultivars. Hybrids are more popular than varieties. Even in 1975/76, the rate of adoption of improved sorghum cultivars in China was 90%. The popularity of different hybrids in China has varied over time. In 1975/76, six sorghum cultivars occupied more than 60% of the total sorghum area. The variety Jin Za No. 5 covered about 19% of the sorghum area of China in 1975/76. In 1980/81, the most popular cultivar was Tie Za No. 6, covering 23% of the total sorghum area. Five other cultivars (Liao Za No.1, Liao Za No. 2, Jin Za

No. 83, Shen Za No. 4 and Tie Za No. 7) together covered about 30%. Only 5% of the sorghum area was under local cultivars in China in 1980/81. In 1985/86, only 2% of the total sorghum area was under local cultivars. Popular cultivars were Shen Za No. 5, Qiao Za No. 2, Liao Za No. 4 and Jin Za No. 94. In 1990/91, the rate of adoption of improved cultivars was 98%, and four cultivars (Long Si Za No. 1, Jin Za No. 12, Tie Za No. 10 and Liao Za No. 5) were popular. By the year 1995/96, 99% of the total sorghum area of China was under improved cultivars. Four improved cultivars (Long Za No. 3, Liao Za No. 6, Liao Za No. 7 and Liao Za No. 10) were popular. About 9% of the sorghum area in China is from ICRISAT parents.

Though Myanmar has 22 recommended improved varieties to cultivate, the rate of adoption remained low. In 1975/76, only local sorghums were in cultivation, while in 1995/96, all areas under improved cultivars (10%) were growing ICRISAT-bred varieties. The popular cultivars were Yezin White Grain 1, Yezin White Grain 2 and Yezin White Grain 3.

Local cultivars in Pakistan have always dominated sorghum cultivation. In 1980/81, only 7% of the total sorghum area in Pakistan was under improved cultivars, which had increased to 21% by 1995/96. However, as of 1995/96, no ICRISAT cultivars have been grown in Pakistan.

The large sorghum-growing area in Egypt is predominantly under local varieties. In 1975/76, only 5% of the area was under improved sorghum cultivars, which had increased to 35% (including 5% area under ICRISAT parents) in 1995/96. In 1975/76, Giza 114 was the only improved cultivar grown in Egypt, while in 1980/81, two cultivars – Giza 114 (4.4%) and Giza 15 (10.6%) – covered 15% of the total sorghum area. By the year 1985/86, Giza 114 was out of cultivation in Egypt, and Giza 15 became the most popular cultivar, covering 15% of the total sorghum area. Giza 15 is still popular in Egypt and covered about 17% of the total sorghum area in 1995/96. Other improved cultivars were Giza 113 and Dorado.

Local varieties are still dominant in Nigeria. Two ICRISAT-bred cultivars (ICSV 111 and ICSV 400) are gaining popularity among farmers of Nigeria. Ogungbile *et al.* (1997) conducted a study to determine the nature and extent and determinants of adoption of ICSV 111 and ICSV 400 in 1996. A survey was conducted in nine villages of Kano state, nine villages in Katsina state, six villages in Kaduna state and three villages in Jigwa state. A total of 219 farmers from 27 villages in four states were interviewed. The rate of adoption of improved cultivars was 28% in Kano, 10% in Katsina, 29% in Kaduna and 3% in Jigwa.

Yapi *et al.* (1998) studied the adoption and benefits of improved sorghum variety S 35 (an ICRISAT-bred variety) in Chad, based on farm survey data collected from 152 farmers from 28 villages in 17 districts for the year 1994/95. The study was conducted in three zones: Guera, Mayo-Kebbi and Chari-Baguirmi, which are located in the Sahelian and the Sahelian-Sudanian zones, where climate affects yield and consequently necessitates short-cycle crop varieties such as S 35. These three zones were target and distribution zones for S 35 in Chad. The adoption rate was higher in Guera (38%) than in Mayo-Kebbi (27%) and Chari-Baguirmi (24%). The lower adoption rates in Chari Baguirmi can be explained by farmers' preference for local red sorghum (*djigari*) rather than to white sorghums such as S 35, along with differences in climate and seed availability.

The same variety, S 35, was also grown extensively in northern Cameroon (Ndjomaha *et al.*, 1998). Ten years after introduction, S 35 was being grown on 50% of the rainfed sorghum area in the Mayo Sava zone.

Yapi *et al.* (1998) studied the adoption of improved sorghum cultivars in three regions – Koulikoro, Ségou and Mopti – of Mali. The area under improved cultivars ranged from 17% to 29% between 1990 and 1995 for all three regions.

Southern and Eastern Africa

Phofu is the most popular improved variety in Botswana. The adoption rate of Phofu in 1997/98 was 21% (Rohrbach and Makhwaje, 1999). The rates of adoption of ICRISAT-bred varieties in Sudan, Malawi, Zambia and Zimbabwe in 1997 were 3%, 10%, 35% and 36%, respectively (Table 9.6). The rate of adoption of improved sorghum cultivars in South Africa in 1997 was 77% but all are under non-ICRISAT cultivars. It should be mentioned here that South Africa came under ICRISAT partnership only in 1994/95, and ICRISAT cultivars are not yet released in South Africa.

Constraints to adoption as reported by the farmers

Ogungbile *et al.* (1999) report that in a survey of Nigerian farmers, low soil fertility was a widely mentioned constraint to adoption of improved varieties. It was the opinion of the farmers that the cultivars (ICSV 111 and ICSV 400) would not do well in marginal land without adequate fertilizer application. Another important constraint mentioned was insect damage. The varieties were reported to be susceptible to stem-borer attack. This was attributed to the sugary nature of the stem. Another problem was die-back, which prevents good crop establishment.

Lack of seeds was another major reason mentioned by most of the respondents in the Nigerian study for not growing the improved varieties. Inadequate supply and high cost of fertilizer also affected the adoption of the cultivars. Credit facilities would be needed to enable the farmers to purchase the necessary inputs (Ogungbile *et al.*, 1999).

Yapi *et al.* (1999) reported constraints to adoption of S 35 in Chad, as mentioned by farmers, to include bird damage, poor soil fertility, seed availability and seed cost.

Reasons for non-adoption in Cameroon were many. The most important reasons cited by farmers include losses due to birds, grain mould, high price of milling, regermination of seed, requirements for soil fertility, poor quality of beer, small stalks for construction, stalks disliked by animals and lack of seed (Ndjomaha *et al.*, 1998).

The most significant constraints to the adoption of improved sorghum cultivars, cited by Mali farmers, are lack of information about the existence and use of new varieties (58%), lack of seed (50%) and poor soil (13%). Lack of information and seed are the most important constraints in all three regions, and poor soil is only a problem in Mopti. In Ségou, there is a strong preference for local varieties. For sorghum in Koulikoro, the need to use fertilizer on improved varieties, bird damage, labour shortages and storage are constraints (Yapi *et al.*, 1998).

Dimensions of Impacts

Impacts on yield

For any crop, it can be difficult to interpret yield levels and changes in yield as measures of research impact. This is particularly true for crops such as sorghum that are customarily grown with few inputs on poor-quality land. Even small changes in the quantities of inputs used or the quality of the land planted to sorghum can have large effects on yield.

Global yield scenario

In the early 1960s, the yield level was very low in most of the developing countries, but by the mid-1990s, yield levels had gone up in Asia. Yield has increased in China, India, Pakistan and North Korea. Per hectare yield in China has increased more than 3 tonnes (3213 kg) and in India by 320 kg (65%). Adoption of hybrids in China is more than 94%. India has about 65% adoption of improved cultivars. Yield increase in Pakistan was by 104 kg (21%). Yield was doubled in North

Korea. However, sorghum yield in Thailand has decreased. In the 1960s, Thailand was growing sorghum for grain purposes, but since the late 1980s a large area under sorghum is now for fodder purpose, and they export dried fodder to Japan. FAO data do not record this fact. FAO data only report area harvested and grain production.

In Africa, yield has increased in South Africa, Egypt, Uganda, Ethiopia, Ghana, Burkina Faso, Lesotho, Nigeria and Namibia to a significant extent. Sorghum yield in South Africa tripled and the adoption rate of improved cultivars in South Africa is 77%. Yield has declined in Niger, Sudan, Mozambique, Rwanda, Kenya and Eritrea to a notable extent. In other African countries there has been no significant change. In many southern African countries, yield was lower in the 1990s than in the early 1960s. Among the explanations for this decline are low fertilizer use and a shift of sorghum cultivation to poorer land. Furthermore, breeders have emphasized developing cultivars with early maturity and yield stability, rather than high yield *per se*. A thorough analysis will be required to identify the reasons for decline in yield in many African countries.

By contrast sorghum yields in European countries have increased substantially. Yield has almost doubled in Italy and France, tripled in Greece and risen fivefold in Spain. In the Americas, yields have doubled in Argentina, Nicaragua, Guatemala and Peru. There were also significant increases in yield in Colombia, Mexico and the USA.

District level impact situation in India

Figure 9.3 portrays the yield scenario in different sorghum growing districts of India for *kharif* sorghum, which was the focus of India's research programme. In Fig. 9.3, we see that yield gains for rainy season (*kharif*) sorghum in Maharashtra and Andhra Pradesh were high where adoption rates were also high. Yield increased at least 750 kg ha^{-1} in these districts and more than 1 tonne in many districts. In the dry season (*rabi*) sorghum, research was less, and fewer improved cultivars were developed, and yield increases were lower (Fig. 9.4).

Yield gain at the farm level

Table 9.7 summarizes the farm-level yield gain information from different studies conducted in Africa. These data are, of course, problematic because they compare yields of improved varieties with yields of local cultivars for farmers who have adopted improved varieties. Obviously this gives a less satisfactory measure of yield advantages for other farmers for whom such yield differences are presumably smaller. None the less, the data offer some useful perspectives. Yields

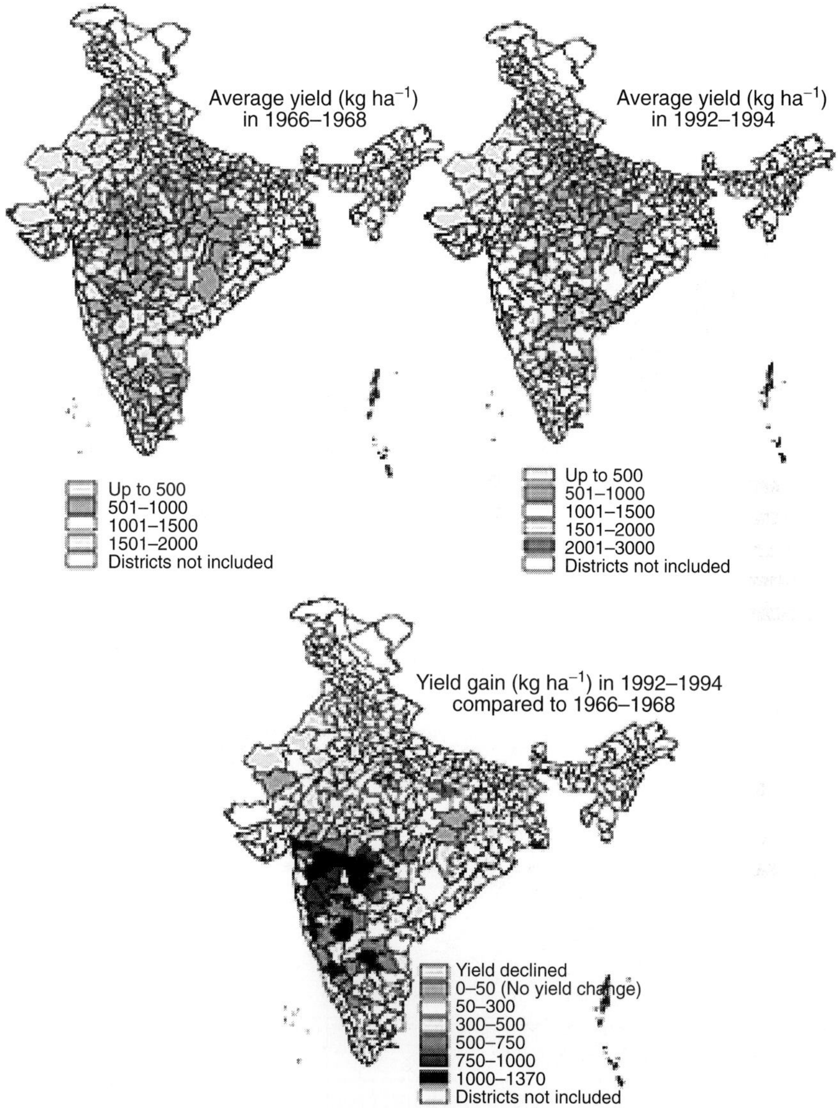

Fig. 9.3. Average yield and yield gain in *kharif* sorghum in India.

of improved cultivars were 7%–63% higher than the best local cultivars in Nigeria. Improved sorghum variety S 35 had 51% yield advantage in Chad and 14% in Cameroon. Ndjomaha *et al.* (1998) reported that during the period 1986–1995, the per hectare difference in productivity between S 35 and the local variety was 432 kg in Mayo Sava, 89 kg in Diamaré, and 52 kg in Mayo Danay regions of Cameroon.

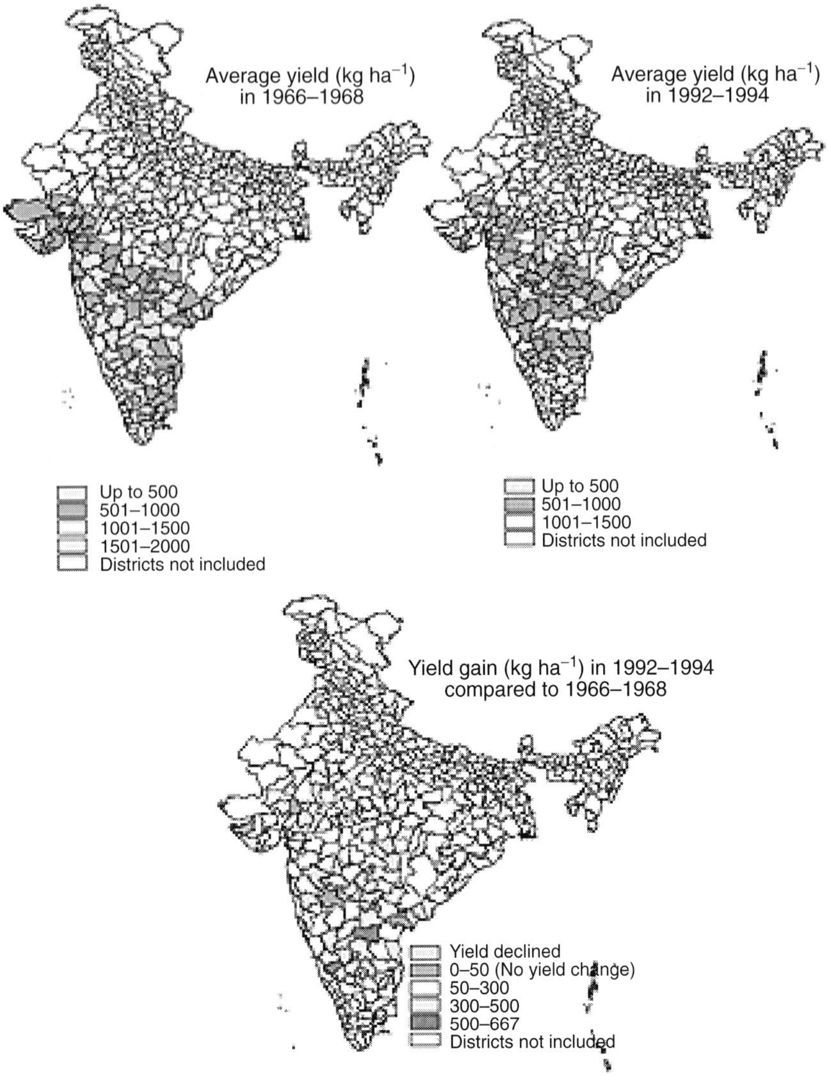

Fig. 9.4. Average yield and yield gain in *rabi* sorghum in India.

These differences indicate a better genetic potential for S 35 in Mayo Sava than in the other two areas, probably because rainfall is more congruent with the 300–800 mm research recommendation. In Mali, sorghum yields increased from 620 kg ha^{-1} with the best local variety to 940 kg ha^{-1} for improved varieties and increases in profits by 51% (Yapi *et al.*, 1998). These yields are consistent with those found in previous studies.

Table 9.7. Impacts of improved sorghum cultivars on yield.

Country	Region	Year	Improved cultivar	Yield (kg ha^{-1}) of Local	Yield (kg ha^{-1}) of Improved	Yield gain (%)
Cameroon	Mayo-Sava	1995	S 35	1220	1650	36
Cameroon	Diamare	1995	S 35	1450	1540	6
Cameroon	Mayo Danay	1995	S 35	1420	1470	4
Cameroon		1995	S 35	1360	1550	14
Chad	Guera	1995	S 35	710	1090	54
Chad	Mayo-Kebbi	1995	S 35	780	1190	53
Chad	Chari-Baguirmi	1995	S 35	810	1180	46
Chad		1995	S 35	760	1150	51
Nigeria	Kano	1996	ICSV 400	875	1165	33
Nigeria	Katsina	1996	ICSV 400	1003	1073	7
Nigeria	Jigawa	1996	ICSV 400	865	1398	62
Nigeria		1996	ICSV 400	914	1212	33
Nigeria	Kano	1996	ICSV 111	875	1221	40
Nigeria	Katsina	1996	ICSV 111	1003	1274	27
Nigeria	Jigawa	1996	ICSV 111	865	1406	63
Nigeria		1996	ICSV 111	914	1300	42

Source: For Cameroon and Chad, Yapi *et al.* (1999) and for Nigeria, Ogungbile *et al.* (1999).

Impacts on cost of production

An alternative measure of productivity gains is the unit cost of production. An analysis in India shows that real cost per tonne of sorghum production decreased in the 1980s and 1990s compared with that of the early 1970s. In Maharashtra, the cost per tonne in the 1990s was 40% below the level in the 1970s. In Rajasthan, this figure was 37% (Table 9.8).

The same measure is available for a few locations in Africa (Table 9.9).[1] S 35 has a cost advantage of 12% in Cameroon and 25% in Chad (Yapi *et al.*, 1999). Using improved varieties of sorghum reduced production costs as much as 25% (US$34 t^{-1}), compared with local varieties in Mali. The absolute production cost per hectare was higher for improved varieties because of additional inputs, but the higher productivity still provided these economies. With this higher productivity, farmers have the opportunity to reduce the area sown to sorghum and diversify their farming to grow other crops either for the market or for their own consumption.

[1] The studies in Africa are specifically for areas of high research impact.

Table 9.8. Impact of improved sorghum cultivars on per tonne production cost in India, 1971–1995.

	Average cost (Rs t^{-1})			Cost reduction (%) compared to early 1970s in	
States	Early 1970s[a]	Early 1980s[b]	Early 1990s[c]	Early 1980s	Early 1990s
Andhra Pradesh	270	NA	286	NA	− 6
Karnataka	224	192	231	14	− 4
Madhya Pradesh	223	169	208	24	7
Maharashtra	253	188	153	25	40
Rajasthan	309	264	195	14	37

Note: All costs are real cost of production. For Rajasthan, real cost is computed on the basis of 1992 prices and all other states based on 1989 prices.
[a]Early 1970s indicate for Andhra Pradesh (average of 1973 and 1974), Karnataka (average of 1972–1974), Madhya Pradesh (1976), Maharashtra (average of 1972–1974) and Rajasthan (average of 1972–1974).
[b]Early 80s indicate for Karnataka (average of 1981–1983), Madhya Pradesh (average of 1981–83), Maharashtra (average of 1982–1983), Rajasthan (average of 1981–1983).
[c]Early 1990s indicate for Andhra Pradesh (average of 1994–1995), Karnataka (1991), Madhya Pradesh (average of 1994–1995), Maharashtra (1995), Rajasthan (1992).
Source: Estimated from cost of cultivation reports (various issues).

Table 9.9. Impacts of improved sorghum cultivar (S 35) on cost of production in Cameroon and Chad, 1995.

		Production costs (CFA francs t^{-1})		
Country	Region	Local	Improved	Unit cost reduction (%)
Cameroon	Mayo-Sava	77,500	57,700	26
Cameroon	Diamare	63,500	58,900	7
Cameroon	Mayo Danay	50,000	49,300	1
Cameroon		63,161	55,607	12
Chad	Guera	89,296	65,825	26
Chad	Mayo-Kebbi	45,994	37,903	18
Chad	Chari-Baguirmi	67,765	49,947	26
Chad		80,805	60,817	25

Source: Yapi *et al.* (1999).

Impact on average yield and instability in yield

Deb *et al.* (1999) conducted a study to quantify the impact of improved sorghum cultivars on yield increase and instability in sorghum yield in India. They measured instability (using the Cuddy-Della Valle index) as well as mean yield of sorghum for two periods: 1966/67 to 1980/81, and 1981/82 to 1993/94. During the first period (1966/67 to 1980/81) percentage of HYV sorghum area to the total sorghum area in India was less than 20%, while in the second period (1981/82 to 1993/94) it was above 20%. The coefficient of genetic diversity among the improved cultivars was also very low in the first period and increased significantly during the second period (Deb and Bantilan, 1998). Therefore, Period 1 can also be treated as a low genetic diversity period while Period 2 can be considered as a genetically diversified sorghum cultivation period. Table 9.10 presents data on average yield and yield variability in sorghum for the two periods. Average yields in India during Period 1 were 582 kg, and 748 kg in Period 2. In all the states except Gujarat, the measured coefficient of variation in yield decreased from the first period to the second period. An implication of this finding is that food security has been strengthened over time through the reduction in year-to-year yield fluctuation.

A more detailed analysis of the changes in yield stability is presented in Deb *et al.* (1999).

Impact on genetic diversity

The genetic diversity in a crop species is related to its stability and improvement. Genetic uniformity increases the possibility that an unexpected pest or disease could cause a major loss in production. Deb and Bantilan (1998) computed genetic diversity in sorghum cultivars for India and measured the relationship between genetic diversity and yield stability.

Table 9.10. Average yield and relative variability in yield of sorghum in different states of India.

	Period I (1966–80)		Period II (1981–93)		Percentage change	
State	Yield (kg ha^{-1})	CV (%)	Yield (kg ha^{-1})	CV (%)	Yield (kg ha^{-1})	CV (%)
Andhra Pradesh	521	23.02	661	21.66	26.84	−5.91
Gujarat	499	31.55	551	42.51	10.38	34.76
Karnataka	985	26.65	957	23.08	−2.91	−13.40
Madhya Pradesh	729	24.08	896	19.52	22.76	−18.96
Maharashtra	609	29.50	902	26.51	17.99	−6.71
Rajasthan	300	58.62	412	50.77	37.47	−13.40
Tamil Nadu	943	28.13	1113	26.24	17.99	−6.71
INDIA	582	10.59	748	13.02	28.47	22.97

Source: Deb *et al.* (1999).

Three types of genetic diversity indices – average, recommended and weighted – were computed following Souza *et al.* (1994). *Average diversity* estimates are based on the average coefficient of diversity (COD) of each variety grown in a given region in a given year. *Weighted diversity* was based on a weighted average of the COD of each variety weighted by the proportion of the area sown to each variety in a given region in a given year. *Recommended diversity* was based on the average COD of each cultivar recommended by either public or private research system or notified by the seed certification agency for a given region in a given year. Deb and Bantilan (1998) explore this topic in detail. All three diversity measures showed increases over time. Interestingly, the level of weighted diversity was much higher in 1994 than in 1966, in all states of India. Analysis shows that sorghum breeders were using different parental materials to develop new improved cultivars rather than depending only on a few parent materials.

The relationship between genetic diversity and yield instability is interesting. Genetic diversity in sorghum cultivation has increased in Andhra Pradesh, Karnataka, Madhya Pradesh and Tamil Nadu, while the index of yield instability has decreased in these states. In Maharashtra, genetic diversity at the end of Period 1 and at the end of Period 2 was almost the same. The variability situation was also similar in these two periods. In Rajasthan, genetic diversity has decreased but relative variability has increased. In other words, yield stability has increased in all the major sorghum-producing states of India with the increase in genetic diversity. Thus, the results suggest that sorghum germplasm research in India has contributed to an increase in genetic diversity and has thereby helped to reduce instability in sorghum yield.

Spillover impacts

Deb and Bantilan (1998) quantified potential spillover impacts of sorghum varieties using International Sorghum Varieties and Hybrid Adaptation Trials (ISVHAT) and All India Coordinated Sorghum Improvement Project (AICSIP) trial data. ISVHAT data used in the study include experimental trials conducted during 1989–1992 in 59 locations of 26 countries of Asia, Africa and Latin America. AICSIP data cover trials conducted at more than 80 locations in India in 1975/76 to 1995/96. They assumed the performance of a variety as a function of environmental variables (location dummy, year dummy) and technology variables (vintage of the variety, origin of the variety). The model was estimated separately for each sorghum domain (see Table 9.4 for description of domains). The results showed that ICRISAT crosses performed well in most sorghum domains, especially in SD1 (rainy season, multipurpose),

SD2 (late maturing, dual purpose), SD3 (early maturing, dual purpose) and SD8(ii) (low latitude). ICRISAT cultivars bred at Patancheru enjoyed a yield advantage of 277 kg ha^{-1} in SD1, 354 kg ha^{-1} in SD3 and 175 kg ha^{-1} in SD2. In other words, yield advantage was as high as 27% in low altitude environments, 15% in SD1 (rainy season, multipurpose), 13% in SD2 (late maturing, dual purpose) and 7% in SD3 (early maturing, dual purpose). The positive yield advantage of ICRISAT-Patancheru bred materials indicates the potential of ICRISAT cultivars to spillover to these test domains. This also indicates that ICRISAT-Patancheru is a suitable location for breeding targeted for wide adaptability.

Brennan and Bantilan (1999) quantified spillover impacts of ICRISAT research on breeding programmes and agricultural production in Australia. They identified ICRISAT germplasm lines released in Australia by breeders and adopted by Australian farmers. For sorghum, the most significant contribution from ICRISAT to Australian agriculture has been the introduction of improved midge resistance combined with desirable white grain and tan plant colour through material such as ICSV 745 and PM 13654. There are several advanced breeding lines that have resistance and other characteristics incorporated from ICRISAT-derived material. As a result, industry experts expect that hybrids with this resistance will be available to growers in the near future, with a significant economic impact on the sorghum industry. On the basis that such resistance is likely to increase yield by 5% in the 50% of the crop affected by midge each year, the expected gains to Australia in terms of yield are estimated at 2.5%. That translates into a cost reduction of US$4.02 per tonne, or an annual cost saving of US$4.69 million at recent average production levels.

Other spillovers can be documented. Macia, a variety released in Mozambique, was also released in Botswana, Tanzania and Namibia (Table 9.11). Similarly, S 35 was developed in India and was adopted in the farmers' fields of Cameroon and Chad. ICSV 111 was developed in India and was released in Burkina Faso, Chad and Nigeria. ICSV 1079 BF was developed in Burkina Faso and cultivated by farmers in Benin, Ghana and Nigeria. SPV 475 was developed for India and is now cultivated in Malawi, Swaziland and Zimbabwe. Seredo was developed for Uganda but is also cultivated by farmers of Ethiopia, Kenya and Tanzania.

These examples show that breeders were successful in generating technology with wide adaptability and technology spillover potential. The results do not sustain this 'location specificity' argument (at least in terms of yields) when the international research system is considered as a source of research spillovers. Sorghum cultivars originating from a collaborative ICRISAT-NARS international research system have proven to be highly transferable within sorghum domains and across different countries around the world.

Table 9.11. Sorghum germplasm spillovers.

Cultivar	Production system* and country where originally selected	Spillover into
5D X 160	Uganda	Rwanda; 20, 21 Burundi
Dinkmash	India	19, 20 Ethiopia
Gambella 1107	Ethiopia	20, 21 Burundi
Ingazi	India	19, 20 Kenya
Macia	Mozambique	19 Botswana, Tanzania, Namibia
Melkamash	India	20 Ethiopia
Seredo	Uganda	19 Ethiopia; 20, 21 Kenya; 20 Tanzania
SPV 475	India	20 Malawi, Swaziland, Zimbabwe
SRN 39	India	19 Sudan, 20 Kenya, 20 Ethiopia
Tegemeo	Uganda	19, 20 Tanzania; 20 Burundi
S 35	India	Cameroon, Chad
CE 151	Senegal	Mauritania
CE 145-66	Senegal	Mauritania
Malisor 84-1	Mali	Côte d'Ivoire
BF 83-3/ 48-2-2	Burkina Faso	Senegal
IRAT	Niger	Burkina Faso, Chad
ICSV 111 IN	India	Benin, Ghana, Nigeria, Senegal, Togo
ICSV 1079 BF	Burkina Faso	Mali, Togo
ICSV 1083 BF	Burkina Faso	Togo
ICSV 1089 BF	Burkina Faso	Senegal, Mali, Togo
ICSV 400	India	Nigeria

Notes:
*Production system 8 (PS 8) is tropical, low rainfall, primarily rainfed, post-rainy season crops are sorghum/oilseed. Western Deccan Plateau of India is the location included in PS 8. Production system 19 (PS 19) covers lowland, rainfed, short-season (less than 100 days) and suitable for sorghum/millet/rangeland, and located in Sahelian eastern Africa, and margins of the Kalahari Desert. Production system 20 (PS 20) covers semi-arid, intermediate season (100–125 days) and suitable for sorghum/maize/rangeland; and located in eastern Africa and parts of southern Africa. Production system 21 is intermediate season (125–150 days) suitable for sorghum/maize/finger millet/legumes and located in eastern and southern Africa. Agro-ecological details of each production system are given in the ICRISAT Annual Report, 1993.
Source: *ICRISAT Southern and Eastern Africa Highlights* (1996), p. 30.

Returns on research

Previous studies have attempted to compute the economic impacts of ICRISAT-NARS research. For example, the net present value (NPV) of benefits from sorghum variety S 35 are estimated at US$15 million in Chad and US$4.6 million in Cameroon, with an internal rate of return

(IRR) of 95% in Chad and 75% in Cameroon (Yapi *et al*., 1999). Improved sorghum cultivars in Mali are estimated to have generated an NPV of US$16 million with an IRR of 69% (Yapi *et al*., 1999). Other estimates have put the rate of returns from sorghum research in Zambia at 11–15% and in Zimbabwe (for ICSV 88060) at 22%.

Summary and Conclusions

Asian NARS have devoted more resources than their African counterparts to sorghum improvement. ICRISAT has worked to develop collaborative strategies appropriate for both sets of parties. ICRISAT has collected and assembled a large quantity of sorghum germplasm from all over the globe. Breeding research effort was large and evolved over time to have complementarities with the growing capacity of some national systems. As a result, a large proportion of released sorghum cultivars has been developed either by ICRISAT or by its partners using ICRISAT materials. In Africa almost all the cultivars released in the 1980s and 1990s are either ICRISAT-bred or acquired through ICRISAT networks. Where improved cultivars were adopted by the farmers, they appear to have increased yield, reduced the cost of production, and decreased yield variability. In general, countries with strong NARS benefited from elite germplasm and parental materials. On the other hand, countries with weak NARS benefited from finished products. Therefore, two distinct breeding strategies are required for strong NARS and weak NARS. For strong NARS, ICRISAT should develop parental materials and elite germplasms. Countries with weak NARS will benefit more if ICRISAT develops more finished products jointly with them.

Acknowledgements

The authors are grateful to Drs C.L.L. Gowda, B.V.S. Reddy, Paula Bramel, N. Kameswara Rao, A.B. Obilana, I. Akintayo, David Rohrbach, Mary Mgonja, Yang Zhen, Wang Liangqun, B.S. Rana, Aziz Fouman Ajirlou, Ms Daw Khin Mar Yee, Ms Kanoktip Lertdprasertrat, Abdul Shakoor, Osman El Obeid Ibrahim, Osman El Nagouly, Clement Kamau, M. Chisi, M. Sadaan, L.T. Mpofu and Peter Setimela for information and help in completing the Impact Monitoring Survey in Asia and Africa; we also acknowledge Dr F.T. Bantilan, Irshad Ahmed and Md. Moinuddin for GIS-based mapping and delineation of sorghum domains. However, the views expressed in this chapter are those of the authors.

References

Brennan, J.P. and Bantilan, M.C.S. (1999) *Impact of ICRISAT Research on Australian Agriculture.* Economic Research Report no. 1, Australian Centre for International Agricultural Research, Canberra, Australia.

Deb, U.K. and Bantilan, M.C.S. (1998) Impacts on genetic diversity. In: Deb, U.K. and Bantilan, M.C.S. (eds) *Spillover Impacts of Sorghum Germplasm Research.* Socio-Economics and Policy Program, ICRISAT, Patancheru, India. Mimeo.

Deb, U.K., Bantilan, M.C.S. and Bantilan, F.T. (1998) Nature, extent and determinants of technology spillover. In: Deb, U.K. and Bantilan, M.C.S. (eds) *Spillover Impacts of Sorghum Germplasm Research.* Socio-Economics and Policy Program, ICRISAT, Patancheru, India. Mimeo.

Deb, U.K., Joshi, P.K. and Bantilan, M.C.S. (1999) Impact of modern cultivars on growth and relative variability in sorghum yields in India. *Agricultural Economics Research Review* 12 (2), 84–106.

FAO (1998) FAO statistical database.

Kamuanga, M.L. and Fobasso, M. (1994) Role of farmers in the evaluation of an improved variety: the case of S35 in northern Cameroon. *Journal for Farming System Research and Extension* 4(2), 93–110.

Ndjomaha, C., Yapi, A. and Adamri, A. (1998) Impact of sorghum S35 in extreme north Cameroon. In: Bantilan, M.C.S. and Joshi, P.K. (eds) *Assessing Joint Impacts: Proceedings of an International Workshop on Joint Impact Assessment of NARS/ICRISAT.* ICRISAT Patancheru, India, pp. 26–27.

Obilana, A.B. (1998) Sorghum improvement. *International Sorghum and Millet News Letter* 39, 4–17.

Obilana, A.B., Monyo, E.S. and Gupta, S.C. (1997) Impact of genetic improvement in sorghum and pearl millet: developing country experiences. In: *Proceedings of the International Conference on Genetic Improvement of Sorghum and Pearl Millet*, 22–27 September 1996, Lubbock, Texas. INTSORMIL and ICRISAT, pp. 119–141.

Ogungbile *et al.* (1999) *Factors Influencing Awareness and Adoption of ICSV 111 and ICSV 400 Sorghum Varieties.* Socio-Economics and Policy Program, ICRISAT, Patancheru, India. Mimeo.

Rana, B.S., Swarnalata Kaul and Rao, M.H. (1997) Impact of genetic improvement on sorghum productivity in India. In: *Proceedings of the International Conference on Genetic Improvement of Sorghum and Pearl Millet*, 22–27 September 1996, Lubbock, Texas. INTSORMIL and ICRISAT, pp. 141–165.

Reddy, B.V.S., Deb, U.K., Stenhouse, J.W. and Prakasha Rao (1998) Sorghum genetic enhancement research process at ICRISAT. In: Deb, U.K. and Bantilan, M.C.S. (eds) *Spillover Impacts of Sorghum Germplasm Research.* Socio-Economics and Policy Program, ICRISAT, Patancheru, India. Mimeo.

Rohrbach, D.D. and Makhwaje, E. (1999) *Adoption and Impact of New Sorghum Varieties in*

Botswana. PO Box 776, Bulawayo, Zimbabwe: Southern African Development Community (SADC)/ International Crops Research Institute for the Semi-Arid Tropics (ICRISAT), Sorghum and Millet Improvement Program (SMIP) (Semiformal publication).

Sanders, J.H., Bezuneh, T. and Schroeder, A.C. (1994) *Impact Assessment of the SAFGRAD Commodity Networks*. Mimeo.

Souza, E., Fox, P.N., Byerlee, D. and Skovmand, B. (1994) Spring wheat diversity in irrigated areas of two developing countries. *Crop Science* 34, 774–783.

Yapi, A., Kergna, A.O., Debrah, S.K., Sidibe, A. and Sango, O. (1998) Impact of sorghum research in Mali. In: Bantilan, M.C.S. and Joshi, P.K. (eds) *Assessing Joint Impacts: Proceedings of an International Workshop on Joint Impact Assessment of NARS/ICRISAT*. ICRISAT, Patancheru, India, pp. 76–93.

Yapi, A., Debrah, S.K., Dehala, G. and Njomaha, C. (1999) *Impact of Germplasm Research Spillovers: The Case of Sorghum Variety S35 in Cameroon and Chad*. Impact Series No. 3. ICRISAT, Patancheru, India.

Impacts of Genetic Enhancement in Pearl Millet 10

M.C.S. BANTILAN AND U.K. DEB

This chapter documents the benefits from pearl millet genetic enhancement research conducted by the International Crops Research Institute for the Semi-Arid Tropics (ICRISAT) in partnership with national agricultural research systems (NARS). ICRISAT-NARS research efforts and the resultant impacts are summarized, recognizing that many improved pearl millet cultivars are the joint products of the partnership. Benefits from pearl millet improvement research are measured in terms of yield gain, reductions in production cost and increases in profitability. This study documents the record of pearl millet germplasm improvement in the form of open-pollinated varieties (OPVs) and hybrids released by national programmes. Data based on farm-level surveys and secondary sources are used to generate productivity and other impact measures. The results indicate that pearl millet farmers adopted improved varieties based on early maturity, yield and profitability gains. Early maturing pearl millet cultivars have proven particularly desirable in drought-prone regions where food security is a severe problem. Lastly, this chapter presents an example of South–South research spillover, where research products developed at ICRISAT found applicability and adaptability across India and sub-Saharan Africa. The results highlight the critical role that an international research institution such as ICRISAT has in enabling research spillovers across national and continental boundaries.

Introduction

Pearl millet (*Pennisetum glaucum*) is a highly drought-tolerant cereal crop and an important food grain. It is generally grown as a rainfed crop on marginal lands with few inputs and little management. Pearl millet provides food for millions of people living in the arid and semi-arid regions of the Indian subcontinent and Africa. It is grown as a food crop in tropical Africa and India, with most production concentrated in Sahelian west Africa and northwestern India. These regions are characterized by high temperatures, short growing season, frequent drought, and sandy and infertile soils. In addition to its use for food, pearl millet has a high feed value for poultry and is a good source of energy and nitrogen in ruminant diets.

Accurate statistics on the area, production and productivity of pearl millet are not available, as pearl millet statistics are often grouped with other minor millets. According to Dendy (1995), pearl millet accounts for only 3.5% of world cereals area and about 1% of the total cereal production. However, it is an extremely important crop in the arid and semi-arid zones, where it is difficult to grow other crops. Harinarayana *et al.* (1999) reported that pearl millet is cultivated in over 30 countries of Asia and Africa on a total area of 24.2 million ha and production is around 16.3 million t. Of this, nearly half is in Asia, with India accounting for 10.4 million ha, or 43% of the total world area. ICRISAT and FAO (1996) provided information on total millet area, production and yield and mentioned the proportion of pearl millet production to total millet production for the period 1992–1994. For the present study, we have compiled information on total millet area, production and yield from FAO (1998) and reported annual averages for 1996–1998 (Table 10.1).

In India, this crop is grown in the drier areas of the central and western regions. Five states (Rajasthan, Maharashtra, Gujarat, Uttar Pradesh and Haryana) account for nearly 90% of the national pearl millet area. Western Africa accounts for nearly 45% of world pearl millet area. The crop is grown in 17 countries in this region, but five countries (Niger, Nigeria, Burkina Faso, Mali and Senegal) account for nearly 90% of the total cultivated area in the region.

Pearl millet grain yields are low, largely because of the harsh environments in which it is grown and also because farmers do not invest in purchased inputs. Yields range from over 1 t ha^{-1} in some countries to as little as 240 kg ha^{-1} in Botswana and Namibia (Table 10.1).

Generally, yield growth has been poor, and production increases have come largely through area expansion rather than higher productivity. However, pearl millet area in India is steadily decreasing (Kelly and Parthasarathy Rao, 1993). Since 1960–1965 about 0.9 million ha have gone out of pearl millet cultivation, particularly in Gujarat, Uttar Pradesh, Haryana, Tamil Nadu, Andhra Pradesh and Punjab.

Table 10.1. Area, production and productivity of millet in Asia and Africa, 1996–1998.

Country	Area (10^3 ha)	Production (10^3 t)	Grain yield (kg ha^{-1})	% share of pearl millet to total millet production[a]
A. Asia				
1 India	13,433.3	10,713.0	797	58
2 Myanmar	224.2	149.5	667	85
3 Pakistan	407.6	189.4	466	97
4 Yemen	97.7	59.6	604	100
Subtotal	14,162.8	11,111.5	784	
B. Western Africa				
5 Benin	36.9	28.3	766	100
6 Burkina Faso	1,203.1	673.0	557	99
7 Côte d' Ivoire	92.7	63.6	686	85
8 Cameroon	70.0	70.9	1,013	100
9 Central African Republic	11.3	11.3	1,000	87
10 Chad	697.3	290.5	414	100
11 Gambia	67.5	64.6	968	97
12 Ghana	170.1	166.3	980	100
13 Guinea	10.5	8.1	775	95
14 Guinea-Bissau	29.6	26.4	890	100
15 Mali	1,052.3	747.7	725	95
16 Mauritania	18.6	3.7	181	100
17 Niger	5,200.0	1,752.7	337	100
18 Nigeria	5,447.3	5,836.3	1,071	98
19 Senegal	871.4	484.8	553	100
20 Sierra Leone	24.0	19.6	818	100
21 Togo	100.1	48.5	485	100
Subtotal	15,102.7	10,296.3	682	
C. Southern and Eastern Africa[b]				
22 Angola	184.6	84.3	453	80
23 Botswana	8.3	2.3	264	100
24 Malawi	37.5	18.6	497	40
25 Mozambique	92.8	46.4	499	80
26 Namibia	268.8	66.2	240	100
27 Sudan	2,465.8	736.7	294	100
28 Tanzania	311.2	287.3	894	70
29 Zambia	79.0	59.4	751	40
30 Zimbabwe	260.7	94.4	356	70
Subtotal	3,708.7	1,395.6	376	
Total[c]	32,974.2	22,803.4	692	

[a]Percentage share of pearl millet to total millet production is taken from ICRISAT/FAO (1996) and relates to 1992–94.
[b]Kenya (1400 t) and Ethiopia (5000 t) are also reported to be producing pearl millet (Harinarayana *et al.*, 1999).
[c]In addition, pearl millet is grown on limited areas in Australia, Korea and USA (estimated around 1 million ha) for forage (Harinarayana *et al.*, 1999).
Source: FAO Statistical Data Base (1998).

Research Methodology

This study draws on a three-pronged approach (Fig. 10.1) to track usage of pearl millet parental lines. First, ICRISAT records provided data on the distribution of elite materials, hybrid parents and released open-pollinated varieties (OPVs) from the ICRISAT breeding programme to NARS seed multiplication agencies (public or private). This analysis was carried out for the period 1986–1998. Second, questionnaires were sent to 160 companies dealing with sorghum and pearl millet seed. A total of 49 companies (of which 37 dealt with pearl millet) responded. They provided information on the nature, extent and importance of ICRISAT breeding materials in their breeding and seed production programmes. Third, on-farm surveys were carried out in India (1683 farmers), Mali (345 farmers), Namibia and Zimbabwe. Information was gathered on adoption of different pearl millet cultivars, farm and farmer characteristics, farmer preferences for specific traits in the improved cultivars, and constraints to the cultivation of improved varieties.

In India, a total of 1683 farmers from 154 villages in five states, namely Maharashtra (360 farmers), Rajasthan (331), Gujarat (419), Haryana (237) and Tamil Nadu (336) were selected. Improved pearl millet cultivars were categorized into five groups: ICRISAT cultivars,

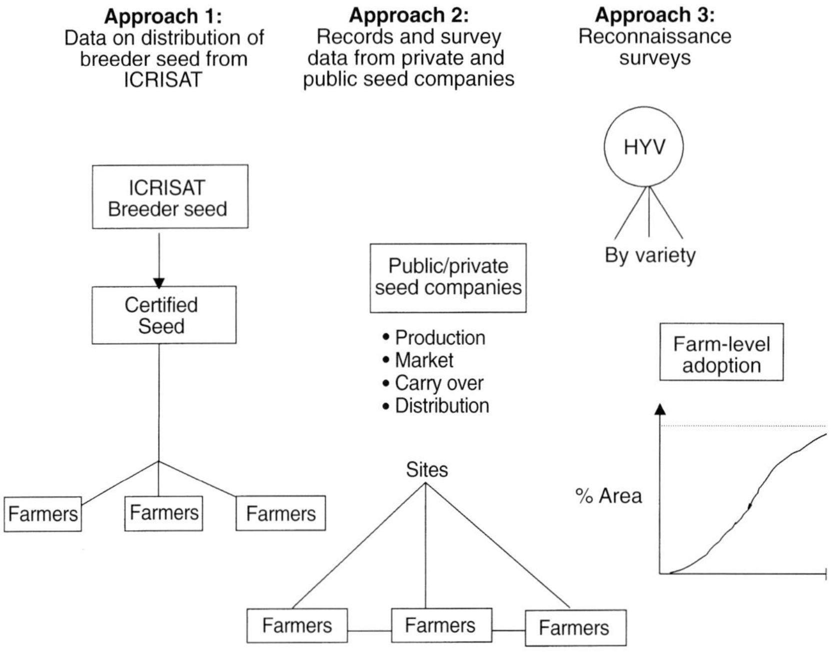

Fig. 10.1. Three approaches to track the adoption and impact of investments in pearl millet research at ICRISAT.

NARS-Public sector cultivars with ICRISAT materials, Private sector cultivars with ICRISAT materials, NARS-Public sector cultivars without ICRISAT materials and Private sector cultivars without ICRISAT materials. Some farmers were not able to name the variety which they grew, but were sure it was an improved cultivar. In such cases we have mentioned the cultivar as unidentified.

Human Resources Involved

Human resources involved in pearl millet genetic enhancement research in Asia and Africa are reported in Table 10.2. At ICRISAT, about five millet breeders are located in Asia and Africa. Fifteen other scientists, including agronomists, crop physiologists, genetic resources specialists, entomologists, pathologists and social scientists are generating information for effective use by the breeders. In India, about 150 pearl millet scientists in the public and private sector are working on this crop. However, in African countries few scientists work on this crop. Many are devoted to more than one crop; often millet and sorghum are combined.

The Research Process

Pre-breeding research

Collection, characterization and maintenance of landraces are essential for crop improvement, and these activities have been a high priority at ICRISAT. As of December 1999, 21,392 pearl millet germplasm accessions from 50 countries conserved at ICRISAT. After collection and assembly, ICRISAT and its NARS partners conduct evaluation trials to identify the useful traits available in the assembled germplasm. This information has been disseminated to researchers worldwide through reports, journal papers and other fora. In response to requests from users, ICRISAT has distributed 94,818 pearl millet germplasm samples to 74 countries: 69% were distributed to Asia, 27% to Africa, and 4% to other continents (Genetic Resources Unit, ICRISAT, 1999, personal communication).

ICRISAT's evolving focus

Research at ICRISAT began in 1972 with greater emphasis on applied rather than basic research. The focus was on grain yield and downy mildew resistance and exploratory research on ergot, smut, and rust resistance and drought tolerance. Equal emphasis was given to the

Table 10.2. Number of pearl millet scientists in different countries (1999 or latest year).

Country	Reference year	BSc	MSc	PhD	Total
Eastern and Central Africa*					
Burundi	1998	1	2		3
Eritrea	1998	4	1		5
Ethiopia	1998	15	26	9	50
Kenya	1998	4	14		18
Rwanda	1998	3			3
Sudan	1998	1	7	20	28
Uganda	1998		4	2	6
Southern Africa*					
Angola	1999				2
Botswana	1999	1	1	2	4
Lesotho	1999	1			1
Malawi	1999		1	2	3
Mozambique	1999	2	1		3
Namibia	1999	1	1		2
Swaziland	1999	1			1
Tanzania	1999	1	3	2	6
Zambia	1999		3	1	4
Zimbabwe	1999	1	2	2	5
West Africa					
Benin	1991				2
Burkina Faso	1991				13
Cameroon	1991				5
Chad	1991				4
Côte d'Ivoire	1991				4
Gambia	1991				4
Ghana	1991				5
Guinea Bissau	1991				19
Mali	1991				10
Mauritania	1991				4
Niger	1991				17
Nigeria	1991				6+11
Senegal	1991				4
Asia					
India	1998				150

Note: *Scientists working in Eastern and Central Africa, and southern Africa are involved in both sorghum and millet research.

Source: For Asia and southern Africa, ICRISAT Impact Monitoring Survey, 1998–2000;
For Eastern and Central Africa, Association for strengthening agricultural research in Eastern and Central Africa, ASARECA (1998);
For West Africa, Anand Kumar (1993).

development of finished products (cultivars) and improved breeding materials/parental lines. Development of improved breeding and screening methodologies was an integral part of applied research (Rai and Hash, 1994).

In the 1970s, breeding of open-pollinated varieties (OPVs), rather than hybrids, was emphasized. This was because ICRISAT had a comparative advantage over NARS, in terms of conducting large-scale inter-population improvement programmes across multiple locations. The Indian NARS had weak or no programmes in OPV breeding in the 1970s. Population breeding products (i.e. improved composites, open-pollinated varieties, early-generation progenies) were perceived to have the additional advantage of strengthening NARS hybrid programmes by providing improved germplasm for deriving hybrid parents. Indian NARS had adequate capacities to develop male-sterile lines. Hence ICRISAT devoted itself to producing restorers, and took to male-sterile lines breeding at a formal project level in the late 1970s.

Since the early 1980s, there has been a considerable improvement in the research capability of NARS in pearl millet research, especially in the Indian subcontinent. This has led to a reordering of ICRISAT's priorities. There was a shift in emphasis towards strategic research followed by continued emphasis on grain yield and downy mildew resistance. Almost all efforts were directed towards the development of improved breeding materials/parental lines (except for a few experimental varieties developed in partnership with NARS). Special effort was made to further refine breeding and screening methodologies, including the application of biotechnology, and relatively greater emphasis than in the past on escaping drought through early maturity (Rai and Hash, 1994).

Breeding for resistance to biotic and abiotic stresses focused mainly on downy mildew resistance. There was very limited research on ergot, smut and rust diseases (Hash, 1997). In India, downy mildew has been the major constraint to production since the 1960s, shortly after hybrids were widely introduced. Since then it has been a major research focus by both ICRISAT and the national programme (Nene and Singh, 1976; Dave, 1987; Rai and Singh, 1987; Shetty, 1987; Singh *et al.*, 1987, 1993). Hash (1997) reviewed the history of downy mildew research. From the published records and from the personal experience initially of D.J. Andrews and Hugh Doggett, it was clear that West African germplasm provided the best sources of genetic diversity for two major yield components (large head volume, large seed size) and high levels of resistance to downy mildew and smut. ICRISAT breeders were successful in incorporating downy mildew resistance genes in new cultivars that have allowed this very serious threat to be brought under control in India – at least for the time being (Hash, 1997).

Breeding for drought resistance received less priority because of the complex nature of the trait and difficulty in assessing the extent of genetic variation for drought resistance, and non-availability of a simple and reliable screening procedure. Another reason was that products arising from a drought-resistance breeding programme at one site were not easily applicable to other drought environments.

In short, in the 1970s the emphasis was on breeding OPVs. In the 1980s the emphasis shifted towards hybrid parents. In the 1990s, the focus was on upstream research in addition to the production of restorers and male-sterile lines, including the development of molecular marker-assisted products.

In the 1990s, ICRISAT made explicit the delineation of six research domains defined in Table 10.3.

Research Products

Intermediate products

ICRISAT has provided parent material to public and private partners since its inception. These include seed parents, i.e. A/B lines as well as pollen parents, i.e. R-lines. A list of varieties/hybrids from ICRISAT parent materials entered into the All India Coordinated Millet Improvement Project (AICMIP) advanced trials is given in Table 10.4.

Table 10.5 lists the quantities of pearl millet breeder seed distributed by ICRISAT to public and private seed multiplication agencies in India during the period 1987–1998. This supply has been substantial, with trends showing an increasing number of requests. Table 10.5 reflects the relatively higher proportion that is supplied to the private sector; for hybrid parents as well as OPVs. Among hybrid parents, 81A, 81B, 841A, 841B, 843A and 843B are the most frequently requested, and therefore supplied, to research agencies, both public and private. The small amounts of hybrid parents, 834A and 834B, that were supplied during 1991–1995 were received by the private sector. The volumes of 841A and 843A supplied in recent years have been increasing, replacing 81A and 81B, which dominated earlier. It is noteworthy that the quantity required remains high, but the responsibility for production and supply of breeder seed of 81A and 81B was turned over to public sector seed corporations in India in 1995.

Table 10.6 shows the extent of distribution of germplasm lines in southern African countries through the SADC/ICRISAT Sorghum and Millet Improvement Programme (SMIP), coordinated by ICRISAT. About 40,000 pearl millet germplasm lines were distributed to eight southern African countries, notably Zimbabwe, Malawi and Botswana.

Table 10.3. Pearl millet (PM) research domains.

Domain	Production system characteristics	Major constraints	Locations
PM I	Sandy, arid zone; early maturing, low-yielding traditional cultivars. Dual-purpose grain and fodder type	Heat and drought; need for reduced photoperiod sensitivity	India (Rajasthan, N. Gujarat, S. Haryana), Pakistan, other W. Asia
PM II	Early-maturing (but later than I). WC-C75 predominant. Dual-purpose grain and fodder type	Downy mildew, smut, and general yield improvement. Need for reduced photoperiod sensitivity	India (N. and E. Haryana, Uttar Pradesh, Madhya Pradesh)
PM IIIa	Medium-maturing hybrids (Asia only) and improved varieties. Grain types. Moderately later-maturing in Africa where traditional cultivars predominate	Downy mildew, drought, and general yield improvement. Photoperiod sensitivity less of a problem than in I and II	Southern Africa (Botswana, Zimbabwe, N. Namibia, S. Angola), India (S. Gujarat, Deccan)
PM IIIb	Medium-late maturing traditional and improved varieties. Grain types	Downy mildew, drought, and general yield improvement. Photoperiod sensitivity less of a problem than in I and II	E. Africa (Ethiopia, high-altitude), southern Africa (Angola, Zambia, Malawi, Mozambique), Latin America (some areas)
PM IIIc	Early- and medium-maturing traditional cultivars with large grain size. Grain types	Downy mildew, drought, and general yield improvement	W. Africa (Ghana, Togo)
PM IV	Post-rainy season/irrigated, improved cultivars. Fodder and dual-purpose types	Rust, downy mildew, general yield improvement	India (Tamil Nadu, Gujarat)
PM V	Sandy, arid zone, rainfed staple cereal, low-yielding traditional cultivars with long panicles. Hill sown. Intercropped with cowpea	Heat and drought, head caterpillars, *Striga*	W. Africa Sahelian zone
PM VI	Semi-arid, rainfed transition zone; low-yielding traditional photoperiod-sensitive cultivars with long panicles. Hill sown	Downy mildew, stem borers, drought, *Striga*	W. Africa Sudanian zone, E. Africa (Kenya, Zaire, Tanzania)

Table 10.4. List of varieties/hybrids entered into the AICMIP advanced trials from ICRISAT parent materials.

Year of first entry	Varieties	Hybrids
1978/79		ICH 154, 165, 105
1985/86		ICMH 423, 451, 83729, 501, 83202, 82601, 83506, 83401, IARI 1, RHRBH 379, 372, 373, 348, HHB 50, 56, 59, AHB 156, 163, PNBH 4
1986/87		ICH 451, HHB 57, 60, 61, 62, 63, ICMH 8370, 84122, 84913, RHRBH 8601, 8602, 8603, 8604, GHB 184, AHB 212, 251, 502
1987/88	ICMS 8010, 8283, 8253, DPBP 851, ICMV 83104, 87402, 84108, 87901, ICMV-F84400, RCB-IC 861, DPBP-IC 862, RCB-IC861, RCB-IC 9	IARI 1, ICMH 85109, 85231, 86217, 87004, RHB 33, 34, 35, 22, 24, 27, 28, 30, RHRBI 8605+B16, 8607, HHB 64, 61, 68, PNBH 6, AHB 615, 619, 623, GHB 179, 181, 205, ICMP 451
1988/89	ICMV 85328, 86104, 86120, 87902, 88907, ICMP 88130	ICMH 87003, 87004, 88088, 85118, 87353, 88951, PHB 122, RHRBH 122, RHRBH 8701, 8702, RHB 50, 54, 58
1991/92	ICMV 87111, 88402, 88908, 88904, 87107, 89410, RCB-IC 891, 892, 901, 902, 911, ECC 6	PUSA 23, HHB 67, 90, 92, ICMH 88735, 89998, 89024, 90952, AHB 838, 840, 919, 1068, 1203, GHB 228, 235, 263, 314, RHB 57, 85, 86, 87, 89, 90, 91, 92, 93, 94, PNBH 11, 14, PHB 133, 136, CZH 859-1
1992/93	CZ-IC 923, 922, 924, GICV 91123, 88921, 92191, PCB-IC 148, RCB-IC 912, 926, 924, 925	HHB 88, 94, 95, 96, 99, 100, CZH 848, PUSA 350, PHB 138, RHB 95, 96, 97, ICMH 91205, AHB 1073, GHB 274, PNBH 17, IBH 5527, 5534
1993/94	AIMP 92901, GICKV 92474, 91773, 92130, CZP-IC-315	PUSA 620, 613, 605, 623, PNBH 18, 19, 20, 22, 23, 25, RHB 98, 99, 100, 101, PHB 141, CZH 921, 922, DBDH 1, HHB 69, 105, 106, 107, 108
1994/95	RCB-IC 224, GICKV 93191, 93471, 93752, 93771, ICMV 93842	
1996/97		ICMH-356, PUSA 322

Source: Bantilan and Deb (2000) prepared from All India Coordinated Millet Improvement Progress Report.

Table 10.5. Supply of pearl millet breeder seed from ICRISAT to seed multiplication agencies in India, 1987–1998.

	Seed supplied samples											
	1987–88		1989–90		1991–92		1993–94		1995–96		1997–98	
Genotype	Public	Private	Public	Private	Public	Private	Public	Private	Public	Private	Public	Private
Hybrid parents												
81A	433	299	118	218	251	338	212	241	33	28	5	4
81B	190	160	111	104	120	163	97	110	13	3	2	2
ICMP 451	209	224	118	139	132	115	99	109	97	146	58	145
834A	44	73	0	3	0	3	0	5	3	3	0	0
834B	25	39	0	2	0	2	0	3	2	2	0	0
ICMP 501	18	10	0	0	0	2	0	4	2	1	42	2
841A	340	97	107	106	137	193	110	146	125	203	59	32
841B	134	59	59	54	67	105	53	74	71	98	27	17
ICMP 423	668	90	57	46	5	51	0	11	2	3	0	0
842A	–	–	–	–	–	–	21	66	152	74	10	37
842B	–	–	–	–	–	–	11	31	35	36	4	21
843A	21	0	84	39	56	118	58	108	133	166	42	62
843B	11	0	44	21	29	61	28	50	56	88	19	54
ICMA 88004	–	–	–	–	–	–	26	72	38	70	85	114
ICMB 88004	–	–	–	–	–	–	12	38	19	30	23	57
ICMR 356	–	–	–	–	–	–	12	41	71	39	48	43
Subtotal	2093	1051	798	732	797	1151	739	1109	852	990	424	590

Continued

Table 10.5. *Continued*

	Seed supplied samples											
	1987–88		1989–90		1991–92		1993–94		1995–96		1997–98	
Genotype	Public	Private	Public	Private	Public	Private	Public	Private	Public	Private	Public	Private
Open-pollinated cultivars												
WC-C75	848	510	149	302	93	59	173	127	14	5	0	0
ICMS 7703	163	202	76	60	21	4	8	15	4	10	0	12
ICMS 7704	193	141	0	0	0	0	0	0	0	0	0	0
ICTP 8203	183	53	238	212	110	295	112	270	40	50	10	10
ICMV 155	–	–	54	86	78	106	180	63	90	42	31	25
RAJ 171	–	–	–	–	–	–	15	55	45	33	47	64
ICMV 221	–	–	–	–	–	–	110	171	222	171	142	97
ICMV 87901	–	–	22	83	0	0	0	0	0	0	0	0
ICMR 501	–	–	15	15	0	0	–	–	–	41	0	21
ICMR 312	–	–	–	–	–	–	–	–	–	–	–	5
Subtotal	1387	906	554	743	553	464	498	744	415	352	230	234
Total	3480	1957	1619	1475	1109	1615	1237	1853	1267	1342	654	824

Table 10.6. Number of pearl millet germplasm lines distributed to the SADC countries.

Country	No. of lines distributed
Angola	97
Botswana	3,000
Lesotho	–
Malawi	5,000
Mozambique	200
Namibia	–
South Africa	100
Swaziland	–
Tanzania	6,000
Zambia	1,200
Zimbabwe	24,000
Total	39,597

Source: SMIP (1999).

In addition to the information gathered from breeder seed records, we also surveyed private companies in India under a study jointly undertaken by ICRISAT and Rutgers University to discover the use of ICRISAT breeding materials by the seed sector in India. Results of the survey are provided in Tables 10.7 and 10.8. Thirty-seven companies involved in pearl millet production responded to the survey questionnaire. Of these 37 companies, 34 are using ICRISAT breeding materials in their programme. About two-thirds use ICRISAT materials directly as hybrid parents, parents in crossing and for selection. One-quarter of the companies are directly producing ICRISAT varieties (Table 10.7). More than half of these companies feel that ICRISAT contributes more than 50% of their material in their breeding programme (Table 10.8). In terms of the level of importance of breeding materials obtained from

Table 10.7. Use of ICRISAT breeding material in the private sector research programme.

Mode of using ICRISAT breeding material	Number of companies	Percentage
By selection from ICRISAT material	24	71
As parents in crossing	22	65
Used directly as parents of hybrids	20	59
Used directly as varieties	8	24

Note: Total number of companies using ICRISAT breeding material = 34.
Source: ICRISAT-Rutgers University Study.

Table 10.8. Contribution of ICRISAT pearl millet breeding material.

Percentage contribution from ICRISAT	Number of companies
Directly released from ICRISAT (100%)	9
76–99%	4
51–75%	4
26–50%	16
Up to 25%	5
No contribution from ICRISAT	6
Details not provided	19

Note: Total number of cultivars released/developed/sold by these companies is 63.
Source: ICRISAT-Rutgers University Study.

different sources, out of 35 companies which responded, 28 mentioned ICRISAT as a very important source while six others mentioned it as one important source.

Varietal production

ICRISAT has also developed OPVs and hybrids. In 1982, an ICRISAT-bred, downy mildew resistant, open-pollinated variety, WC-C75, produced grain and stover yields equal to the best available hybrid at that time (BJ 104) and was released in India. This variety provided a timely alternative to the susceptible BJ 104, and to low-yielding local landraces. The rapid multiplication of WC-C75 and its adoption by farmers helped to prevent a decline in pearl millet production. In 1986, an ICRISAT downy mildew resistant hybrid, ICMH 451 (also known as MH 179) was released. It outyielded all other varieties and hybrids released earlier, and its seed production was relatively easy and profitable.

Table 10.9 shows the temporal distribution of pearl millet cultivar releases by origin in different countries. The average number of released varieties per annum has increased over time, especially in India. In southern Africa, most releases came only after the SADC/ICRISAT SMIP was launched in 1983.

Released cultivars, according to their pedigrees, are classified as ICRISAT cross, ICRISAT parent and ICRISAT network (i.e. cultivars developed by national programme or germplasm materials released as superior varieties through ICRISAT network trials). There was no release based on ICRISAT material prior to 1982. Out of 49 releases worldwide in the 1980s, 23 releases were of ICRISAT origin; out of 59 releases in the

Table 10.9. Number of released pearl millet varieties with ICRISAT content by period.

		1981–90					1991–98						Total	
Country	1965–80 Others	ICRISAT cross	ICRISAT parent	ICRISAT network	Others	All sources	ICRISAT cross	ICRISAT parent	ICRISAT network	Others	All sources	Unknown parent	ICRISAT derived	All sources
Asia														
India	15	5	8	1	23	37	4	24		7	35	2	42	87
Pakistan							1	1			2		2	2
Africa														
Botswana							1						1	1
Burkina Faso								3					3	3
Chad								2					2	2
Malawi							1		1		22		2	2
Mali			1			1		2			22		3	3
Mauritania								2			2		2	2
Namibia				1		2		-			2		2	4
Niger		3				3							3	3
Senegal		1				2	1				1	1	2	3
Sudan				1		1							1	1
Tanzania								2			2		2	2
Zambia		1				1	1	2	2		3		4	4
Zimbabwe				1		1	1	2			3		4	4
Total	15	10	9	4	23	48	10	39	3	7	54	3	75	123

1990s, 52 were of ICRISAT origin. Two points can be noted. First, particularly in Africa, many of the released varieties were developed by ICRISAT. Second, as NARS breeding programmes grew stronger in India, ICRISAT parents (rather than finished material) grew more in importance.

Adoption of Improved Cultivars

Adoption rates of improved pearl millet cultivars in different countries are provided in Table 10.10. Inter-country comparison of adoption shows that adoption rates vary from 5% to 65%. In India, the adoption rate is 65% and in Namibia, it is nearly 50%. Adoption rates are around 20–30% in Zambia, Mali, Zimbabwe and Botswana.

Figure 10.2 shows adoption trends in different districts of India for the period 1966–1994, based on district-level data obtained from published sources. Adoption of improved pearl millet cultivars has increased significantly over time, starting from very low adoption levels in the late 1960s. In 1992–1994, adoption was over 80% in most districts in Maharashtra (central India), Gujarat (western India) and Tamil Nadu (southern India). About 40 districts of India had attained more than 80% adoption rates. Increasing adoption over time was influenced by the development of downy mildew-resistant varieties at 4- to 5-year intervals. Widespread adoption has led to major yield gains, as discussed later.

Indian farmers were asked to rank the traits they liked in the improved cultivars they are growing. High grain yield ranked first in all states, while high fodder yield ranked second in Maharashtra, Haryana and Gujarat (Table 10.11). Other farmer-preferred traits were short duration, disease (downy mildew) resistance, drought resistance, good taste and large grain size (Bantilan *et al.*, 1999a,b). Tamil Nadu farmers cited 18 different factors that influenced them to adopt improved cultivars, but about 60% cited high yield, 10% cited resistance to drought and 9% cited seed availability (Ramasamy *et al.*, 1999).

Adoption of improved cultivars in three regions in Mali rose from 12% in 1990 to 23% in 1995 (Yapi *et al.*, 1998). Across the three study regions in Mali, the main reasons for adoption of new millet varieties are earliness (91%), productivity (72%) and food quality (33%). These reasons vary in order of importance in the three regions, perhaps due to rainfall differences.

About 50% of the total pearl millet area in Namibia is under one pearl millet variety, Okashana 1, developed by ICRISAT (Rohrbach *et al.*, 1999). Reasons for the high adoption were: (i) strong assistance from an international research centre such as ICRISAT; (ii) close collaboration of researchers with farmers; and (iii) complementary investments in seed production.

Table 10.10. Adoption of improved pearl millet cultivars.

Country	Region	Source	Year	Percentage of area planted to ICRISAT cross	ICRISAT parent	ICRISAT network	Others	All improved
Asia								
India	National	Bantilan and Deb (2000)	1990	18	15		27	60
India	National	Bantilan and Deb (2000)	1996	21	17		27	65
India	Eastern Rajasthan	Bantilan *et al.* (1999a)	1996	12	9.6		35	66.6
India	Haryana	Bantilan *et al.* (2000a)	1996	2	66		18	86
India	Gujarat	Bantilan *et al.* (2000b)	1995	31	47		21	99
India	Maharashtra	Bantilan *et al.* (1999b)	1994	36	43		15	94
India	Tamil Nadu	Ramasamy *et al.* (1999)	1994	22.6	6.6		48	77
Africa								
Angola	National	SMIP (1999)	1997					10
Botswana	National							30
Mali	Segou	Yapi *et al.* (1998)	1995					29
Mali	Koulikoro		1995					20
Mali	Mopti		1995					17
Malawi	National	SMIP (1999)	1997					7
Mozambique	National	SMIP (1999)	1997					11
Namibia	National	Rohrbach *et al.* (1999)	1997	49				49
Tanzania	National	SMIP (1999)	1997					1
Zambia	National	SMIP (1999)	1997					19
Zambia	Southern Province	Obilana *et al.* (1997)	1995	19				19
Zambia	Western Province	Obilana *et al.* (1997)	1995		62			62
Zimbabwe	National		1996	16	11			27
Zimbabwe	Southern Zimbabwe	Obilana *et al.* (1997)	1995	14				14

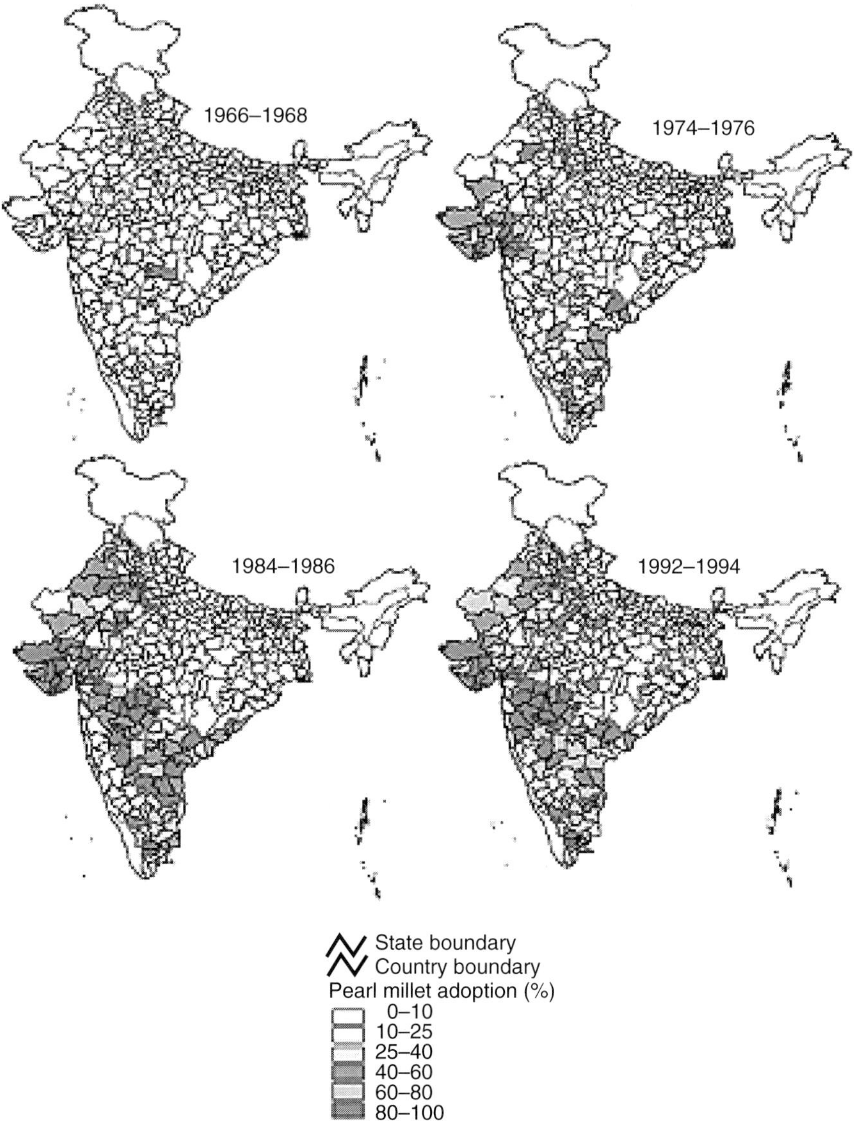

Fig. 10.2. Rate of adoption (%) of improved pearl millet cultivars in India.

In brief, reasons for high adoption of improved pearl millet cultivars are: high yield, short duration, reduced farmer risk due to early maturity and downy mildew resistance (India), and availability of seeds through private and public seed sector.

Table 10.11. Traits of improved pearl millet cultivars liked by farmers of selected states of India.

	Ranks provided by the farmers of				
Traits	Maharashtra	Rajasthan	Gujarat	Haryana	Tamil Nadu
Reference year	1994	1996	1995	1996	1994
Grain yield	1	1	1	1	1
Fodder yield	2	4	2	2	
Short duration		2	6	3	
Disease resistance	3	5	3	4	3
Drought resistance	2	3	5	5	2
Better taste	4	7	4		
Bold grain size	5	6			4

Source: Bantilan *et al.* (1999a,b, 2000a,b) and Ramasamy *et al.* (1999).

Constraints to adoption as reported by farmers

Indian farmers were asked to cite and rank the constraints they face in adopting improved pearl millet cultivars. The major constraints were non-availability of seed, low fodder yield of existing cultivars, lack of awareness, high water requirement for improved cultivars, poor extension service, and poor grain and fodder quality (Bantilan *et al.*, 1999a,b).

The most significant constraints to adoption cited by Mali farmers are lack of information about the existence of new varieties (49%), lack of seed (33%) and poor soil (26%) (Yapi *et al.*, 1998). Lack of information and seed are the most important constraints in all three regions, while poor soil is only a problem in Mopti. In Ségou, there is a strong preference for local varieties. The need for fertilizer is the most important constraint in Koulikoro.

Dimensions of Impacts

Improvement in efficiency in NARS research

As already noted, progress in the release of new varieties has increased significantly as a result of ICRISAT support to NARS. Use of ICRISAT-developed material that can be tested by NARS has reduced research lags – for example, the variety Okashana 1, earlier developed and tested by ICRISAT in India, underwent only 3 years of adaptive testing before being released in Namibia, thus greatly reducing the time and expense of developing a new variety from scratch. Another major factor in improving NARS research efficiency has been large-scale training and capacity building efforts by ICRISAT. For example, in southern Africa, which lacked trained research staff, over 650 scientists and technicians have undergone training programmes or received scholarships for higher education.

Impacts on yield

District-level yields data for 1992–1994 and 1966–1968 from 238 districts in India were compared in order to estimate the impacts on yield. Yield has increased in almost all the districts. For example, in the late 1960s, most districts of Maharashtra and Gujarat recorded yields less than 500 kg ha^{-1} and slightly higher than 500 kg in Tamil Nadu and Haryana. However, in the 1990s, this had increased by 500–1000 kg ha^{-1} in Gujarat, Maharashtra and Haryana (Table 10.12). Yield increases have been particularly large in some districts where adoption levels are high.

Results of farm surveys show that in all Indian states, improved cultivars give higher grain and fodder yields than local varieties. The percentage increase is higher for grain yield than for fodder yield.

Adoption of new millet varieties in Mali increased pearl millet yields from 570 kg ha^{-1} with the best local variety to 930 kg ha^{-1} for improved varieties (Yapi *et al.*, 1998). These yields are consistent with those found in previous studies. Shetty *et al.* (1991) noted that, in Mali millet, yields vary from 300 kg ha^{-1} in the Sahelian zone to 700 kg ha^{-1} in the zone with most rainfall in the south, compared with on-station yields of 1500–2000 kg ha^{-1}. On-farm yield estimates by Yapi *et al.* (1998) seem consistent with these data. With production at these levels, farmers are able to feed their families and have surplus grain to market. Growing improved varieties assures food security and reduces production risks linked to late season drought (Yapi *et al.*, 1998).

Table 10.12. Impact of improved pearl millet cultivars on pearl millet yield in different states of India, 1971–1994.

	Average yield level (kg ha^{-1})			Yield gain (%) compared to 1971–74	
State	1972–74	1981–83	1992–94	1981–83	1992–94
Gujarat	641	1380	1534	115	139
Haryana	578	725	1309	25	126
Rajasthan	265	373	557	41	110

Source: Deb *et al.* (2000).

Impacts on cost of production and farm profit

Results from cost of cultivation data showed that the average cost of pearl millet production per tonne, in 1992–1994 compared with 1972–1974, has declined by 35%, 42% and 59% in Gujarat, Haryana and Rajasthan in India, respectively (Table 10.13). Farm-level surveys in India showed that improved cultivars have more than 40% lower costs of production estimated on a full-cost basis (Table 10.14).

Yapi *et al.* (1998) reported that improved varieties reduced production costs in Mali by 38% (US\$38 t^{-1}), compared with local millet varieties. The absolute production cost per hectare was higher for improved varieties because of additional inputs, but the higher productivity still provided economies. Improved cultivars have increased farm profit in Mali by 63%. The net income of Indian farmers, computed on a variable cost basis, increased by up to five times (Table 10.14).

Returns on research

Several studies have estimated the returns from pearl millet research in Mali, Namibia and Zimbabwe. Considering research and extension costs, the net present value of benefits from research on improved varieties of millet in Mali was estimated at US\$25 million, representing an internal rate of return (IRR) of 50% (Yapi *et al.*, 1998). Internal rates of return for pearl millet research in Zimbabwe from SDMV 89004 were estimated at 44%. Farm-level studies in Namibia showed that the internal rate of return to pearl millet research was 50%, with a net present value (NPV) of this research of more than US\$10 million in 1998 (Rohrbach *et al.*, 1999).

Spillover impacts

An important objective of international agricultural research institutions is to determine the extent to which research undertaken at one location

Table 10.13. Impact of improved pearl millet cultivars on cost of production of pearl millet in India, 1971–1995.

	Average cost (Rs t^{-1})			Cost reduction (%) compared to 1972–74	
State	1972–74	1981–83	1992–94	1981–83	1992–94
Gujarat	3814	2665	2464	30	35
Haryana	4277	2881	2488	33	42
Rajasthan	3898	1676	1593	57	59

Table 10.14. Impacts of adoption of improved pearl millet cultivars: results of farm-level studies.

			Impacts on						
			Yield gain (%)			Increase in labour use (%)			
Country/region	Year	Cultivars	Grain	Fodder	Reduction in unit cost (%)	All	Female	Per hectare net farm income	Remarks
India									
Eastern Rajasthan	1996	Improved	228	12	47	60	140	Rs. 1134	
Haryana	1996	Improved	182	68	47	44	44	Rs. 2062	
Gujarat (*kharif*)	1995	MH 179	247	72	54	133	170	Rs. 2818	Wide adaptability due to disease resistance, short duration, high grain and fodder yield
Gujarat (summer)	1995	MH 179	462	119	59	261	306	Rs. 5557	
Maharashtra	1994	Improved	95	7	43	25	16		
Tamil Nadu	1994	ICMS 7703	108		18	59	45	Rs. 3567	
Mali									
Segou	1995	Improved	63		38				Stable yield, improved food security. Generated NPV of US$25 million with an IRR of 50%
Koulikoro	1995	Improved	65						
Mopti	1995	Improved	52						
Namibia	1997	Okashana 1	24						Broadly accepted for early maturity, bold grain; basis for start of national seed industry. Provided NPV US$11.7 million with an IRR of 50%
Zimbabwe	1996	SDMV 89004							Widely accepted for early maturity and bold grain. Estimated IRR is 44%.

Source: Deb *et al.* (2000) for eastern Rajasthan, Haryana, Gujarat and Maharashtra; Ramasamy *et al.* (2000) for Tamil Nadu; Rohrbach *et al.* (1999) for Namibia and Yapi *et al.* (1998) for Mali.

may impact on other regions of interest. ICRISAT has, as a policy, distributed a wide range of parental materials to breeding programmes in the NARS and private seed industries throughout the semi-arid tropics. This has contributed to enhanced technology spillover. For example, ICMV 221, Okashana 1 and WC-C75 were originally bred for India, but ICMV 221 was also released in Kenya and Uganda.

An open-pollinated variety (ICTP 8203), developed at ICRISAT-India from Togo populations, was introduced to Namibian farmers through the SADC/ICRISAT Sorghum and Millet Improvement Programme (SMIP) and the efforts of the Rossing Foundation during 1986/87 and 1987/88 along with a total of 50 varieties on demonstration trial. Farmers liked this variety when they saw it in the demonstration field. In 1989, the Rossing Foundation distributed large quantities of seed of ICTP 8203 under the name of Okashana 1. Okashana 1 now occupies about 50% of the pearl millet area in Namibia (Rohrbach *et al.*, 1999). Okashana 1 (ICMV 88908) was released in Malawi, Namibia and Botswana. In Zimbabwe, private seed companies produce and market Okashana 1, though it is yet to be formally released.

Similarly, WC C75 was released in Zambia. Kaufela was developed for Zambia but also released in Botswana, Tanzania and Mozambique. Okoa and Shibe were originally selected for Zimbabwe but Okoa was also released in Botswana and Shibe in Tanzania (ICRISAT, 1996, p. 30; Monyo, 1998). These indicate that the genetic material used in the development of these cultivars has wide adaptation, suggesting that there are important spillovers from ICRISAT genetic enhancement research in pearl millet.

Conclusions

This chapter documents the benefits generated from genetic enhancement research in pearl millet in sub-Saharan Africa and Asia. The pearl millet breeding programme at ICRISAT, in partnership with NARS, has released 75 new varieties and hybrids during 1981–1998. ICRISAT had also developed hybrid parents and supplied materials to its public- and private-sector partners throughout India and Africa. During 1981–1998, all released cultivars in the study countries (except India) were ICRISAT-derived (either ICRISAT bred, or developed from ICRISAT parents or obtained through ICRISAT networks). The increased dominance of ICRISAT parent-material-based releases indicates the importance of ICRISAT's role in the development of parent materials and other intermediate breeding products. The availability of high quality ICRISAT-developed parental materials and their use by private and public sector pearl millet breeders have substantially shortened the research and development time and resulted in increased efficiency of NARS breeding programmes. Among the improved cultivars adopted in farmers' fields, a significant proportion are ICRISAT-bred or based on ICRISAT materials. Adoption of private-sector cultivars with ICRISAT parentage is also increasing.

ICRISAT research has helped to increase yield, reduce production costs, and improve the efficiency of breeding programmes throughout the world. Countries with less well-endowed research facilities, especially in Africa, have benefited most from ICRISAT-bred cultivars and through research spillovers.

Acknowledgements

The authors are grateful to Drs K.N. Rai, K. Anand Kumar, C.T. Hash, E.S. Monyo, Paula Bramel, N. Kameswara Rao, David Rohrbach, H.A. Freeman, John Brennan and Jeff Davis for their useful comments and insightful interactions during early versions of this chapter; Ajay Vadrachary is thanked for editorial assistance; and Dr F.T. Bantilan, Irshad Ahmed and Md. Moinuddin for GIS-based mapping and delineation of pearl millet domains. However, the views expressed in this chapter are those of the authors.

References

Anand Kumar, K. (1993) Pearl millet in Western Africa. In: Byth, D.E. (ed.) *Sorghum and Millets: Commodity and Research Environments.* International Crops Research Institute for the Semi-Arid Tropics (ICRISAT), Patancheru, India.

Bantilan, M.C.S. and Deb, U.K. (2000) *Joint impacts of partnership of ICRISAT-NARS pearl millet genetic improvement.* Paper presented at the American Association for Advancement of Science (AAAS) conference held at Washington, DC, USA.

Bantilan, M.C.S., Chopde, V.K., Gupta, S. and Deb, U.K. (1999a) *Factors influencing adoption of improved pearl millet cultivars in Rajasthan, India.* Poster paper presented at the workshop on Summer Training in Socio-economics and Policy Research on 27 July 1999 at the ICRISAT.

Bantilan, M.C.S., Rao, Y.M., Gujral, H.K. and Deb, U.K. (1999b) *Critical factors influencing adoption of improved pearl millet cultivars in Maharashtra, India.* Poster paper presented at the workshop on Summer Training in Socio-economics and Policy Research of ICRISAT, 27 July 1999 at the ICRISAT.

Bantilan, M.C.S., Deb, U.K., Gupta, S. and Chopde, V.K. (2000a) Improved pearl millet cultivars in Haryana: adoption and impacts. Unpublished manuscript.

Bantilan, M.C.S., Deb, U.K. and Rao, Y.M. (2000b) Improved pearl millet cultivars in Gujarat: adoption and impacts. Unpublished manuscript.

Dave, H.R. (1987) Pearl millet hybrids. In: *Proceedings of the International Pearl Millet Workshop*, 7–11 April 1986. ICRISAT, India. Patancheru, India, pp. 121–126.

Deb, U.K., Bantilan, M.C.S. and Rai, K.N. (2000) *Impacts of improved pearl millet cultivars in India.* Paper presented at the workshop on Impacts of Agricultural Research in India, 10–11 February, ICRISAT, Patancheru, India.

Dendy, D.A.V. (1995) Sorghum and millets: production and importance. In: Dandey, D.A.V. (ed.) *Sorghum and Millets: Chemistry and Technology*. American Association of Cereal Chemists, St Paul, Minnesota, pp. 11–26.

FAO (1998) *FAO Statistical Data Base 1998*.

Harinarayana, G., Anand Kumar, K. and Andrews, D.J. (1999) Pearl millet in global agriculture. In: Khairlwal, I.S., Rai, K.N., Andrews, D.J. and Harinarayana, G. (eds) *Pearl Millet Breeding*. Oxford & IBH Publishing, New Delhi, pp. 480–506.

Hash, C.T. (1997) Research on downy mildew of pearl millet. In: Bantilan, M.C.S. and Joshi, P.K. (eds) *Integrating Research Evaluation Efforts*. ICRISAT, Patancheru, India, pp. 121–128.

ICRISAT (1996) *Southern and Eastern Africa Highlights, 1996*. ICRISAT, Patancheru, India, p. 30.

ICRISAT/FAO (1996) *The World Sorghum and Millet Economies: Facts, Trends and Outlook*. Joint publication by ICRISAT, India and the FAO, Rome.

Kelly, T.G. and Parthasarathy Rao, P. (1993) Sorghum and millet in Asia. In: Byth, D.E. (ed.) *Sorghum and Millets – Commodity and Research Environment*. ICRISAT, Patancheru, India, pp. 93–117.

Monyo, E.S. (1998) 15 years of pearl millet improvement in the SADC region. *International Sorghum and Millets Newsletter* 39, 17–33.

Nene, Y.L. and Singh, S.D. (1976) Downy mildew and ergot of pearl millet. *PANS* 22, 366–385.

Obilana, A.B., Monyo, E.S. and Gupta, S.C. (1997) Impact of genetic improvement in sorghum and pearl millet: developing country experiences. In: *Proceedings of the International Conference on Genetic Improvement of Sorghum and Pearl Millet, 22–27 September 1996, Lubbock, Texas.* INTSORMIL/ICRISAT Publication No. 97-5, pp. 119–141.

Rai, K.N. and Hash, C.T. (1994) Genetic enhancement of pearl millet at ICRISAT. In: Bantilan, M.C.S. and Joshi, P.K. (eds) *Evaluating ICRISAT Research Impact.* ICRISAT, Patancheru, India, pp. 19–25.

Rai, K.N. and Singh, N.B. (1987) Breeding pearl millet male-sterile lines. In: *Proceedings of the International Pearl Millet Workshop*, 7–11 April 1986. ICRISAT, Patancheru, India, pp. 127–137.

Ramasamy, C., Bantilan, M.C.S., Elongovan, S. and Asokan, M. (1999) Perceptions and adoption decisions of farmers in cultivation of improved pearl millet cultivars – a study in Tamil Nadu. *Indian Journal of Agricultural Economics* 54 (2), 139–154.

Ramasamy, C., Bantilan, M.C.S., Elongovan, S. and Asokan, M. (2000) *Improved Pearl Millet Cultivars in Tamil Nadu: Adoption, Impact, and Returns to Research Investment.* Impact Series No. 7. ICRISAT, Patancheru, India.

Rohrbach, D.D., Lechner, W.R., Ipinge, S.A. and Monyo, E.S. (1999) *Impact from Investment in Crop Breeding: the Case of Okashana 1 in Namibia*. Impact Series No. 4. ICRISAT, Patancheru, India.

Shetty, H.S. (1987) Biology and epidemiology of downy mildew of pearl millet. In: *Proceedings of the International Pearl Millet Workshop, 7–11 April 1986*. ICRISAT, Patancheru, India, pp. 147–160.

Shetty, S.V.R., Beninati, N.F. and Beckerman (1991) *Strengthening Sorghum and Pearl Millet Research in Mali*. ICRISAT, Patancheru, India.

Singh, S.D., Ball, S.L. and Thakur, D.P. (1987) Problems and strategies in the control of downy mildew. In: *Proceedings of the International Pearl Millet Workshop, 7–11 April 1986*. ICRISAT, India, pp. 161–172.

Singh, S.D., King, S.B. and Werder, J. (1993) *Downy Mildew Disease of Pearl Millet*. Information Bulletin No. 37. ICRISAT, Patancheru, India.

SMIP (1999) *Collaboration and Achievements in Southern Africa, 1983–1999*. Matopos Research Station, Bulawayo, Zimbabwe.

Yapi, A., Kergna, A.O., Debrah, S.K., Sidibe, A. and Sanogo, O. (1998) Impact of sorghum and millet research in Mali. In: Bantilan, M.C.S. and Joshi, P.K. (eds) *Assessing Joint Research Impacts: Proceedings of an International Workshop on Joint Impact Assessment of NARS/ICRISAT Technologies for the Semi-Arid Tropics, 2–4 December 1996*. ICRISAT, Patancheru, India, pp. 76–93.

11 The Impact of International and National Investment in Barley Germplasm Improvement in the Developing Countries

A. Aw-Hassan and K. Shideed
in collaboration with S. Ceccarelli, W. Erskine, S. Grando and R. Tutwiler

Barley grain is used for animal feed, malt and food for human consumption. Archaeological evidence shows that barley was used in human food several thousand years ago (Bhatty, 1992). Although replaced by wheat and rice in modern times, barley still remains an important food grain in some developing countries, particularly in marginal areas where it may be the only viable crop. The annual per capita consumption of barley for the 1995–1997 period was estimated to be 41.0 kg in Morocco, 20.2 kg in Algeria, 16.2 kg in Iraq, 14.3 kg in Ethiopia, 9.4 kg in Tunisia and 6.1 kg in Kazakhstan (FAO, 2001).

The most important use of barley grain is for animal feed. Barley straw is used as animal feed in West Asia, North Africa, Ethiopia, Eritrea, Yemen, in the Andean region of South America and in the Far East. Barley stubble is grazed in summer in large areas of West Asia and North Africa. Barley is also used as animal feed at the vegetative stage (green grazing) or is cut before maturity and either directly fed to the animals or used for silage. Barley straw is also used for animal bedding and as cover material for hut roofs. Malting barley, the second largest use after feed, is grown as a cash crop in a number of developing countries.

The average area, production and value of barley production for 1994–1998 are given in Table 11.1. Developing countries grow about 19 million ha of barley: 72% is grown in West Asia and North Africa, 19% is grown in Central Asian countries, and about 6% is grown in Latin America. In West Asia and North Africa, the major producers are Turkey, Morocco, Syria, Iran, Iraq, Algeria and Ethiopia. Most of the

Table 11.1. Average area, production and value of barley production for 1994–1998.

	Area harvested (million ha)	Production (million t)	Value (million $)
World	66.8	150.3	23,257
Developing countries	19.0	27.4	4,017
Central Asia	4.4	3.4	
Kazakhstan	3.9	2.8	251
West Asia/North Africa	13.7	19.1	
Turkey	3.6	7.9	1,162
Morocco	2.2	2.3	299
Syria	1.7	1.3	214
Iran	1.8	2.9	321
Iraq	1.4	1.0	145
Ethiopia	1.0	1.0	341
Algeria	1.0	0.7	115
Latin America and Caribbean	1.0	1.8	363

barley (88%) in Central Asia is grown in Kazakhstan. The annual average barley production in the developing world is about 27 million t, with a value of perhaps US$4.0 billion at 1997 prices.

International barley research dates back only 25 years. The International Centre for Agricultural Research in the Dry Areas (ICARDA) was established in 1977 with the mission to improve the welfare of people in the dry areas of the developing world by increasing the production and nutritional quality of food while preserving and enhancing the resource base. ICARDA has a mandate to develop improved barley varieties for all developing countries. Because barley is mainly grown under harsh environments with low rainfall and because it is mainly cultivated by small resource-poor farmers in areas where no other crop could grow, barley improvement research should benefit those small resource-poor farmers the most.

Data and Methodology

To assess ICARDA's contributions to barley improvement, data were collected from collaborating national agricultural research programmes in eight countries: Algeria, Ecuador, Egypt, Iraq, Jordan, Morocco, Syria and Tunisia. The data include human capital in barley breeding, NARS varietal releases, estimates of area planted by variety, yield advantage of improved germplasm and the composition of NARS breeding pools.

Human Resources and Expenditures on Barley Improvement Research

ICARDA

ICARDA's barley improvement programme started with the establishment of the centre in 1977. ICARDA expenditure on barley improvement research, in terms of human and financial resources, is presented in Table 11.2 for 1980, 1990 and 1997. The human capital investment in barley improvement research increased from 3.5 scientists in 1980 to 4.85 in 1990 and to 5.55 in 1997. The total research expenditure figures were estimated at about US$0.9 million, 1.2 million and 1.4 million for the 3 years, respectively.

NARS

The human capital investment in barley breeding programmes for nine developing countries is given in Table 11.3. There are 46 scientists in these programmes of which 19 are involved in breeding. The rest are involved in support sciences, including pathology (24%), entomology (6%), and agronomy, physiology and quality assessment (26%). Approximately 46% of all scientists have PhD degrees, while the rest have Master's or lower degrees. Most of these scientists also work on other crops. The time spent on barley improvement is about 55% of the total time endowment of the above mentioned individuals. For breeders, there is greater concentration on barley. About 76% of the breeders' time is devoted to barley improvement, and the remainder is allocated to other crops.

Table 11.2. Human resources and expenditure in barley germplasm improvement programme at ICARDA.

	Scientist-years		
	1980	1990	1997
Breeding	2.50	3.50	4.00
Entomology	0.00	0.25	0.20
Pathology	1.00	0.50	0.35
Biotechnology	0.00	0.00	0.40
Total	3.50	4.25	4.95
GRU	0.00	0.60	0.60
Grand total	3.50	4.85	5.55
Million US$	0.875	1.212	1.387

GRU, Genetic Resource Unit.

Table 11.3. Human resources endowments in barley breeding research in developing countries.

	Breeding	Pathology	Entomology	Agronomy/ physiology	Quality	Total
	Number of scientists (PhDs in parentheses)					
Algeria	2 (0)	1 (0)	0	0	0	3 (0)
Ecuador	1 (1)	0	0	0	0	1 (1)
Egypt	4 (3)	1 (1)	1 (1)	1 (1)	0	7 (6)
Ethiopia	3 (1)	1 (1)	1 (0)	3 (0)	0	8 (2)
Iraq	1 (1)	0	0	2 (0)	0	3 (1)
Jordan	2 (0)	0	0	2 (0)	0	4 (0)
Morocco	2 (1)	2 (0)	1 (1)	1 (1)	1 (0)	7 (3)
Syria	1 (0)	1 (0)	0	1 (0)	0	3 (0)
Tunisia	3 (3)	5 (3)	0	2 (2)	0	10 (8)
Total	19 (10)	11 (5)	3 (2)	12 (4)	1 (0)	46 (21)
Full-time equivalent (scientist-years)						
Algeria	1.4	0.8	0.0	0.0	0.0	2.2
Ecuador	0.8	0.0	0.0	0.0	0.0	0.8
Egypt	3.8	0.8	0.8	0.8	0.0	6.2
Ethiopia	3.0	0.3	0.5	1.0	0.0	4.8
Iraq	1.0	0.0	0.0	0.6	0.0	1.6
Jordan	1.2	0.0	0.0	0.9	0.0	2.1
Morocco	0.8	1.0	0.5	0.2	0.3	2.8
Syria	1.0	0.5	0.0	0.5	0.0	2.0
Tunisia	1.4	1.0	0.0	0.5	0.0	2.9
Total	14.4	4.4	1.8	4.5	0.3	25.4

Among the national barley breeding programmes, the larger producers, such as Morocco, Syria and Iraq, have less than one full-time breeder per million hectares of cultivated barley area. Algeria and Tunisia, which produce somewhat less barley, have higher manpower, with between one and three full-time breeders per million hectares. Some small producers, such as Ecuador, Egypt and Jordan, which grow less than 100,000 ha of barley, have the highest human capital concentration with more than ten full-time breeders per million hectares. Another indicator of investment in barley improvement research is the ratio of breeders holding PhD degrees to the total number of breeders. Syria, Jordan and Algeria have no PhD-level breeders, Morocco has one PhD out of two breeders, and Egypt has three PhD out of four breeders.

The conclusion of this analysis is that human resources in national barley breeding programmes are uneven in terms of education level and in terms of number of breeders per cultivated area. On the whole, the human capital investment in barley improvement research in developing countries is relatively low, with fewer than two full-time breeders for every million hectares of barley. A large proportion (about half) of the scientists in these programmes hold Master's or lower degrees, and all the breeders work less than full-time on barley improvement.

Impact on Production and Release of New Varieties

Production and release of new varieties

A total of 111 barley varieties have been released in 23 developing countries during the 1980–1999 period. The total number of varieties released and annual average releases in 5-year periods are given in Table 11.4.

Table 11.4. Total and average annual number of barley varietal releases by country for 1980–2000.

Country	1981–85	1986–90	1991–95	1996–2000	Total
Egypt	0	2	2	2	6
Algeria	0	6	4	0	10
Morocco	5	5	2	2	14
Tunisia	0	4	1	1	6
Ecuador	0	1	2	0	3
Jordan	2	1	0	0	3
Syria	0	1	2	0	3
Iraq	0	0	4	0	4
Chile	0	2	0	0	2
China	0	4	0	2	6
Cyprus	1	1	2	3	7
Ethiopia	4	2	1	2	9
Iran	0	1	2	5	8
Lebanon	0	1	0	2	3
Libya	0	0	2	4	6
Mexico	0	1	0	1	2
Nepal	0	1	0	0	1
Pakistan	0	3	1	4	8
Qatar	2	1	0	0	3
S. Arabia	0	1	0	0	1
Turkey	0	0	2	1	3
Vietnam	0	1	0	0	1
Yemen	0	2	0	0	2
Total	14	41	27	29	111
Average annual release	2.8	8.2	5.4	5.8	5.55

Varietal releases increased from an average of three per year during the 1980–1984 period to about eight per year during the 1985–1989 period, when it peaked. After that, varietal releases stabilized at a rate of approximately six per year. It is clear that varietal production tripled during the late 1980s but remained constant during the 1990s. Varietal production is affected by the investment in research capacity at ICARDA and national programmes. ICARDA's research expenditure in barley improvement research increased by 38% during the 1980s. The increased varietal production could be attributed to increased access to germplasm by national programmes and to the increased capacity of national programmes. ICARDA's barley breeding programme has increased NARS access to germplasm through its international nurseries and germplasm exchange programmes. ICARDA also contributes to NARS capacity building through training, collaborative research projects, and various exchanges.

ICARDA contribution to varietal release

ICARDA's contribution to varietal releases can be determined by the origin of released varieties using pedigree analysis. The released varieties were classified into six categories: (i) ICARDA crossed and selected varieties; (ii) ICARDA crosses selected by NARS; (iii) NARS crosses with ICARDA parent; (iv) NARS crosses with NARS parent; (v) ICARDA germplasm accession; and (vi) material from other international sources.

The summary results of this analysis are shown in Table 11.5. About 78% of all barley varieties released by the 23 developing countries during the 1980–1999 period were ICARDA-related material; 52% were ICARDA crosses, of which 38% were selected at ICARDA and 14% were selected by NARS. In addition, 11% of the varietal releases were NARS crosses with at least one parent from ICARDA, while an additional 16% of barley released varieties are direct releases received from ICARDA germplasm accessions. These data show that ICARDA barley breeding was an important factor in the varietal releases in the developing world. ICARDA, like other CGIAR centres, has played a dual role: it produces varieties that are being directly transferred to farmers and is also an important source of breeding material for NARS breeding programmes.

Trends in varietal contents and evolution of ICARDA contribution

ICARDA's contribution to NARS variety development has evolved over time as a result of the changing needs and capacities of NARS. This

Table 11.5. The proportion of the types and content of released barley varieties in the developing countries.

	Total	ICARDA-cross/ ICARDA-selection	ICARDA-cross/ NARS-selection	NARS-cross/ ICARDA-parent	ICARDA germplasm accession	Total ICARDA-related	NARS-cross/ NARS-parent	Other international sources
Algeria	10	0.5	0.0	0.0	0.0	0.5	0.5	0.0
Chile	2	0.5	0.0	0.0	0.0	0.5	0.0	0.5
China	6	0.3	0.7	0.0	0.0	1.0	0.0	0.0
Cyprus	7	0.0	0.1	0.0	0.0	0.1	0.9	0.0
Ecuador	3	1.0	0.0	0.0	0.0	1.0	0.0	0.0
Egypt	6	0.0	0.3	0.0	0.0	0.3	0.7	0.0
Ethiopia	9	0.0	0.0	0.1	0.1	0.2	0.8	0.0
Iran	8	0.6	0.0	0.1	0.0	0.8	0.0	0.3
Iraq	4	0.3	0.3	0.0	0.0	0.5	0.5	0.0
Jordan	3	0.0	0.0	1.0	0.0	1.0	0.0	0.0
Lebanon	3	0.7	0.0	0.0	0.3	1.0	0.0	0.0
Libya	6	0.5	0.5	0.0	0.0	1.0	0.0	0.0
Mexico	2	1.0	0.0	0.0	0.0	1.0	0.0	0.0
Morocco	14	0.5	0.0	0.3	0.0	0.8	0.2	0.0
Nepal	1	0.0	0.0	0.0	1.0	1.0	0.0	0.0
Pakistan	8	0.1	0.6	0.0	0.1	0.9	0.0	0.1
Qatar	3	0.0	0.3	0.0	0.0	0.3	0.0	0.7
S. Arabia	1	0.0	0.0	0.0	1.0	1.0	0.0	0.0
Syria	3	1.0	0.0	0.0	0.0	1.0	0.0	0.0
Tunisia	6	0.7	0.3	0.0	0.0	1.0	0.0	0.0
Turkey	3	0.0	0.0	1.0	0.0	1.0	0.0	0.0
Vietnam	1	1.0	0.0	0.0	0.0	1.0	0.0	0.0
Yemen	2	0.0	0.0	0.0	1.0	1.0	0.0	0.0
Average*		0.38	0.14	0.11	0.16	0.78	0.15	0.07

**Note*: The total may not add up to 1 due to rounding up.

evolution can be observed in Fig. 11.1. The clearest trend is the noticeable rise in the proportion of released varieties that are selected by NARS. This trend is also reflected in the decline in the proportion from material selected at ICARDA in the late 1980s.

Another clear trend is the decline over time of the proportion of material released from NARS crosses. This is in spite of the fact that the number of crosses made by NARS has jumped for the last 10 years. This implies an increasing demand by NARS for ICARDA crosses, which may be due to the fact that ICARDA has access to larger pools of genetic material than NARS. The third clear trend is that the release of varieties that were directly selected from ICARDA germplasm accessions represented an important proportion of released varieties in the early development phase of the programme.

These trends suggest two points. First, the increased selection activities by NARS indicate increased research capacity which is related to a decline in ICARDA's selection activity. This is consistent with the decentralization of barley breeding in which varietal selection is increasingly devolved to NARS. The second point is that the drop in NARS crosses suggests that selection, rather than crossing, has become the most widespread strategy for NARS operating with limited human and financial resources. This strategy depends on a continuous flow of improved germplasm from ICARDA.

The increasing NARS self-reliance for varietal selection, and the more general decentralization of barley breeding have implications for resource allocation within ICARDA's barley breeding programme

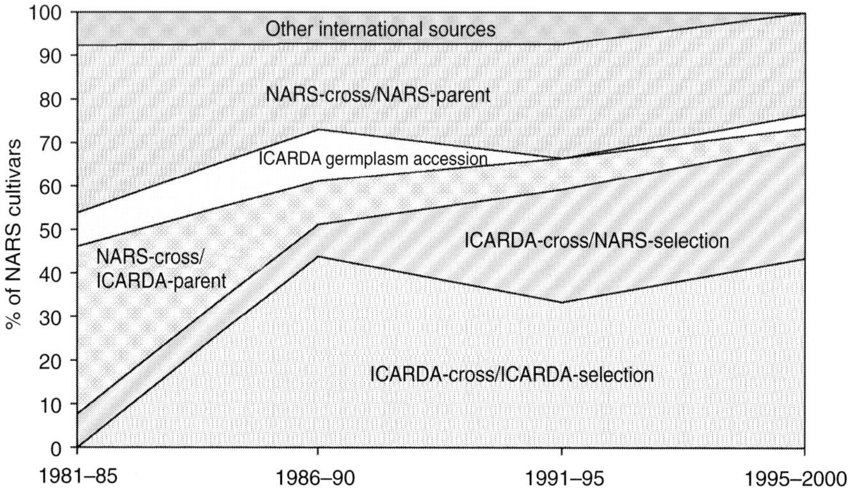

Fig. 11.1. Contribution of ICARDA content indicators in NARS released cultivars of barley.

as well as NARS. The implication is that NARS should allocate more of their time for cultivar development, while ICARDA should allocate more resources for pre-breeding and for genetic resource conservation and management. The strong NARS can do some or all of their own crossing as well as selection. These NARS, however, benefit from their access to the large pool of germplasm material through their collaboration with ICARDA and the regional networks coordinated by ICARDA. This also affects the types of germplasm demanded by NARS. For example, the distribution of material for specific adaptation to NARS, in the form of segregating populations, has increased since 1990. This is expected to increase the number of releases for different agro-ecological conditions and their acceptability and adoption by farmers. Thus, the role of ICARDA as a provider of genetic diversity is important for the germplasm improvement in the developing countries.

ICARDA Contribution to NARS Pre-breeding Research

Analysis of NARS breeding pools and crossing blocks is required in order to study the impact of ICARDA pre-breeding work. NARS breeding pools provide information on cultivars to be released over the next 5–10 years and, thus, the likely impact of ICARDA in the immediate future. Data in Table 11.6 indicate that ICARDA advanced lines and released cultivars represent the major part in the composition of NARS crossing blocks. In fact, the contribution of this content has increased from 46% in 1987 to about 54% in 1997, suggesting that ICARDA's

Table 11.6. Composition of NARS barley crossing blocks and parental contributions of crosses for 1987 and 1997.

	Crossing blocks		Crosses	
	1987	1997	1987	1997
Total	420	718	430	595
Proportions by type of germplasm source:			Parents	
1- NARS	0.12	0.23	0.49	0.41
2- ICARDA	0.46	0.54	0.35	0.36
3- Other countries	0.38	0.14	0.13	0.09
4- Local landraces	0.02	0.07	0.03	0.09
5- ICARDA landraces	0.01	0.02	0.00	0.05
Total	1.00	1.00	1.00	1.00

Countries: Iraq, Tunisia, Algeria, Ecuador, Egypt, Jordan, Morocco, Syria, Ethiopia.

likely impact on variety releases will be substantial in the immediate future. The importance of NARS' own advanced lines and released cultivars have also increased between 1987 and 1997, while the advanced lines and released cultivars from other countries have declined. The contribution of landraces has slightly increased.

Another measure for assessing the contribution of ICARDA to NARS pre-breeding research is the source of parents for the NARS crosses. The sources of parents for NARS crosses in 1987 and 1997 are shown in Table 11.6. The number of crosses by NARS has increased by 38% for the 10-year period. As a whole, over one-third of the parents of NARS crosses were ICARDA advanced lines and released cultivars. The contribution of ICARDA parents to NARS crosses has remained around 35–36% in those years. Parents from NARS' own advanced and released lines account for 49% of the crosses made in 1987 and 41% in 1997. Interestingly, the content of landraces in the NARS crosses has increased from 3% in 1987 to 14% in 1997. This trend is again consistent with NARS efforts to develop locally adopted modern varieties.

Economic Impact

The economic surplus (ES) model has often been used to estimate the benefits of agricultural research (Evenson, 1974; Akino and Hayami, 1975; Hertford and Schmitz, 1977; Byerlee and Traxler, 1995). Alston *et al.* (1999) discussed several variants of the ES model and provided procedures for estimating research benefits for different scenarios in both *ex ante* and *ex post* analysis. The ES model can be used to compute annual flows of research benefits and costs. The general procedure requires data on yield advantage of new varieties, adoption path, the change in cost of production due to the use of new varieties, producer prices, and demand and supply price elasticities. When the distribution of benefits between producers and consumers is not a concern, gross annual research benefit (GARB) can be calculated, which does not require information on elasticities.

The ES model for a small open economy can be expressed as:

$$\Delta ES = P_t Q_t k_t (1+0.5 k_t \varepsilon) \quad (1)$$

where P_t is prices in year t, Q_t is quantity in year t, and k_t is the supply shift down in year t as proportion of initial price and ε is supply price elasticity. This model is used to estimate gross research benefits for nine countries (Algeria, Ecuador, Egypt, Ethiopia, Iraq, Jordan, Morocco, Syria and Tunisia) for 24 years, 1980–2000. The parameters used in the analysis are discussed below.

Yield gain

Data on yield advantage of new barley varieties were collected from national programmes. Yield improvement estimates were also obtained from annual reports of on-farm trials. The data for the above nine countries are reported in Table 11.7. The yield advantage data refer only to dominant varieties in terms of cultivated area for each country. The highest yield advantage of 25% was recorded in Algeria and the lowest at 10% in Ecuador. However, the actual production increase from new varieties depends on their adoption.

Diffusion profile of new varieties

The data collected from NARS indicate that adoption of improved varieties of barley is growing in several countries. High adoption levels were reported in 1997 for Ecuador (55%), Egypt (50%), Jordan (50%) and Tunisia (40%). Relatively low adoption levels were reported in Morocco (19%), Iraq (14%) and Ethiopia (11%). Algeria and Syria, two large producers, had the lowest adoption levels of 5% or less of the total barley area. Figure 11.2 shows a steady increase, since the 1980s, of the area cultivated with improved varieties for four countries. The overall average of the diffusion level of improved barley varieties, weighted by area, for the eight countries in Table 11.7 is 14%, of which 10% are ICARDA crosses or have ICARDA parents, and 4% are entirely NARS varieties.

Table 11.7. Barley varietal releases, diffusion, yield advantage and average yield in selected countries.

	Proportions of area by type of germplasm			
Country	ICARDA-related (%)	NARS (%)	Yield average (million t ha^{-1})	Yield advantage (%)
Algeria	0.00	0.03	0.96	25
Ecuador	0.55	0.00	0.65	10
Egypt	0.50	0.00	2.16	24
Ethiopia	0.00	0.11	1.00	12
Iraq	0.14	0.00	0.69	32
Jordan	0.50	0.00	0.81	15
Morocco	0.12	0.07	1.04	24
Syria	0.05	0.00	0.78	23
Tunisia	0.40	0.00	0.93	25
Area weighted average	0.10	0.04	0.91	

Source: our survey.

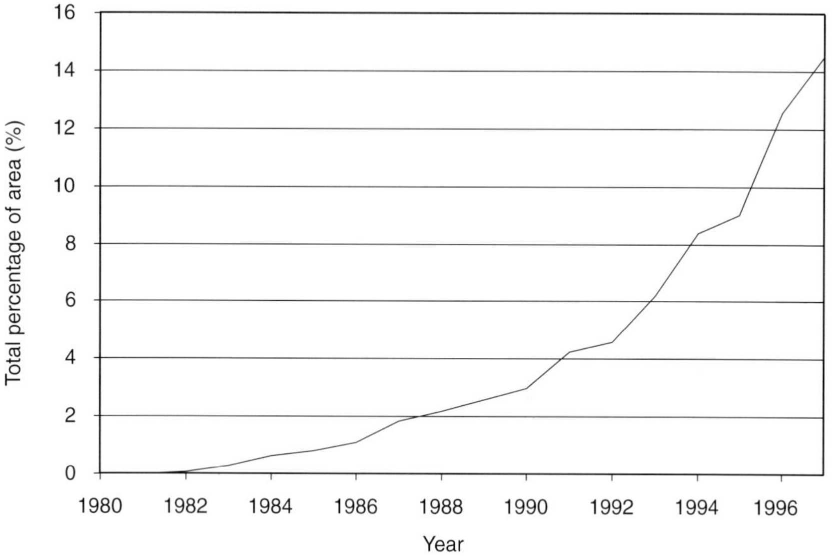

Fig. 11.2. Adoption of barley cultivars in West Asia and North Africa.

Estimating the *k*-factor

The *k*-factor in equation (1) is essential for estimating the benefits of research. The parameter k_t is estimated as $g/e\ A_t$, where g is the yield improvement ratio

$\left(1 = \frac{Yield T_V}{Yield New}\right)$ and A_t is the proportion of area under improved varieties

in year *t*. The price supply elasticities vary for each country depending on the potential for expansion of barley production and elasticities of other competing crops. Price supply elasticities for most agricultural commodities vary from 0.1 to 1.0 (Alston *et al.*, 1995). Estimates of supply elasticities of barley for three countries (Jordan, Tunisia and Morocco) ranged from 0.14 in the low potential areas of Jordan to 0.944 in the high potential areas of Morocco.[1] The supply elasticity depends, for example, on the supply elasticity of the factors of production accountable. Barley is cultivated in marginal areas and, on the aggregate, its supply is relatively inelastic because of the inelastic supply of suitable land. A single price supply elasticity of 0.3 was used for all countries.

[1] These estimates were provided by Nabil Chaherli from a study presented at IFPRI-ICARDA Policy and Property Rights Research Workshop in Hammamet, Tunisia, 26–29 November 1998.

Estimating cost of barley germplasm improvement research

Annual data on the total budget, the total number of scientists employed, and the number of scientists (in full-time equivalent) involved in barley improvement are available for ICARDA. The expenditure series for ICARDA barley improvement programmes was constructed by multiplying ICARDA's average cost per scientist by the number of the full-time scientists in barley improvement. Only part of total ICARDA barley expenditures was allocated to the nine countries included in the analysis. The ICARDA contents in released barley varieties in these countries were used as an approximation for the proportion of ICARDA barley expenditure allocated to each country.

NARS barley expenditures were estimated based on data provided by Pardey *et al.* (1989). They reported in the appendix (pp. 414–421) total agricultural research expenditures (in 1980 purchasing power parity, PPP) and total number of researcher by country for 1961–1965 to 1981–1985. These two series were used to calculate average cost per NARS scientist in 1980 PPP dollars. This cost per scientist was converted to 1997 PPP dollars using a US consumer price index and multiplied by the 1997 survey data on the number of NARS barley improvement scientists in each country to obtain 1997 barley improvement expenditures. Total NARS expenditures were then projected forward to 1999 and back to 1977 using Pardey *et al.*'s data on growth rates of research expenditures and total number of researchers by region/country, assuming that barley research expenditures have increased at the same rate as research expenditures on all crops. This method of estimating investment in barley research is based on a number of assumptions. These are, that the cost of supporting a barley researcher is the same as the average cost per researcher for other crops, and that barley research expenditures have commanded a constant share of total expenditures. Regardless of these simplifying assumptions, the estimated expenditures are a reasonable approximation (Byerlee and Traxler, 1995).

In 1997, the nine NARS invested approximately US$6.9 million at 1990 prices in barley improvement research and they employed 46 scientists. ICARDA's global expenditure on barley improvement research in 1997 was about US$1.4 million. About 52% or US$0.724 million was spent in the nine countries. This represents about 9% of the total investments in barley improvement research in the nine developing countries.

Estimating returns to research investment

Gross research benefits were calculated using the economic surplus model for the nine countries for 24 years (1977–2000), using the small open-economy model in equation (1). Average (1980–1998) c.i.f prices for each country were calculated from the *FAO Trade Year Book*. The differences in country c.i.f prices reflect the differences in transportation costs. These prices approximate the social value of barley production for each country. These prices were deflated with price indexes for cereals using 1990 as constant. Production was kept constant for the whole analysis period at the 1980–2000 average level. A lag period of 10 years for ICARDA and 5 years for NARS was assumed. NARS have shorter breeding cycles to release varieties once they receive advanced material from ICARDA. The gross annual research benefits for 1997 are given in Table 11.8. These nine countries had a combined benefit of about US$92.5 million from barley improvement research in 1997. This is about 13 times the amount that these countries have spent in barley improvement research including ICARDA's contribution, which was estimated at approximately US$7 million. These estimates are indicative of gross annual research benefits. The internal rates of return to research investment are computed for the nine countries and reported in Table 11.8. With the exception of Morocco, a large country, which has an IRR (internal rate of return) of about 51%, all the countries have returns to research investment lower than 50%. Iraq and Tunisia have attained an IRR of 38% for their research investment, while Egypt and Jordan had similar IRR of 32% and 31%, respectively. The other four countries had estimated IRR lower than 30%.

The impact of improved germplasm is not limited to the yield advantage. In addition, released varieties have important traits such as resistance to disease and drought. For example, the three barley cultivars released in Ecuador are resistant to yellow and/or leaf rusts. Similarly, the improved varieties of Tunisia and Jordan are tolerant to disease and/or drought. Breeding for drought and disease resistance has an important impact in reducing the inherent risk associated with rainfed farming and thus helps the partner countries in bridging their food and feed deficit.

Conclusion

Formal barley breeding research for non-malt improvement is only about 20 years old and is relatively new in developing countries. Barley research in the developing countries, in spite of ICARDA's effort to stimulate national investment for the last 20 years, still remains underfunded, particularly in the large producing countries. The production of new germplasm started in the early 1980s. ICARDA has significantly contributed to the productivity of national programmes in terms of barley varietal production. ICARDA increases the capacity of NARS through

Table 11.8. Estimated research benefit due to barley improvement for selected countries.

	Algeria	Ecuador	Egypt	Ethiopia	Iraq	Jordan	Morocco	Syria	Tunisia	Total
Area 1997 (1000 ha)*	677	56	58	897	1173	50	1996	1572	311	6791
Production 1997 (1000 t)	644	35	126	953	778	43	1324	983	160	4885
Research expenditure 1997 (US$m)	0.505	0.153	0.207	1.237	0.539	0.283	1.440	0.250	2.282	6.90
Research expenditure per 1000 ha 1997 (US$)	746	2738	3587	1378	460	5621	721	159	7338	2527
Gross annual research benefits 1997 (US$m)	1.9	1.4	8.0	8.5	23.6	0.9	32.0	5.6	10.6	92.5
Internal rate of returns (%)	22	29	32	22	38	31	51	27	38	32

Area and production are for the year 1997, except Algeria which recorded exceptionally small area that year. So, the 5-year figure (1993–1997) is used for that country.

Production level is 1980–2000 average for the period of the analysis.

Discount rate of 3% is used.

training and through increasing their access to larger pools of genetic diversity and providing a flow of improved germplasm. About 70% of the released barley varieties in 24 developing countries had either been selected at ICARDA or had an ICARDA parent or came directly from ICARDA's accessions. Since the 1980s, the proportion of material released from NARS crosses has declined, while the proportion of material selected by NARS has increased, though mainly from ICARDA crosses. This indicates complementarity between ICARDA and NARS; i.e. ICARDA produces crosses, while NARS programmes either select them for specific conditions or use them as parents for their own crosses.

References

Ahmed, M.M. and Sanders, J.H. (1991) The economic impacts of Hageen Dura-I in the Gezira scheme, Sudan. In: *Proceedings of the International Sorghum and Millet CRSP Conference, 8–12 July 1991, Corpus Christi, Texas*. INTSORMIL Publication No. 92-1.

Akino, M. and Hayami, Y. (1975) Efficiency and equity in public research: rice breeding in Japan's economic development. *American Journal of Agricultural Economics* 57 (February), 1–10.

Alston, J.M., Norton, G.W. and Pardey, P.G. (1999) *Science under Scarcity: Principles and Practice for Agricultural Research Evaluation and Priority Setting.* CAB International, Wallingford, UK.

Bhatty, R.S. (1992) Dietary and nutritional aspects of barley in human food. In: Munck, L. (ed.) *Barley Genetics VI.* Munksgaard International Publishers, Copenhagen, Denmark, pp. 913–923.

Byerlee, D. and Traxler, G. (1995) National and international wheat improvement research in the post-green revolution period: evolution and impacts. *American Journal of Agricultural Economics* 77 (May), 268–278.

Evenson, R.E. (1974) The green revolution in recent development experience. *American Journal of Agricultural Economics* 56 (May), 387–394.

FAO (2001) Food sheet balance at www.fao.org

Hertford, R. and Schmitz, A. (1977) Measuring economic returns to agricultural research. In: Arndt, T.M., Dalrymple, D.G. and Ruttan, V.W. (eds) *Resource Allocation and Productivity in National and International Agricultural Research.* University of Minnesota Press, Minneapolis.

Pardey, P.G., Roseboom, J. and Anderson, J.R. (eds) (1989) *Agricultural Research Policy.* Cambridge University Press, Cambridge.

12 The Impact of CIAT's Genetic Improvement Research on Beans

N.L. Johnson, D. Pachico and C.S. Wortmann

The common bean (*Phaseolus vulgaris*) is the world's most important food legume, a group which also includes chickpeas, lentils and faba beans. Beans are grown under diverse conditions, from sea-level to over 3000 m (van Schoonhoven and Pachico, 1998). Traditionally they have been produced by small farmers for subsistence and income generation. Beans are a major source of food for all income levels, and are particularly important as a protein source for the poor in Latin America and Africa.

The common bean's centre of origin is Latin America, and the region continues to be the world's most important bean-growing area, accounting for half of global output. Brazil is the largest bean producer in the world, producing over 2 million t in 1998 (Table 12.1).[1]

Eastern and Southern Africa are also important bean-producing regions. Three countries from this region are among the world's largest producing countries: the DR Congo, Burundi and Tanzania. Kenya may well be the largest producer in Africa; however, its production is not well documented (S. Beebe, 2000, personal communication).

Official production estimates for many countries may also underestimate total production, since they often omit beans intercropped with other crops. A further shortcoming of official statistics on common bean production is that they often fail to distinguish between *Phaseolus vulgaris* and other leguminous species referred to as 'beans' – for example, mung beans or adzuki beans – some of which may be of a completely

[1] Statistics are from FAO unless otherwise cited.

Table 12.1. Production and yields in selected bean-producing countries, 1998.

Country	Production 1998 (t)	Yield (t ha^{-1})
Brazil	2,183,767	0.6559
China	1,513,174	1.2536
Mexico	1,475,282	0.7413
USA	1,398,300	1.8054
Myanmar	1,077,570	0.8112
Ethiopia	410,000	0.9762
Argentina	290,000	1.0861
Burundi	274,902	1.0182
Korea	270,000	0.8438
Tanzania	50,000	0.6944

Source: FAO and S. Beebe, 2000, personal communication.

different genus. According to FAO, India is the world's largest bean producer, yet in fact India produces very little common bean (S. Beebe, 2000, personal communication). Care must therefore be used when interpreting data on bean production, and alternative data sources, where available, may be more reliable than official government or FAO statistics (see Wortmann *et al.*, 1998)

Genetic Improvement in Beans

The international agricultural research centres

CIAT (International Centre for Tropical Agriculture) began working on common beans in 1973 at its headquarters in Cali, Colombia. At that time, the majority of bean production was from small farmers, and this group became the target for CIAT's research. The main bean-producing countries, such as Brazil and Mexico, had significant large farm sectors at that time; however, CIAT focused on small farmers because they accounted for the majority of production and also for equity reasons (Pachico, 1986).

The selection of this target group had clear implications for CIAT's activities. Since the vast majority of small farmers produced under rainfed conditions with virtually no purchased inputs, scientists in CIAT's bean programme focused on breeding low-input, disease-resistant varieties (Pachico, 1986).

In 1979, the first of several scientists was outposted to Central America, and the following year the first regional programme was established there. This work focused specifically on finding varieties resistant to the bean golden mosaic virus (BGMV), which was causing severe crop damage throughout the region.

In 1983–1984, a second regional programme was established in the Great Lakes Region of Africa, focusing on pest and disease resistance along with soil fertility.

Table 12.2 shows CIAT's human and financial investment in bean research over time. Total resources devoted to bean improvement grew steadily from the 1970s until the mid-1990s, when a centre-wide restructuring led to a reduction in funding for all crop improvement research at CIAT. A similar trend can be observed in the number of breeders in CIAT's bean programme. The programme started with one breeder in 1970 and grew to seven in 1990. After 1985, nearly half the breeders were stationed in Africa.

National programmes[2]

Bean breeding as a systematic and organized activity appears to have begun in Latin America in the 1930s, when both Mexico and Brazil began to conduct bean variety trials. Bean breeding was consolidated during the 1940s with the establishment of Rockefeller Foundation-supported programmes in Mexico and Colombia as well as national programmes in Colombia, Peru and Chile. Much of the breeding work at this time consisted of evaluation and selection of local varieties.

In 1962, the Central American Cooperative Network for Bean Improvement (PCCMF) was founded in San José, Costa Rica. This net-

Table 12.2. CIAT human and financial investment in bean improvement.

	1970	1975	1980	1985	1990	1997/8
Total programme costs (thousands of 1990 US$)	356	2,855	5,633	8,089	13,858	7,678
Principal scientists	2	14	17	20	26	18.5
Breeders	1	2	4	5	7	5
Breeders in Africa				2	3	3
Breeders as % of programme	50	14	24	25	27	27
Breeders in Africa as % of breeders	0	0	0	40	43	60

Source: CIAT Annual Reports and expert opinion.

[2] This section draws heavily on Voysest (1983).

work, which began by testing varieties from Mexico and Central America, later became part of a larger bean improvement effort coordinated by the Inter-American Institute for Agricultural Cooperation (IICA). By the late 1960s, it was testing varieties from throughout the Americas and had distributed ten improved lines, including three that actually became released varieties.

Because the national programmes in Latin America and the Caribbean (LAC) were relatively well developed by the time CIAT was established, CIAT and national programmes generally worked as partners in the production of improved varieties. Breeders and pathologists in CIAT would identify promising materials. Recombinations of these materials were made and then sent out in the early stages of selection (F3 and F4) to national programmes to be selected in the regions where they would ultimately be released (Pachico, 1986).

Tables 12.3 and 12.4 present two measures of NARS (national agricultural research systems) investment in bean breeding in LAC

Table 12.3. Breeders in national programmes and breeding intensities in LAC for selected countries and years.

	Breeders in national programmes			Breeders/million tonnes of production		
Country	1979	1989	1998	1979	1989	1998
Argentina	2	4	3	8.51	32.05	10.34
Bolivia	2	5	3	547.95	1022.49	236.22
Brazil	10	**10**	10	4.57	4.33	4.58
Colombia	7	8	6	34.40	54.78	72.12
Costa Rica	2	5	**1**	93.71	80.24	43.05
Cuba	2	0.5	**2**	176.66	146.84	71.43
Chile	4	4	4	250.00	21.74	127.43
Dominican Republic	3	0	**1**	60.37	0.00	41.23
Ecuador	4	6	4	172.44	188.19	96.57
El Salvador	1	2	**0**	21.50	44.87	
Guatemala	4	4	**1**	63.36	44.15	11.96
Haiti	1	2	0	19.28	34.48	
Honduras	4	5	**1**	115.97	87.65	13.17
Nicaragua	3	4	**1**	103.90	63.98	11.82
Panama	0	3	0		589.51	
Peru	4	5	4	72.91	85.32	56.85
Venezuela	2	4	0	64.94	87.77	
LAC total	74	88.5	41	18.20	22.51	12.67

Bold indicates expert opinion or estimate.
Sources: Expert opinion and an international directory of bean researchers in Latin America.

Table 12.4. Breeders in national programmes and breeding intensities in Africa, selected years.

	Breeders in national programmes			Breeders/million tonnes of production		
Country	1980	1990	1998	1980	1990	1998
Burundi		2	1		6.05	3.64
DR Congo		5	5		37.04	35.71
Ethiopia		4	10.5		48.21	25.61
Kenya	1	4	4			
Madagascar		1	1		22.75	13.89
Malawi	1	1	1	16.13	11.76	11.11
Rwanda		1	1		5.13	8.33
S. Africa		3	3		22.06	57.69
Sudan		4	4		1333.33	333.33
Tanzania		5	6		20.00	24.00
Uganda		1	2		2.53	9.09
Zimbabwe		2	2		42.55	44.44
Africa total*	2	33	40.5	0.833	17.02	21.65

*Excluding Kenya, where production data not available.

and Africa: the number of breeders working in the NARS[3] and breeding intensity, which is defined here as the number of breeders per million tonnes of bean production. It is important to note here that the NARS includes not only national-level government programmes but also state-level institutions, universities, and even the private sector. This private sector is not currently a major force in bean breeding, though the private sector is involved in seed production and distribution.

Looking first at Latin America (Table 12.3), what stands out is that there has been a significant decline in both the number of breeders and in breeding intensity in many countries since 1990.[4] Many countries in Central America and the Caribbean had significant reductions in the number of breeders in their programmes. The countries of South America, especially the Andean region and the southern cone, remained relatively stable over the period in terms of numbers of breeders.

[3] The data come from expert opinion and from various editions of a directory of bean improvement scientists (see Johnson, 1999, for more details).

[4] Data on number of breeders are from expert opinion and, in the case of LAC, from various editions of a directory of an international bean improvement association.

The breeding intensity measure allows us to look at how much a country is investing in beans as a function of its production.[5] On average, the breeding intensity in LAC fell sharply between 1989 and 1998, from 18.2 breeders per million tonnes to 12.67. Again, this average masks a great deal of variation, and it is strongly affected by very high intensities in a few countries, namely Panama and Bolivia. In the case of Bolivia, the high breeding intensity corresponds to a specific effort to introduce bean production as a new crop for export, an initiative supported by a local university and CIAT (PROFRIZA, 2000).

Brazil, the largest producer in the region, has had a constant breeding intensity of about four-and-a-half scientists per million tonnes of production, half the regional average and the lowest of any country that actually has a breeding programme. This is consistent with the results of previous studies that found that larger producing countries tend to invest relatively less in breeding than smaller countries do, suggesting that there are scale effects in breeding research (Walker *et al.*, 1999). However, it is also the case that the improvement efforts of the Brazilian programme focus on a narrow range of bean types, namely carioca and black beans.

Recent studies report the number of scientists per million tonnes of potato, wheat and cassava in the late 1990s to be about 5.5, 5 and 0.5, respectively (Walker *et al.*, 1999; Johnson and Manyong, 2000). Compared to these, the regional average of 12.67 seems high; however, one must also consider the fact that the price per tonne of beans is generally much higher than the price of these other commodities. Given the high value of beans compared with other commodities, LAC may actually be under-investing in bean research relative to research on other commodities.

In Africa (Table 12.4), the picture is quite different. African investment in bean breeding is increasing; both the number of breeders and the intensity of breeding have increased since 1990. This rapid growth is in large part due to the extremely low levels of investment in the 1980s. In 1980, only two countries, Malawi and Kenya, had bean breeders. In 1998, 11 countries had breeders. The most dramatic case may be that of Ethiopia, where the number of breeders went from four in 1990 to 10.5 in 1998. In terms of breeding intensity, Africa is today close to where LAC was in 1989.

[5] Number of breeders is clearly a proxy for total investment, since it doesn't say anything about operational budgets. Another shortcoming of using breeders as a proxy is that changes in the structure of research funding in much of LAC has resulted in a shift in breeders' activities away from research towards crop promotion and fund-raising.

Germplasm Exchange and Production of Improved Bean Varieties

Table 12.5 shows the countries that have released CIAT-related bean varieties. Since CIAT does not release varieties directly, the term 'CIAT-related' is used to identify a variety that was released by a national programme but had significant input from CIAT. CIAT-related varieties include germplasm accessions from CIAT gene banks (sometimes referred to as genetic resource units or GRUs), crosses made in CIAT, crosses made in NARS using CIAT parents, and varieties transferred through CIAT-supported networks. The data show wide coverage in both LAC and Eastern/Southern Africa, the main bean-producing areas in CIAT's global mandate for the developing world. Brazil has released the most CIAT-related varieties (44), followed by Argentina (30), Cuba (23), Bolivia and Rwanda (19 each).

Composition of released varieties

Not all varieties released by national programmes are CIAT-related. National breeding programmes often exchange germplasm directly with other national programmes and agricultural research organizations. The materials obtained may be released directly as varieties, or used as sources of genetic diversity in countries' own breeding programmes. Figures 12.1 and 12.2 show how the proportion of CIAT-related varieties in total released varieties has changed over time. The data show that CIAT varieties are a majority of total varieties, and that the two trend lines move together, suggesting a complementary rather than competitive relationship between CIAT and the NARS.

The fact that CIAT-related varieties make up a large percentage of released varieties, while CIAT breeders constitute a very small percentage of total breeders, appears to suggest that CIAT's breeding programme is very productive. CIAT benefits from having high levels of human and financial capital. CIAT breeders generally have larger research budgets than their NARS counterparts, and are also more likely to have PhDs. More importantly, this observation probably reinforces the complementarity between CIAT and NARS breeders, in which the former tend to work further upstream and handle more materials, while the NARS partners spend more time in on-farm testing.

Composition of CIAT-related varieties

As mentioned earlier, a CIAT-related variety may take one of several forms. Figures 12.3 and 12.4 show how the composition of CIAT-related

Table 12.5. Countries that have released CIAT-related bean varieties.

Country	Number of varieties released
Argentina	30
Australia	2
Bolivia	19
Brazil	44
Burundi	12
Canada	1
Chile	6
Colombia	11
DR Congo	12
Costa Rica	18
Cuba	23
Cyprus	1
Dominican Republic	3
Ecuador	9
El Salvador	5
Ethiopia	13
Guatemala	11
Haiti	1
Honduras	12
Kenya	11
Madagascar	1
Malawi	7
Mexico	4
Mozambique	4
Nicaragua	14
Panama	8
Peru	12
Philippines	1
Rwanda	19
South Africa	5
Spain	3
Swaziland	3
Tanzania	7
Turkey	3
Uganda	11
USA	2
Venezuela	8
Zambia	3
Zimbabwe	3

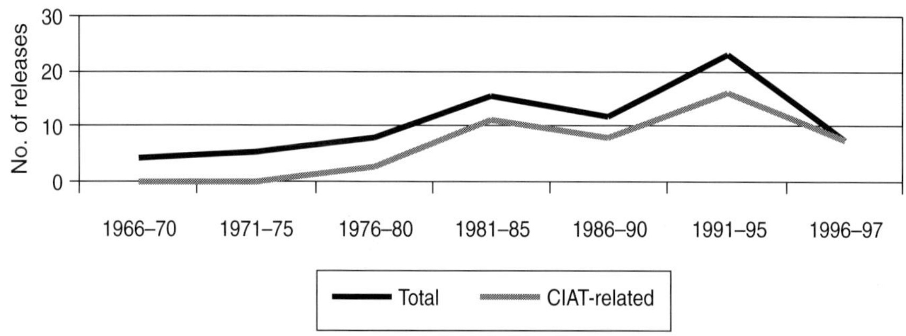

Fig. 12.1. Average annual bean variety releases (LAC).

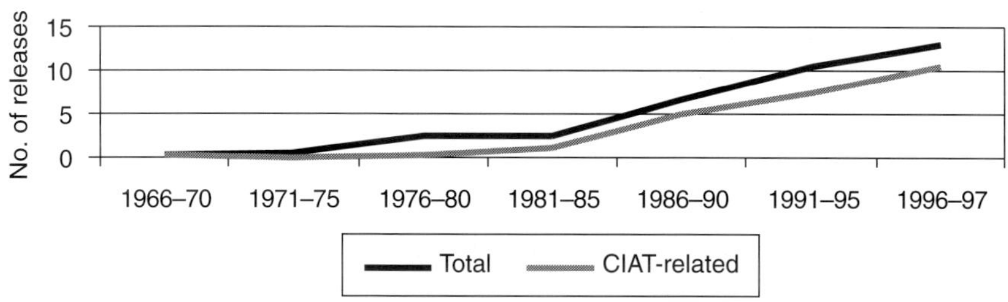

Fig. 12.2. Average annual bean variety releases (Africa).

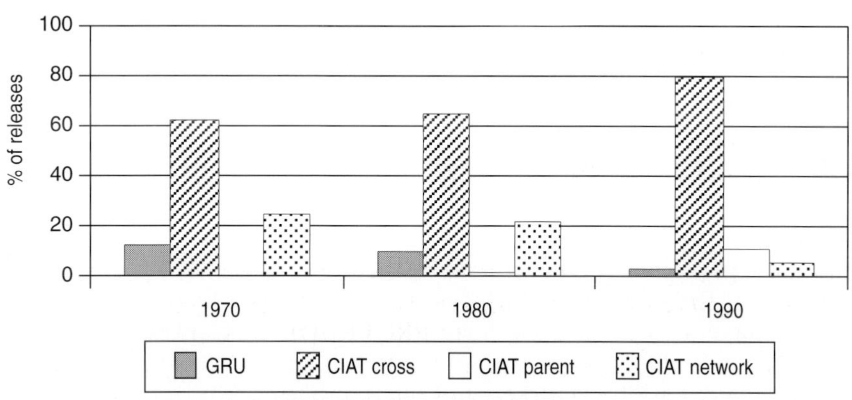

Fig. 12.3. Composition of CIAT-related bean releases (LAC).

varieties has changed over time. Several trends are visible. The first is that the proportion of germplasm accessions from CIAT germplasm banks that are released as varieties has declined between the 1970s and the 1990s. Germplasm accessions – essentially selected landraces collected by CIAT – were never a major source of CIAT-related varieties in

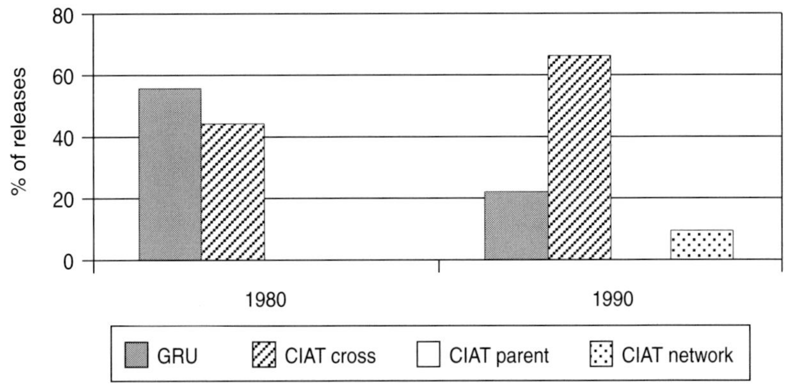

Fig. 12.4. Composition of CIAT-related bean releases (Africa).

LAC, reflecting the relatively advanced stage of bean breeding in the region when CIAT was established. In Africa, by contrast, germplasm accessions were over half of all CIAT-related varieties in the 1980s, as CIAT's newly established breeding programme introduced these selected landraces from LAC and other regions into Africa. In the 1990s, the data suggest that the usefulness of germplasm accessions as a source for varieties was beginning to decline.

The decline in landrace releases has been accompanied by an increase in the release of crosses – both CIAT crosses and, in the case of LAC, crosses made in NARS using CIAT parents. The latter category was non-existent in the 1970s, but accounted for 11% of total releases in the 1990s. In Africa, South Africa has recently released a locally bred variety with CIAT parents, the first one on the continent.

Figures 12.3 and 12.4 also present data on CIAT network varieties. These are varieties to which CIAT's only contribution was to support the germplasm exchange network through which the germplasm arrived in the country of release. CIAT supports bean germplasm exchange via networks such as PROFRIJOL in Central America, PROFRIZA in the Andean region, the Eastern and Central Africa Bean Research Network (ECABREN) and the Southern Africa Bean Research Network (SABRN).

According to the data, in LAC network varieties declined from 25% of CIAT-related varieties in the 1970s to only 6% of varieties in the 1990s. In Africa, the proportion of CIAT-related varieties that were network varieties grew from 0 to 9% between the 1980s and the 1990s. To conclude that this trend implies a declining importance of bean networks in LAC would be a misunderstanding. Rather, the decline in importance over time of network varieties is a function of how they are defined.

A variety is classified as being a CIAT network variety if CIAT's *only* contribution to the variety was its support of the network through which the germplasm was transferred. If a CIAT cross or a GRU from a CIAT collection is transferred through a CIAT-supported network, it will not be categorized as a network variety. In the early years, CIAT networks played an important role in transferring germplasm from one national programme to another. Over time, as the gains from this type of transfer declined and as CIAT's work in germplasm collection and improvement grew, we would expect to see fewer and fewer varieties passing through the networks that have no other connection to CIAT. This is, in fact, what the data show for LAC. Because the African networks were recently established, it is too soon to tell if the same pattern will emerge.

Germplasm exchange networks continue to be very active in LAC. While past data are not available for comparisons, a recent review of the Central American network PROFRIJOL reported the transfer of 18,444 materials over the past 7 years (1987–1996) (Viana, 1998). Of these, 11,433 were from the member countries and 7861 were from CIAT. An important piece of further research would be to look at the role and importance of the networks in the production of varieties.

Adoption and Impact of Improved Bean Varieties

Tables 12.6 and 12.7 show data on the adoption and impact of CIAT-related bean varieties in 1998 by country for selected countries in LAC and Africa. Adoption data for LAC (Table 12.6) were drawn from existing empirical impact studies carried out during the 1980s and 1990s.[6] Data for Africa (Table 12.7) come from empirical studies and estimates of the Pan African Bean Research Alliance (PABRA; see Wortmann, 1999). The total area planted in each country comes from FAO data, as do most of the prices. It is important to note that this Table does not include all countries in which CIAT-related varieties have been released and/or are being grown. Only countries for which reliable estimates were available are included, and some of these country estimates only include certain regions or certain bean types.

[6] See the CIAT impact assessment webpage for abstracts of many of the studies. Where no reliable data or estimates were available for a country, it was left out of the analysis. This was the case for both Cuba and Mexico. CIAT varieties have been released in the country, but there is simply no basis on which to estimate adoption and impact.

Table 12.6. Impact of CIAT-related bean varieties in Latin America, 1998.

	Area planted to beans (ha)	Percentage in CIAT-related varieties (%)	Yield gain over local varieties (t ha^{-1})	Incremental production (t)	Value of increased production (US$)
Costa Rica	39,000	85	0.10	3,290	1,447,432
Guatemala	122,780	40	0.23	11,272	7,434,601
Nicaragua	153,720	30	0.23	10,399	3,529,816
Honduras	83,000	45	0.11	4,152	1,409,295
El Salvador	85,000	25	0.29	6,059	2,325,588
Panama	11,000	40	0.25	1,105	727,021
Brazil	3,307,760	51	0.17	279,211	125,873,960
Argentina	285,000	77	0.24	53,270	28,818,153
Colombia	138,022	10	0.20	2,787	1,198,827
Bolivia	11,640	82	0.25	2,375	920,799
Ecuador	61,520	20	0.13	1,683	889,363
Peru	73,334	16	0.35	4,001	2,878,131
LAC total	4,371,776	49	0.21	379,604	177,452,986

Source: adapted from Johnson (1999).

Table 12.7. Adoption and impact of CIAT-related bean varieties in Africa, 1998.

	Area planted to beans (ha)	Percentage in CIAT-related varieties (%)	Yield gain over local varieties (t ha^{-1})	Incremental production (t)	Value of increased production (US$)
Uganda	360,000	15	0.2	8,830	2,648,976
Ethiopia	420,000	8	0.4	11,508	2,877,119
Kenya			0.6	188	65,905
Tanzania	360,000	4	0.2	3,264	1,142,436
Rwanda	190,000	16	0.9	28,888	8,666,480
DR Congo	248,000	48	0.3	35,000	10,500,000
Malawi	160,000	1	0.2	353	105,797
Africa total	1,738,000	15	0.4	88,032	26,006,712

Source: adapted from Johnson (1999).

According to the data, a total of 49% of bean area in LAC was planted to CIAT-related varieties in 1998, ranging from 10% in Colombia to 85% in Costa Rica. The yield gains associated with improved varieties average 210 kg ha^{-1}, ranging from 100 kg ha^{-1} in

Costa Rica to 350 kg ha^{-1} in Peru.[7] Average yields in the LAC region are generally between 500 and 1000 kg ha^{-1}. In 1998, the gross annual value of increased production was US$177.5 million, and the cumulative value since the 1970s is over a billion dollars.[8]

These yield and production data are consistent with aggregate yield data from FAO, which reports yield gains of 16% for LAC as a whole between 1970 and 1998. The Andean countries had average yield gains over the period of 54%, while in Central America and the Caribbean, yields increased by 23%.

Not all of these yield gains can, of course, be attributed to CIAT's research. Since many of these varieties have been developed in collaboration with NARS, it is difficult to disentangle CIAT's contributions from those of NARS collaborations.

Perhaps the most dramatic case of CIAT impact is that of Brazil, where the data show that the world's largest bean producer has half of its area planted to CIAT-related varieties. Until the early 1990s, CIAT-related varieties were planted on about 10% of area (Janssen *et al.*, 1992). Since the mid-1990s, two CIAT-related varieties, Perola and Apore, have reportedly become very popular in Brazil, in some cases even replacing the dominant local variety, Carioca. This trend was first identified in seed sales data, and was later confirmed by expert opinion (EMBRAPA). Empirical adoption and impact studies will need to be done to document this very significant impact.

Another case of significant impact is that of Argentina, the second largest bean producer in LAC. According to the data, 77% of area is planted to CIAT-related varieties. Unlike in other bean-producing countries, Argentinians consume relatively few beans. Most of the production is for export. Argentina has traditionally produced a white bean called Alubia for the European market, and more recently it has began producing small black beans for export to other countries within LAC. Virtually all of Argentina's black bean varieties come from CIAT (Vizgarra, 1999). Argentina's production went from 39,000 to 290,000 t between 1970 and 1998, placing it among the world's top ten producers (FAO).

The Central American countries have about 40% of area planted to CIAT-related varieties; however, this average masks wide variations within countries. One of CIAT's major successes in the region was in resistance to bean golden mosaic virus (BGMV). In BGMV-prone areas, resistant varieties are found on a high proportion of bean land (Johnson

[7] Reported yield gains are net gains associated with improved varieties. Yield gains associated with increased input or other changes in management practices are not included.

[8] In 1990 US dollars.

and Klass, 1999). In areas that are not at high risk for BGMV, adoption of improved varieties is often low. Lack of reliable seed production and distribution systems are part of the problem, not only in Central America but also in many parts of Africa.

In Africa, bean improvement work began much later than in LAC, and the diffusions curves are estimated to be in very early stages in many countries. None the less, several countries report significant impact; on average, 15% of bean area in the study countries was planted to CIAT-related beans in 1998. Yield gains associated with these new varieties were 400 kg ha^{-1}. In 1998, the gross value of increased production associated with CIAT-related varieties in the study countries was US$26 million, with a cumulative impact of US$116 million.[9]

According to FAO, average yields in the region have gone from 438 to 558 kg ha^{-1} from 1984 to 1998, an increase of 17%. One reason that the yield gain associated with CIAT-related varieties is so high compared with regional yield changes is that adoption of CIAT-related varieties may involve a shift from bush beans to climbing beans. Climbing beans can yield 1 to 2 t ha^{-1} more than bush beans; however, they are much more input-intensive, a fact which must be considered in assessing their overall productivity impacts and economic impacts.

Climbing beans were introduced into the Great Lakes region in the early 1980s, and by 1998 Rwanda and the eastern areas of the DR Congo were estimated to have 16 and 48%, respectively, of their bean area planted to these varieties. Similar figures are likely for Burundi, but no data are available from that country. It is estimated that in Rwanda alone by 1994, climbing beans were generating net benefits of US$8–15 million annually (Sperling *et al.*, 1994).

Returns on Research on Bean Genetic Improvement

To calculate the return on bean-breeding work, we need to compare the stream of benefits associated with improved bean varieties with the investments made in breeding over the years. The impact data provide a rough estimate of the benefit streams associated with new varieties, though it is likely to underestimate total impact, since not all countries which produce CIAT-related varieties are included. In terms of research costs, to accurately estimate the internal rate of return (IRR) of improved varieties, we would need to include both

[9] Values in 1990 US dollars.

CIAT and NARS costs. Since cost data are not available for NARS, we will make the standard assumption that NARS investment is equal to CIAT investment. This is the same as assuming that CIAT and the NARS are each responsible for 50% of the benefits from the CIAT-related improved varieties.

There are two options for calculating CIAT costs. The first is to use data on total bean research investment. These data were presented in Table 12.2. The problem with using these data is that they include all bean research, not just breeding research. Work on improved agronomic and management practices, on networks,[10] on training, and other activities not directly associated with bean genetic improvement is all included in this cost figure. Given that the impact estimates presented in the previous section include only those benefits that can be directly attributed to genetic improvement, these estimates will underestimate the total benefits associated with bean research. Using this measure of total investment in bean research, the net benefits to bean research became positive in 1988, and as of 1998 the internal rate of return for bean research was 18%.

The other alternative is to use only the portion of research costs devoted to breeding. While CIAT cost data are not broken down this way, one way to estimate the portion of resources devoted to breeding research is to assume that it is the same as the proportion of total scientists in the bean programme who are breeders. These percentages are also shown in Table 12.2. This method is likely to underestimate the total amount of resources devoted to crop improvement because in many cases scientists such as pathologists, virologists, geneticists or agronomists work in support of the genetic improvement programmes. Using this estimate of total investment breeding research, net benefits became positive in 1984, and as of 1998 the internal rate of return was 33%.

Given the shortcomings of each method, it is plausible that the true rate of return for bean genetic improvement lies somewhere within the 18–33% range. The rate of return may also differ according to where the research is done. While it was not possible to get data on total bean research expenses for Africa and LAC, we can estimate breeding research costs for each continent using the proportion of breeders stationed in each region.

[10] The impact of CIAT network varieties, varieties whose genetic material is not related to CIAT but which were transferred through CIAT networks, were not included in the adoption and impact estimates. This decision was made in order to be consistent with other commodities studies that form part of the IAEG Impact Project.

Using this data, it was estimated that as of 1998 the IRR for research in LAC was 32%, while in Africa it was 60%. The difference is because the breeding programme in Africa was able to benefit from the past work in LAC and was able to reduce the lag time between the establishment of a programme and the release of a variety. Net benefits to research in Africa became positive after just 4 years, compared with 14 years in LAC. While it is not appropriate to compare the two IRRs directly, since one is a marginal rate of return and the other a total rate of return, it is useful to calculate the two returns because they show the value of extending research done at headquarters to other parts of the world.

Summary, Conclusions and Thoughts on the Future

Over the past three decades, bean improvement research has had substantial impact on bean production in many parts of the world. As of 1999, 39 countries had released 362 bean varieties with some connection to CIAT. The increased production associated with these varieties over the years has an estimated value of over US$1.2 billion. The rate of return to bean breeding is estimated to be between 18 and 33%, suggesting that it is a highly profitable activity.

Intensifying breeding work on abiotic as well as biotic stresses is expected to extend the benefits of improved varieties to more marginal areas that have not yet seen significant benefits. Increasing genetic improvement work on climbing beans – which have traditionally received only about 2% of research investment – also has potential for increasing output. Breeding for micronutrient content also shows promise as a way to increase the contribution of beans to human health, which could have significant impact among the poor. Biotechnology will play an important role in bean breeding, where it is seen as a way to improve the efficiency of conventional breeding. Biotechnology will contribute to more accurate manipulation of resistance genes, for example.

Despite the globalization of bean markets, many consumers retain their preferences for certain bean types, and the diversity of these preferences means that it is often not cost-effective for major exporters to supply these markets. This leaves an opportunity for local farmers to specialize in the production of these high-value bean types. Targeting some of these varieties in genetic improvement activities could have both production and poverty impacts.

Critical to the success of genetic improvement programmes that target varieties to particular production and market environments is the flow of information between producers, marketers, consumers, extension workers, researchers and other stakeholders in the innovation process. Improved information technology such as GIS clearly has a role to play (see for example Wortmann *et al.*, 1998), and it will also be locally important that seed production and distribution systems be extended and strengthened.

One way that CIAT and NARS are attempting to improve the efficiency of the varietal development process is through increasing the participation of different stakeholders. In participatory breeding, farmers are involved in choosing criteria for making crosses as well as identifying promising selections. By incorporating farmers into the process, it is hoped that resulting varieties will be better suited to farmers' needs and will, therefore, be more quickly and widely adopted.

In an era of decreased funding for agricultural research, using the limited resources as wisely as possible will be critical to ensuring that bean producers and consumers receive at least as much impact from genetic improvement research over the next 30 years as they have during the past 30.

Acknowledgements

The CIAT Bean Programme, specifically Oswaldo Voysest, Steve Beebe, Matthew Blair, Roger Kirkby, Howard Gridley and Michael Thung.

References

FAO. Agricultural Database, various years.

Janssen, W., Teixeira, S. and Thung, M.I. (1992) The adoption of improved bean varieties in Brazil. *Trends in CIAT Commodities, Working Document 111.*

Johnson, N. (1999) *Global impacts of CIAT germplasm improvement programs, 1967–1997: progress report on the IAEG study.* Paper presented at 1999 Annual Review. CIAT, Cali, Colombia.

Johnson, N. and Klass, J. (1999) *The impact of crop improvement research on rural poverty: a spatial analysis of BGMV-resistant bean varieties in Honduras.* Presented at the International Conference on the Impact of Agricultural Reseach on Poverty, CIAT/IICA/CATIE, San José, Costa Rica, September 1999.

Johnson, N. and Manyong, V. (2000) *The impact of IARC genetic improvement programs on cassava and beans.* Presentation at the American Association for the Advancement of Science (AAAS), 21 February 2000, Washington, DC.

Pachico, D. (1986) *After the Green Revolution: Technical Change in Bean Production in Colombia, Costa Rica and Guatemala.* CIAT (mimeo).

PROFRIZA (2000) Un cultivo ancestral avanza a la modernidad, tiempo de transación: 1988–1998. In: Voysest, O. (ed.) *Final Report of the Regional Bean Program for the Andean Zone (PROFRIZA).*

Sperling, L., Scheidegger, U., Buruchara, R., Nyabyenda, P. and Munyanesa, S. (1994) *Intensifying Production Among Smallholder Farmers: The Impact of Improved Climbing Beans in Rwanda.* CIAT working document.

van Schoonhoven, A. and Pachico, D. (1998) *Rice and Beans in Latin America: A Summary Report on the Economic Impact of Improved Varieties.* CIAT.

Viana, A. (1998) *The Flow of Germplasm and the Impact of PROFRIJOL in Central America, 1987–1996.* PROFRIJOL, Guatemala [in Spanish].

Vizgarra, O. (1999) Common bean production in Argentina. *Michigan Dry Bean Digest* 23(2), 16–19.

Voysest, O. (1983) *Bean Varieties in Latin America and their Origins.* CIAT, Cali, Colombia [in Spanish].

Walker, T. *et al.* (1999) *Potato Genetic Improvement in Developing Countries and CIP's Role in Varietal Change.* CIP (mimeo).

Wortmann, C. (1999) *Estimates of Impact of CIAT-related Varieties.* CIAT, Uganda (mimeo).

Wortmann, C., Kirkby, R., Eledu, C. and Allen, D. (1998) *Atlas of Common Bean Production in Africa.* CIAT and the Pan Africa Bean Alliance (PABRA).

Economic Impact of International and National Lentil Improvement Research in Developing Countries 13

A. AW-HASSAN AND K. SHIDEED
IN COLLABORATION WITH A. SARKER, R. TUTWILER AND W. ERSKINE

Lentils, one of humanity's oldest food crops, originated in the Fertile Crescent of the Near East (Webb and Hawtin, 1981). As a food, lentils provide valuable protein and, unlike several other food legumes, few anti-nutritional or toxic factors have been reported in lentils. They also require a comparatively short cooking time and are one of the most easily digested of pulses. Lentils may be consumed whole, decorticated and split, or ground into flour. Although lentils are mainly human food, they may occasionally be used to feed animals, particularly poultry. The straw and pod walls, residues from threshing, have a high feed value. The seed coats left after decortication are also considered a valuable feed and may contain up to 13% protein. Lentils are sometimes grown as a fodder with the whole plants being grazed green or cut and fed to livestock. They may also be ploughed in as a green manure. Although lentils are not a major food crop on a world scale, they are nevertheless important in certain countries. The wide range of uses of lentils and their by-products, coupled with their value in many farming systems, and ability to thrive on relatively poor soils and under adverse environmental conditions, has ensured their continued role as crop species.

The average area, production and value of lentils for the 1994–1998 period are presented in Table 13.1. Total area planted worldwide to lentils in 1998 was 3.35 million ha, among which 2.86 million ha was planted in developing countries, representing 85% of the total.

Table 13.1. Average area, production and value of lentil production for 1994–1998.

	Area harvested (1000 ha)	Production (1000 t)	Production (million $)
World	3348	2818	1367
Developing countries	2859	2237	1100
Latin America and Caribbean	43	39	16
Bangladesh	207	168	88
China	93	110	26
WANA	1188	995	479
Turkey	603	604	290
Iran	233	120	63
Syria	130	131	56
Ethiopia	62	34	18
Pakistan	62	32	14
Morocco	50	26	13
Iraq	20	15	10
Egypt	5	7	4
Jordan	4	3	1

Source: FAO agricultural statistics except Iraq which is provided by IPA: Agricultural Research Centre in Iraq.

The annual average lentil production in the developing countries was about 2.2 million tonnes, with a value of US$1.1 billion at the average prices of 1994–1998.

ICARDA (International Centre for Agricultural Research in the Dry Areas) has a mandate for the improvement of lentils for all developing countries, which it has pursued through plant breeding, germplasm exchange with national programmes, and capacity building of NARS (national agricultural research systems). The immediate objective of this report is to document and quantify the impact of international and national lentil breeding research on the production of this crop, and then to document the effect of ICARDA on NARS productivity.

Data and Methodology

Identification of ICARDA's contributions to genetic resource management, pre-breeding research and cultivar development research of NARS lentil breeding programmes required collection of appropriate data sets. To this end, data were collected for China, Syria, Iraq, Pakistan, Bangladesh, Egypt, Jordan and Sudan. The collected data sets include NARS varietal releases, estimates of area planted by variety, yield advantage of ICARDA germplasm, the composition of NARS breeding pools, and human capital in lentil breeding.

Survey data were supplemented by ICARDA release records and pedigree information, and these are confirmed by ICARDA breeders. Type and contents of released varieties by institutions (ICARDA, NARS, other institutions) were determined by pedigree analysis. An economic surplus (ES) approach is used to compute the returns on international and national investment in lentil germplasm improvement research.

Lentil Improvement Research

Lentil is an under-exploited and under-researched annual legume and is often treated as orphan crop even in major producing countries. It is an autogamous species with very little outcrossing. This, coupled with the difficulty of making crosses by hand and the fact that male sterility has yet to be identified, limits the choice of breeding methods to those which have been developed for self-pollinated species. Local types of lentils are frequently characterized by narrow adaptability.

The breeding strategies used for lentil have evolved with time. In the first stage, the variation in the ICARDA lentil germplasm collection was directly exploited, with selection made between and within landraces. Selection pressure for an appropriate phenology, which was the major factor in the domestication of lentil, was the driving force behind ICARDA's breeding strategy for lentil in this stage (Robertson and Erskine, 1997). These selections were distributed to national programmes through an international nursery network to test for local adaptation. Most of the cultivars developed at this stage were derived from selection within heterogeneous populations, and were not the result of hybridization (Muehlbauer and Slinkard, 1981). The germplasm on which this selection was based came largely from West Asia and North Africa (WANA), which is the centre of origin.

The particular combinations of characters required for specific regions were often not found 'on the shelf' in the collection. Consequently, in the second stage, ICARDA started hybridization and selections from segregating populations. Stable lines were then distributed to national programmes for testing in their respective agroclimatic conditions. Research in this stage resulted in the release of a number of cultivars in different regions.

However, lentil lines developed from selection at ICARDA in West Asia are mostly limited in adaptation to the home region. As a result, the breeding programme has decentralized to work closely with national programmes having different agro-ecological conditions. In this third stage of lentil improvement, potential crosses are identified with NARS partners and subsequently made at ICARDA's Tel Hadya station. Then country-specific segregating populations are shipped to national cooperators for local selection. More than 200 crosses are made annually at ICARDA, targeted to address different stresses in specific agro-ecological zones. Selections made by national programmes are fed back into the international nursery network for wider distribution. Increasingly, national programmes use ICARDA-derived material in their own hybridization programmes.

Separate programmes target improvements for the diverse environments in which lentil is grown. Abiotic and biotic stresses affect lentil, and sources of resistance are being identified. In addition, multiple resistance is often needed, and several accessions have resistance to two or more diseases. Rust is the most important foliar disease. ICARDA screens for rust resistance through joint research with the national programmes of Ethiopia, Morocco and Pakistan. As a result of this effort, rust-resistant cultivars have been released in Chile, Ecuador, Ethiopia, Morocco and Pakistan. An international nursery for rust resistance was initiated with national programmes in 1990 to clarify the host–pathogen relationships in different regions and to assist in identifying variation in the fungus. Vascular wilt is the most important soilborne disease of lentil in the Mediterranean region and also causes major yield losses in the Indian subcontinent. ICARDA has developed an efficient screening method for vascular wilt in lentil and has identified several useful sources of resistance. Ascochyta blight causes losses in productivity in WANA, parts of the Indian subcontinent and Canada. Good sources of resistance to ascochyta blight have been identified in cooperation with NARS.

Although lentils are considered to be one of the most cold-tolerant of the food legume crops, they are unable to withstand the very severe winters of the higher elevations in the Mediterranean region, such as on the Anatolian plateau in Turkey or the high plateaux in the Maghreb countries. In these areas the crop is normally sown during early spring. The Indian subcontinent is the largest lentil-producing region in the world. The crop is grown as a winter (*rabi*) crop, and is normally sown after the end of the summer monsoon rains, from October to December, and raised on conserved soil moisture. If water is available, the crop may be given one or two irrigations. In the Nile valley of Egypt and northern Sudan, lentils are sown in the early winter but, under the extremely arid conditions of this region, almost the entire water requirement of the crop is met by irrigation. Thus, both cold tolerance and drought tolerance are desirable traits.

Other important traits for lentil improvement include biomass (seed and straw) and seed yield, attributes for mechanical harvesting, earliness and response to irrigation.

Human Resources and Expenditures in Lentil Improvement Research

Human capital and expenditure at ICARDA

Data on total budget, total number of scientists employed, and the number of scientists (in full-time equivalents) involved in lentil improvement are available for ICARDA. The expenditure series for ICARDA lentil improvement programmes was constructed by multiplying ICARDA's average cost per scientist by the number of full-time scientists in lentil improvement. ICARDA expenditure on lentil improvement research in terms of human and financial resources is presented in Table 13.2 for 1980, 1990 and 1997. The human capital investment in lentil improvement research increased from 2.55 scientists in 1980 to 3.05 in 1997. This pattern in human resource investment is consistent with changes in financial expenditures. The total research expenditures were estimated at US$0.64 million and US$0.76 million for these 2 years, respectively.

Human capital and expenditures at NARS

NARS lentil expenditures were estimated, based on data provided by Pardey *et al.* (1991). They reported in their appendix (pp. 414–421) total agricultural research expenditures (in 1980 purchasing power parity, PPP) and total number of researchers by country for 1961–1965 to 1981–1985. These two series were used to calculate average cost per NARS scientist in 1980 PPP dollars. This cost per scientist was multiplied by the 1997 survey data on the number of NARS lentil improvement scientists in each country to obtain 1997 lentil improvement expenditures for NARS.

Table 13.2. Human resources and expenditure in the lentil germplasm improvement programme at ICARDA.

	Units	1980	1990	1997
Breeding	(SY)	2.25	2.00	2.00
Entomology	(SY)	0.30	0.25	0.20
Pathology	(SY)	0.00	0.00	0.30
Biotechnology	(SY)	0.00	0.00	0.30
Total	(SY)	2.55	2.25	2.80
GRU	(SY)	0.00	0.25	0.25
Grand total	(SY)	2.55	2.50	3.05
Expenditure	(million US$)	0.64	0.63	0.76

Notes: SY, scientist-years; GRU, genetic resource unit.

Total NARS expenditures were then projected forward to 1999 and back to 1980 using Pardey *et al.*'s data on growth rates of research expenditures and total number of researchers by region/country, and assuming that lentil research expenditures have increased at the same rate as research expenditures on all crops. This method of estimating investment in lentil research is based on a number of assumptions: that the cost of supporting a lentil researcher is the same as the average cost per researcher for other crops, and that lentil research expenditure has commanded a constant share of total expenditures.

The human capital investment in lentil breeding programmes for the six countries (China, Egypt, Iraq, Jordan, Pakistan and Sudan) is given in Table 13.3. There are 34 scientists in these NARS, of which 14 (41%) are breeders, and the rest are supporting scientists including pathologists (21%), entomologists (6%) and agronomists (29%). About 38% of all scientists have PhD degrees and the rest have Master's degrees or lower. On the whole, the time spent in lentil improvement is about 67% of the number of scientists. About 90% of breeders' time is invested in lentil breeding activities.

Table 13.3. Human resources endowments in lentil breeding research in developing countries.

	Number of scientists (PhDs in parentheses)					
	Breeding	Pathology	Entomology	Agronomy/ physiology	Quality	Total
Bangladesh	2(1)	2(1)	1(0)		1(0)	6(2)
China	4(0)	0	0	0	0	4(0)
Egypt	3(2)	1(1)	1(1)	2(1)	0	7(5)
Iraq	0	0	0	3(1)	0	3(1)
Jordan	0	0	0	4(0)	0	4(0)
Pakistan	2(1)	2(2)	0	0	0	4(3)
Sudan	2(0)	1(1)	0	1(1)	0	4(2)
Syria	1(0)	1(0)	0	0	0	2(0)
Total	14(4)	7(5)	2(1)	10(3)	1(0)	34(13)
Full-time equivalent (scientist-years)						
Bangladesh	1.5	1.2	0.2	0.0	0.5	3.4
China	3.7	0.0	0.0	0.0	0.0	3.7
Egypt	3.0	0.2	0.1	2.0	0.0	5.3
Iraq	0.0	0.0	0.0	3.0	0.0	3.0
Jordan	0.0	0.0	0.0	0.0	2.0	2.0
Pakistan	2.0	0.4	0.0	0.0	0.0	2.4
Sudan	1.3	0.0	0.0	0.3	0.0	1.6
Syria	1.0	0.5	0.0	0.0	0.0	1.5
Total	12.5	2.3	0.3	5.3	2.5	22.9

There is an uneven distribution of human resources in the national lentil breeding programmes. The larger producers, such as China and Pakistan, have relatively lower investments in their lentil improvement research. The smaller sized producers, such as Egypt, Iraq and Jordan, have higher staffing per unit of output. Another indicator of investment in lentil improvement research is the proportion of breeders holding PhD degrees to the total number of breeders. Only Egypt and Pakistan have PhD holders in lentil breeding. Egypt has two PhDs out of three breeders, and Pakistan has one PhD holder out of two breeders. The other four countries have no PhD holders in lentil breeding.

NARS financial expenditures are presented in Table 13.4. It shows that NARS lentil expenditures have increased substantially, from US$0.160 million in 1980 to US$0.76 million in 1990. The rate of increase was slower between 1990 and 1997, from US$0.76 million to US$1.67 million. This sharp increase in lentil expenditures of these six countries could be interpreted as an indirect impact of ICARDA's lentil improvement programme on NARS investment in crop breeding. With this expansion in NARS expenditures, ICARDA's share of total lentil expenditures reduced sharply, from 63% in 1980 to 26% in 1990 and to 15% in 1997.

The conclusion of this analysis is that human resources in lentil breeding research are uneven in terms of education level and number of breeders per cultivated area. On the whole, the human capital investment in lentil improvement is relatively low in the developing countries.

Table 13.4. NARS expenditures in lentil improvement research (million US$).

	1980	1990	1997
Egypt	0.057	0.092	0.154
Iraq	0.070	0.229	0.524
Jordan	0.043	0.080	0.158
Syria	0.025	0.058	0.124
China	0.054	0.127	0.259
Pakistan	0.052	0.095	0.156
Bangladesh	0.057	0.238	0.461
Sudan	0.073	0.110	0.156
Total NARS	0.159	0.760	1.670
ICARDA's contribution	0.272	0.268	0.323
Total	0.431	1.027	1.993
ICARDA share as % of total ICARDA NARS expenditures	63	26	16

Impact of Lentil Improvement Research on Germplasm Production and Release of Varieties

Production and release of new varieties

Data in Table 13.5 show total numbers of released lentil varieties for 22 countries. A total of 54 varieties have been released by the 22 countries during the 1980–1999 period. Separation of total releases into sub-periods with a 5-year interval shows that the largest number of releases, 20 varieties, occurred during the 1995–1999 period, with an annual average of about four varieties. Previously, the number of releases was stable during the 1985–1989 and 1990–1994 periods. Only three varieties were released during the 1980–1984 period, when Ethiopia was the only country releasing improved lentil varieties.

Table 13.5. Total and average annual number of lentil varietal releases by country for 1980–1999.

Country	1980–84	1985–89	1990–94	1995–99	Total
China	0	1	1	2	4
Egypt	0	0	1	4	5
Iraq	0	0	1	1	2
Jordan	0	2	1	0	3
Pakistan	0	1	1	1	3
Sudan	0	0	2	1	3
Syria	0	1	0	0	1
Algeria	0	3	0	0	3
Argentina	0	0	1	0	1
Bangladesh	0	0	2	2	4
Chile	0	1	0	0	1
Ecuador	0	1	0	0	1
Ethiopia	3	0	1	2	6
Iran	0	0	0	1	1
Lebanon	0	1	0	1	2
Lesotho	0	0	0	2	2
Libya	0	0	1	0	1
Morocco	0	0	1	0	1
Nepal	0	1	0	0	1
Portugal	0	0	0	2	2
Tunisia	0	2	0	0	2
Turkey	0	1	3	1	5
Total	3	15	16	20	54
Average annual release	1	3	3	4	3

ICARDA contribution to varietal release

Cultivar development comprises two main plant breeding activities. These are crossing lines to create new varieties and screening varieties developed elsewhere. ICARDA's contribution to varietal releases can be determined by the origin of released varieties using pedigree analysis. The released varieties were classified into five categories: (i) ICARDA crossed and selected varieties; (ii) ICARDA crosses selected by NARS; (iii) NARS crosses with ICARDA parents; (iv) NARS crosses with NARS parent; and (v) ICARDA germplasm accessions selected and released as varieties.

The impact of ICARDA lentil breeding research can be easily shown from data in Fig. 13.1 which presents the origin of released varieties. The table shows that ICARDA made the crosses for 50% of all lentil varieties released by the 22 countries during the 1980–1999 period. Information on the origin of parents of NARS releases adds to ICARDA impact by showing that an additional 2% of the varieties released from NARSs crosses had ICARDA parents. These data suggest that ICARDA, like other CGIAR centres, is playing a dual role; it is producing varieties that are being directly transferred to farmers, and is also a source of breeding material for NARS breeding programmes. An additional 30% of lentil released varieties is attributed to ICARDA germplasm accession. Thus, ICARDA has contributed to 81% of the released varieties.

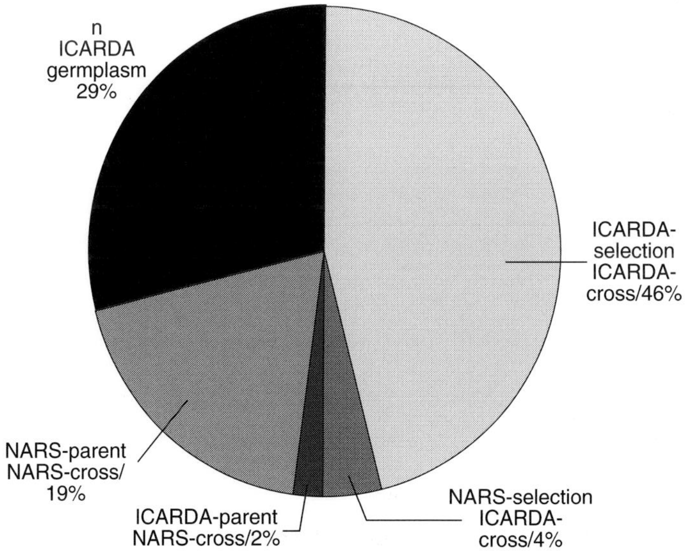

Fig. 13.1. The proportion of the types and content of released lentil varieties.

Only ten released varieties were originated from NARS crosses and parents. Thus, the ICARDA lentil breeding programme is an important factor in the release of new varieties in developing countries.

Figure 13.2 shows the evolution of ICARDA's contribution over time. The clearest trend is the noticeable increase in the proportion of released varieties which are crossed at ICARDA. Overall, the share of ICARDA crosses in released varieties reached its highest level in recent years.

Another important trend is the decline over time of the proportion of releases from NARS crosses of any source. The sharpest decline is in the releases of NARS crosses from ICARDA parents. This implies an increasing demand by NARS for ICARDA crosses. The increased demand for ICARDA crosses may be attributed to the fact that ICARDA has access to a larger pool of genetic material than NARS do. The third important trend is that ICARDA germplasm accessions represent an important proportion of released varieties. These trends suggest that the increased crossing and selection activities by ICARDA are still the main source for varietal production in the developing countries. This is consistent with the strategy of the lentil breeding programme where cultivars of specific traits are developed and selection is made on the agro-ecological conditions of the

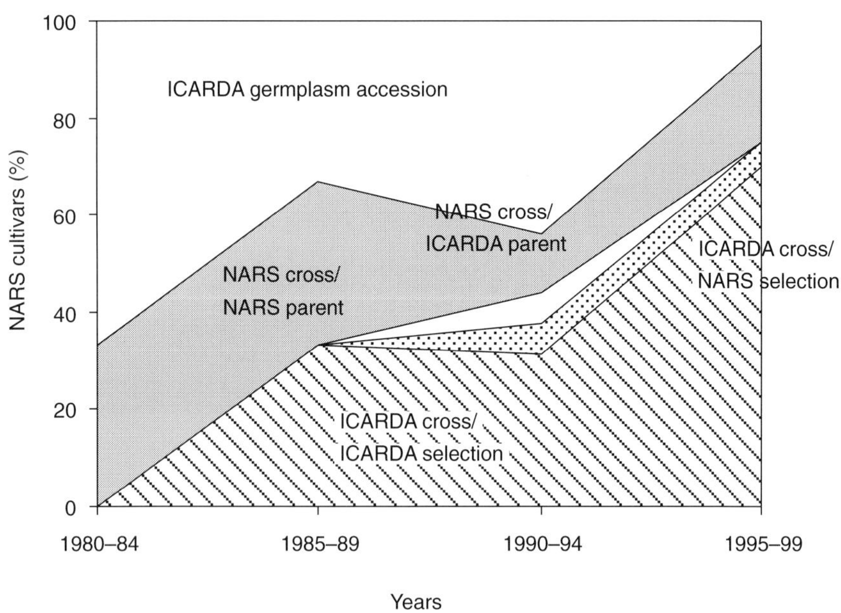

Fig. 13.2. Contribution of ICARDA content measures to NARS-released cultivars of lentil.

target region. Meanwhile, the decline in NARS crosses suggests that selection rather than crosses is the most efficient use of NARS human resources, while they have access to a continuous flow of improved germplasm through their collaboration with ICARDA. However, as NARS capacity increases, the crosses from their own parents are expected to increase substantially.

ICARDA Contribution to NARS Pre-breeding Research

Analysis of NARS breeding pools and crossing blocks is required in order to study the impact of ICARDA pre-breeding work. NARS breeding pools provide information on cultivars to be released over the next 5–10 years and, thus, the likely impact of ICARDA in the immediate future. The data in Table 13.6 indicate that ICARDA advanced lines and released cultivars represent the major part of the composition of NARS crossing blocks in 1997. The contribution of this content has increased substantially, from 22% in 1987 to 57% in 1997, suggesting that ICARDA's likely impact on variety releases is substantial in the immediate future. The importance of NARS' own advanced lines and released cultivars has decreased from 45% to 22% between 1987 and 1997, as NARS have increased their reliance on ICARDA in their search for high yielding improved lentil varieties. The contribution of landraces has also decreased from 23% to 16% between the same 2 years.

Another measure for assessing the contribution of ICARDA to NARS pre-breeding research is the source of parents for the NARS crosses. The number of crosses made by NARS has increased sharply

Table 13.6. Composition of NARS lentil crossing blocks and parental contributions of crosses for 1987 and 1997.

	Crossing blocks		Crosses	
	1987	1997	1987	1997
Total	114	379	50	119
Proportions by type of germplasm source				
1- NARS	0.45	0.22	0.25	0.35
2- ICARDA	0.22	0.57	0.40	0.41
3- Other countries	0.10	0.05	0.00	0.03
4- Local landraces	0.21	0.12	0.25	0.15
5- ICARDA landraces	0.02	0.04	0.10	0.06

Countries: China, Egypt, Iraq, Jordan, Pakistan and Sudan.

during the 10-year period. The data in Table 13.6 show that parents from ICARDA advanced lines and released cultivars contributed greatly to the crosses made by NARS in 1987 and 1997. The contribution of ICARDA parents accounts for 40% in 1987 and 1997. Parents from NARS' own advanced lines accounted for 25% of the crosses made in 1987. The contribution of this category of material increased sharply in 1997, accounting for 35% of the 1997 crosses. The share of local landraces in NARS crosses decreased from 25% in 1987 to 15% in 1997.

Economic Impact

Estimating research benefits

Economic surplus (ES) models have been widely used in previous studies (Evenson, 1974; Hertford and Schmitz, 1977; Byerlee and Traxler, 1995) to estimate the benefits and costs of agricultural research.

The total benefits for a small open economy can be expressed as:

$$\Delta ES = P_t.Q_t.k_t\,(1+0.5k_t.\,\varepsilon)$$

where P_t is prices in year t, Q_t is quantity in year t, and k_t is the downward shift in supply curve in year t as proportion of initial price and ε is the price elasticity of supply.

The size of the research-induced supply shift, the k-factor, is a crucial determinant of the total benefits from research. The accuracy in estimating k and its path over time, reflecting adopting lags, will determine the accuracy and validity of the estimates of research benefits (Alston *et al.*, 1999).

The economic surplus model can be used to compute annual flows of research benefits and costs. The general procedure requires data on yield advantage of new varieties, adoption path, the change in cost of production due to the use of new varieties, producer prices, and demand and supply price elasticities. When the distribution of benefits between producers and consumers is not a concern, gross annual research benefits (GARB) can be calculated, which does not require information on elasticities (Traxler, 1998). Schwartz *et al.* (1993) used this approach in estimating the economic returns on cowpea research in Senegal.

Given that releases of lentil varieties are relatively new in the sense that most varietal release took place in the late 1980s and diffusion has picked up in the 1990s, an *ex post* estimation of the economic surplus will not reflect the actual benefits of the lentil improvement programme. Therefore, the GARB from production increase due to improved lentil yields for 1997 (or $GARB_{1997}$) were estimated. The parameters used in this analysis are discussed below.

Yield gains

Yield improvement is the most important impact of any commodity improvement programme. For farmers to adopt the improved varieties, new releases must outperform local cultivars in grain yield and/or other traits, such as drought- and disease-resistance. Only Type I varietal change, and thus yield advantage, is observed for lentil.[1] Type II varietal change is not yet evident in lentil production under rainfed conditions. Estimation of benefits from research requires data on yield improvement on farmers' fields with and without new varieties. Yield improvement information was collected from national programmes. These were primarily based on the results of on-farm trials.

The immediate yield advantage of switching from unimproved to improved varieties is as high as 45% in China. The lowest yield advantage of 15% was obtained in Syria. Other countries have a yield advantage of 30–40% compared with the base yield of traditional varieties (Table 13.7). The overall reported yield gain from improved lentil germplasm for five countries (China, Egypt, Iraq, Pakistan and Jordan) was estimated at 41%. However, national average yields still remain low due to low adoption of improved varieties and high climatic variability.

The impact of improved germplasm is not limited to yield advantage. In addition, released varieties have important traits, such as resistance to disease and drought. These traits may reduce costs of production. For example, two released cultivars in Egypt display a high response to water, one other variety is resistant to root wilt, and another cultivar is resistant to drought. Similarly, one of the two improved varieties released in Iraq is suitable for mechanical harvesting, whereas the other one is red so as to meet consumers' preferences. Breeding for drought- and disease-resistance has an important impact in reducing the inherent risk associated with rainfed farming. For example, three cultivars originated from ICARDA (Adaa, Gudo and Chalen) were the only unaffected varieties in research centres and farmers' fields after a

[1] Byerlee and Traxler (1995) described two types of varietal technical change. Type I change occurs in areas where modern varieties are replacing traditional varieties resulting in a sharp increase in crop yield. Type II change occurs in areas where farmers are adopting newer generation modern varieties to replace older generation modern varieties. These changes assure the maintenance of yield stability in the face of evolving pest biotypes.

Table 13.7. Average lentil production, yield, and estimated yield gains due to improved germplasm use for selected countries.

Country	Area[1] (1000 ha)	Production[1] (1000 t)	Yield average[1] ($t\ ha^{-1}$)	Estimated yield gain[2] (%)	Adoption (%)	Production increase (Mt)	GARB 1997 (US$ million)
Bangladesh	207	171	0.83	30	12	6,158	2.85
China	93	110	1.19	45	05	2,484	0.42
Egypt	5	7	1.54	40	50	1,396	0.70
Iraq	20	15	0.77	25	25	967	0.29
Jordan	4	3	0.64	23	21	133	0.07
Pakistan	62	32	0.52	30	32	3,114	1.46
Syria	130	116	0.89	16	25	4,628	1.86
Total/average	521	455	0.87	29	17	18,881	7.67

Source: (1) FAOSTAT; (2) estimated from national surveys, on-farm trials reports and expert estimates. Average yield and adoption are weighted by the cultivated area.

rust epidemic wiped out most of the local landraces in Ethiopia in the winter of 1997 (Bejiga *et al.*, 1998).

Diffusion of new varieties

Formal lentil breeding research is relatively new in the developing countries. The production of new germplasm started in the early 1980s. The adoption of this material is affected by many factors. The most important of these is the performance of the new varieties, which varies with environment and management factors. Another important issue is the availability of improved seed. State seed agencies do not always serve lentil growers, because lentils are not seen as a strategic food crop; and without hybrids, there is little incentive for seed companies to market improved varieties. Regardless of these difficulties, the data collected from NARS indicate that the adoption of improved lentil varieties is growing in several countries.

Improved lentil varieties were disseminated in Syria in 1987 with about 2% of lentil area being planted to improved cultivars. By the year 1990, the adoption rate of improved lentil cultivars was 9%. With continuous cooperation between ICARDA and Syria's NARS, more improved varieties were released, and the adoption rate increased to 25% in 1997. Similar increases occurred in other countries of the study, where adoption rose during the 1990s. The national research programme of Pakistan reported that about 32% of lentil area in the targeted region of that country is now planted with improved lentil varieties. Similarly, about 25% of lentil area in Iraq is planted with improved varieties. In Bangladesh, the area cultivated with improved lentil cultivars increased from 12% in 1997 to 30% in 1999 because of an effective technology transfer project named 'mission pulses'. The overall average adoption rate for six countries (Bangladesh, China, Egypt, Iraq, Jordan and Pakistan) is estimated at 17% in 1997. Hence, the area planted with improved lentil cultivars in these countries is about 90,000 ha. ICARDA-based varieties account for about 80% of releases and might thus account for a proportional share of the area under improved varieties. Lentil production is expected to increase as the diffusion of improved varieties expands. This conclusion is supported by the fact that current adoption of improved lentil varieties is in its early stages and has not reached its ceiling in these countries. Increased food production is an important policy objective for these food-deficit countries.

Estimating gross annual research benefits from lentil improvement

The gross annual research benefits for 1997 can be expressed as:

$$GARB_{1997} = \Delta Q_{1997} \times P_{1997}, \text{ and } \Delta Q_{1997} = Q_{1997} \times g \times A_{1997},$$

where g is the yield improvement ratio ($\text{Yield}_{\text{New Vars}}/\text{Yield}_{\text{Trad Vars}}{}^{-1}$). A_{1997} is the proportion of area under improved varieties in 1997, and Q_{1997} is lentil production. Information on lentil yield increase due to improved varieties and adoption rates was collected through surveys from national programmes. The 1997 c.i.f. prices (P_{1997}) for each country were calculated from the *FAO Trade Yearbook*. These prices approximate the social value of lentil production for each country. With a weighted average yield improvement of 29%, and adoption rate of 17%, a production increase of about 19,000 t is attributed to the adoption of improved cultivars in these countries for 1997. The total GARB for 1997 was estimated at about US$7.7 million for the seven countries (Bangladesh, China, Egypt, Iraq, Pakistan, Jordan and Syria). Large producing countries such as Bangladesh, China and Pakistan benefited most from this technology. Bangladesh realized the highest benefit of US$2.85 million in 1997, followed by Syria with US$1.9 million and Pakistan with US$1.47 million. Other small producing countries did not realize as much benefit. This result is particularly due to low adoption of improved varieties. Innovative ways of enhancing varietal diffusion at a much faster rate is a necessary condition for realizing greater returns from lentil research.

Conclusion

ICARDA, after its establishment in 1978, initiated a global lentil improvement research programme for the developing countries. There was hardly any lentil improvement research before that time. Thus ICARDA played a crucial role in strengthening lentil improvement research in the developing countries, through training, networking and collaborative research programmes. Fifty-two varieties have been released in 22 developing countries since 1980. Currently ICARDA accounts for not more than 15% of the total investment in lentil improvement research, yet it contributed to 81% of the varietal releases either by developing and transferring cultivars, or by providing parents for national crosses. In addition, ICARDA facilitated flows of germplasm through its exchange and network programmes. ICARDA also affected national pre-breeding work by providing advanced lines and specific traits used for NARS crosses.

Data reported by a number of national programmes indicate that the diffusion of improved lentil varieties is in its early stages, but is steadily increasing. The diffusion of improved varieties to larger areas is slowed down by external factors. These factors include: (i) the difficult nature of the production environment; (ii) high climate and soil variability, which limits the likelihood of a single variety being able to dominate large areas; and (iii) the inability of seed systems to get high quality seed to small farmers in marginal environments. One way to achieve higher varietal diffusion is to decentralize breeding programmes, with the strategy of adapting germplasm to the specific agro-ecological conditions of different sub-regions. Increased lentil varietal diffusion would also require innovative ways to involve extension and seed systems in the technology development and transfer process.

References

Alston, J.M., Norton, G.W. and Pardy, P. (1999) *Science under Scarcity: Principles and Practice for Agricultural Research Evaluation and Priority Setting*. CAB International, Wallingford, UK.

Bejiga, G., Tadesse, N. and Erskine, W. (1998) Ethiopia fights back lentil diseases. In: *ICARDA Caravan*, no. 9.

Byerlee, D. and Traxler, G. (1995) National and international wheat improvement research in the post-Green Revolution period: evolution and impacts. *American Journal of Agricultural Economics* 77 (May), 268–278.

Evenson, R.E. (1974) The Green Revolution in recent development experience. *American Journal of Agricultural Economics* 56 (May), 387–394.

Hertford, R. and Schmitz, A. (1977) Measuring economic returns to agricultural research. In: Arndt, T.M., Dalrymple, D.G. and Ruttan, V.W. (eds) *Resource Allocation and Productivity in National and International Agricultural Research*.University of Minnesota Press, Minneapolis.

Muehlbauer, F.J. and Sinkard, A.E. (1981) Genetics and breeding methodology. In: Webb, C. and Hawtin, G. (eds) *Lentils*. Commonwealth Agricultural Bureaux and ICARDA, pp. 69–90.

Pardey, P.G., Roseboom, J. and Anderson, J.R. (1991) *Agricultural Research Policy – International Quantitative Perspective*. Cambridge University Press, Cambridge.

Robertson, L.D. and Erskine, W. (1997) Lentil. In: Fuccillo, D., Sears, L. and Stapleton, P. (eds) *Biodiversity in Trust: Conservation and Use of Plant Genetic Resources in CGIAR Centers*. Cambridge University Press, Cambridge.

Schwartz, L.A., Sterns, J.A. and Oehmke, J.F. (1993) Economic returns to cowpea research, extension, and input distribution in Senegal. *Agricultural Economics* 8, 161–171.

Traxler, G. (1998) *Impact of CGIAR germplasm improvement on food production: design proposal*. Unpublished report prepared on request by the IAEG of the CGIAR.

Webb, C. and Hawtin, G. (eds) (1981) *Lentils*. Commonwealth Agricultural Bureaux and ICARDA.

Impacts of Genetic Improvement in Groundnut 14

M.C.S. Bantilan, U.K. Deb and S.N. Nigam

Introduction

This chapter documents groundnut genetic enhancement research conducted by ICRISAT in partnership with NARS in Asia and Africa, and quantifies its impact. Impacts of research in groundnut are measured in terms of variety release, yield gain and reduction in production cost per tonne. Estimates of the impacts are measured at both the farm and aggregate level. This chapter also investigates whether ICRISAT content in improved varieties affects varietal impacts. The extent of adoption of improved groundnut varieties in Africa and Asia is presented. Levels of adoption are high in some countries such as China, South Africa, Swaziland, Mozambique, Argentina and Brazil. There is a notable increase in yield in most of the groundnut growing countries, and per tonne production costs of improved varieties were less than that of local varieties. Increase in yield was associated with uptake of improved groundnut varieties. An important policy implication arising from this study is the need for breeding by ICRISAT in partnership with NARS. For promotion of improved groundnut varieties countries need to ensure the availability of seed and exchange of information.

Groundnut is a major oilseed and food crop worldwide. In 1999, 31.96 million tonnes were produced from 23.57 million ha (Table 14.1). Groundnut is important for its oil and protein, and is a valuable commodity for both human use and livestock feed. The plant originated in South America but is now widely distributed throughout

Table 14.1. Groundnut (in-shell) area, production and yield in different countries, 1979–1999.

Country	Area ('000 ha)			Yield (t ha^{-1})			Prod ('000 t)		
	1979–81	1989–91	1997–99	1979–81	1989–91	1997–99	1979–81	1989–91	1997–99
Africa									
Benin	90	97	122	0.7	0.7	0.8	60	69	102
Botswana	4	1	1	0.3	0.5	0.5	1	1	0
Burkina Faso	129	182	221	0.5	0.7	0.9	70	121	194
Chad	168	186	412	0.6	1.0	1.0	93	164	432
Congo, DRC	477	628	528	0.7	0.8	0.8	334	513	403
Egypt	12	13	49	2.1	2.2	3.0	26	27	146
Gambia	72	85	84	1.1	1.1	1.1	79	96	92
Ghana	98	129	188	1.3	0.9	1.0	125	127	193
Guinea	128	104	166	0.7	0.8	1.0	83	80	168
Kenya	11	20	18	0.6	0.6	0.6	7	12	11
Malawi	250	86	113	0.7	0.7	0.8	176	52	92
Mali	166	188	157	0.9	1.0	0.9	141	174	148
Morocco	28	22	28	1.2	0.9	1.6	34	20	44
Mozambique	350	342	284	0.4	0.3	0.5	131	113	138
Niger	181	86	237	0.6	0.3	0.4	105	30	103
Nigeria	572	878	2,506	0.9	1.4	1.0	503	1,181	2,616
Senegal	1,053	857	722	0.7	0.9	0.9	690	757	651
Sierra Leone	15	23	37	0.8	0.9	0.9	12	20	34
Sudan	960	332	1,461	0.8	0.6	0.7	769	174	976
Swaziland	2	3	7	0.5	1.5	1.4	1	4	9
Tanzania	91	110	116	0.6	0.6	0.6	54	62	73
Uganda	109	185	200	0.7	0.8	0.7	80	149	137
Zambia	28	66	125	0.6	0.4	0.4	18	28	51
Zimbabwe	183	192	207	0.6	0.6	0.5	101	108	109

Asia									
Bangladesh	25	38	35	1.1	1.1	1.1	28	41	40
China	2,346	2,947	4,041	1.5	2.1	2.8	3,501	6,082	11,465
India	7,132	8,562	7,203	0.8	0.9	1.0	5,999	7,570	7,440
Indonesia	496	633	643	1.6	1.8	1.5	806	1,124	968
Malaysia	7	1	1	3.3	3.7	3.8	25	5	6
Myanmar	489	523	472	0.8	0.9	1.2	390	456	554
Pakistan	49	84	101	1.2	1.1	1.1	60	89	107
Philippines	52	45	25	0.9	0.8	1.0	47	35	25
Sri Lanka	12	10	10	0.6	0.6	0.6	7	6	6
Syria	10	11	13	1.8	2.0	2.3	18	22	29
Thailand	103	116	89	1.2	1.4	1.5	128	160	135
Vietnam	106	207	257	0.9	1.1	1.4	94	218	352
Latin America and the Caribbean									
Argentina	289	166	337	1.3	2.1	1.7	403	350	594
Brazil	282	85	95	1.5	1.7	1.8	433	142	168
Haiti	47	42	27	0.7	0.8	0.8	35	34	22
Jamaica	2	2	3	1.2	1.1	1.2	2	2	3
Mexico	67	87	93	1.1	1.3	1.4	73	110	134
Paraguay	29	36	28	1.0	1.1	1.1	28	39	30
Developed countries									
Australia	32	20	25	1.5	1.6	1.6	48	31	39
Greece	4	2	1	2.8	3.9	3.7	11	7	3
Israel	5	3	4	4.4	6.6	6.1	22	20	24
Japan	33	18	12	1.9	2.0	2.3	61	36	27
South Africa	317	129	83	1.0	1.2	1.7	304	150	143
USA	595	739	582	2.6	2.6	2.9	1,550	1,893	1,713
World	18,750	2,0274	23,191	1.0	1.2	1.4	18,546	23,458	31,805

Note: Each figure is a 3-year average for the respective period, e.g. 1979–1981.
Source: Estimated from FAO data.

tropical, subtropical, and warm temperate areas in Asia, Africa, Oceania, North and South America, and Europe. In semi-arid areas, it is largely grown by smallholder farmers under rainfed conditions. Developing countries accounted for over 95% of world groundnut area and about 94% of total production during 1997–1999. Production is concentrated in Asia and Africa, with Asia accounting for about 60% of global area and 70% of production. Major groundnut producers in Asia are India (accounting for 44% of global area and 33% of total output) and China (17% of area and 36% of output). Africa accounts for 35% of global groundnut area and 21% of groundnut production; major producers in Africa are Nigeria, Sudan and Senegal. The USA and Argentina are also major groundnut-growing countries.

Groundnut Genetic Enhancement Research at ICRISAT

Compared to the cereal crops, groundnut has received relatively little research attention. Groundnut research to date has generally been conducted by government or public institutions, apart from some private-sector efforts in the USA (Freeman *et al.*, 1999, p. 30). In general, one or two scientists per country are involved in groundnut research in most of the African countries that produce groundnuts. By comparison, in India, more than 150 scientists are involved in groundnut research. At ICRISAT, five or six breeders and an equal number of other scientists are involved in the improvement of the groundnut crop. Research and development activities carried out by groundnut scientists include collection, evaluation and conservation of germplasm, breeding with specific targets, and distribution of germplasm and enhanced genetic materials to the national programmes. ICRISAT has also collaborated with the national agricultural research systems (NARS) in Africa and Asia and has conducted various international trials and observation nurseries. ICRISAT has also played a major role in training groundnut scientists.

Pre-breeding research

Collection, characterization and maintenance of landraces are essential for crop improvement, and these activities have been a high priority at ICRISAT. As of December 1999, 15,342 groundnut germplasm accessions from 92 countries were conserved in ICRISAT's collections. Forty-eight traits are included in the characterization descriptors.

Once genetic resources have been collected and assembled, ICRISAT and its NARS partners conduct evaluation trials to identify the useful traits available in the germplasm. Economically important traits sought in the evaluation are high oil content, early maturity, large seed and high shelling ratio. So far, researchers have identified sources of resistance for disease (335 accessions), insects (31), drought tolerance (46) and high protein (100). This information has been disseminated to researchers worldwide through reports, journal papers and other fora. In response to requests from users, ICRISAT has distributed 86,088 groundnut germplasm samples to 93 countries up to December 1999. Generally, groundnut germplasm is supplied as pods to scientists in India. It is always supplied as seed to scientists outside India, after complying with all the necessary phytosanitary requirements (Kameswara Rao, 2000).

Breeding research domains

ICRISAT groundnut breeders focus on eight breeding research domains. A 'research domain' is delineated as a homogeneous eco-region defined in terms of soil and climatic conditions, and spreads beyond the geographical boundaries of one country. Table 14.2 summarizes the location and characteristics of each of the eight groundnut research domains identified by groundnut scientists.

Research focus through time

The genetic enhancement approach in groundnut at ICRISAT has differed between subsistence farming systems and high-input farming systems. Resistance breeding was the main focus of the efforts to improve groundnut productivity in subsistence systems, while yield and quality breeding were the targets for high-input farming systems. Breeding methods for both systems have included mass selection, pedigree methods, modified pedigree selection (single seed descent; SSD) and backcross breeding. Backcross breeding is increasing in usefulness as simple traits of economic importance are being discovered. In recent years, molecular genetic research has been adopted and molecular marker technology and genetic transformation work are in use.

The emphasis of groundnut breeding research focus at ICRISAT has changed over time, in keeping with research achievements at ICRISAT and with the increased capability and research infrastructure of NARS. Table 14.3 summarizes the groundnut research portfolio at ICRISAT since 1976. A summary of achievements in different research activities is mentioned below.

Table 14.2. Groundnut research domains.

Domain	Production system characteristics	Major constraints	Locations
GN I	Rainy season, 90–100 days duration, rainfed. Oil and confectionery use	Drought, late leaf spot, rust, rosette, aflatoxin	Mid tier of Sahel (Senegal, Mali, Burkina Faso, Niger), India (Gujarat)
GN II	Rainy season, 100–120 days duration, rainfed. Mostly oil use	Late leaf spot, rust, drought, aflatoxin, rosette	E. Africa (Sudan), India (N. Maharashtra, Madhya Pradesh)
GN III	Rainy season, 90–130 days duration, rainfed. Oil and confectionery use	Late leaf spot, rust, rosette, millipedes, pod rots	Southern tier of Sahel (Nigeria, Gambia, Cameroon, Ghana), India (N. coastal Andhra Pradesh, Orissa, W. Bengal), Bangladesh, N. Vietnam, Indonesia
GN IV	Rainy season, 100–120 days duration, rainfed. Mostly oil use	Late leaf rust, drought, leaf miner, *Spodoptera*	India (S. Maharashtra, Andhra Pradesh, Tamil Nadu, Karnataka), Myanmar, Thailand, S. Vietnam
GN V	Summer season, 110–120 days duration, full irrigation. Mostly oil use	No major constraint, but iron chlorosis and bud necrosis disease could be important	India (Gujarat, N. Maharashtra, Madhya Pradesh)
GN VI	Post-rainy season, 100–120 days duration, full irrigation. Mostly oil use	Late leaf spot, bud necrosis disease, leaf miner, *Spodoptera*, white grubs	India (W. & S. Maharashtra, Andhra Pradesh, Tamil Nadu, Orissa, Karnataka, Kerala), N. India, Pakistan, Nepal (rainy season)
GN VIIa	Rainy season, mostly monocropping, 120–140 days rainfed (generally well distributed). Mostly confectionery use. Large seeded varieties preferred	Early leaf spot, rust, rosette, aphids, jassids	Southern Africa (N. Mozambique, N. Zimbabwe, C. Malawi, E. Zambia, S. Tanzania, Zaire)
GN VIIb	Rainy season, mono- and intercropping, 90–110 days duration, rainfed. Mostly oil use	Late leaf spot, rust, rosette, drought, aflatoxin	Southern Africa (S. Mozambique, S. Zimbabwe)
GN VIII	Rainy season, rainfed (bi-modal rainfall). Mono- and intercropping, 90–120 days duration. Oil and confectionery use. Three-seeded Valencia types preferred	Early leaf spot, late leaf spot, rust, rosette, pod rots	Central Africa (N. Tanzania, N. Zaire, Uganda, Rwanda, Burundi, W. Kenya)

Source: ICRISAT medium term plan, 1994–1998.

Table 14.3. Groundnut research portfolio at ICRISAT (Asia, SEA, WCA).

Research area	1976	1986	1996	2000
Foliar diseases	✓	✓	✓	✓
Rust	✓	✓	✓	✓
Late leaf spot	✓	✓	✓	✓
Early leaf spot	✓	✓	✓	✓
Aflatoxin	✓	✓	✓	✓
Insect pests				
Sucking pests	–	✓	✓	–
Defoliators	–	✓	✓	–
Soil pests	–	✓	✓	–
Virus diseases				
Rosette	–	✓	✓	✓
Bud necrosis	✓	✓	✓	–
Peanut clump	–	✓	✓	✓
Drought	✓	✓	✓	✓
Yield and adaptation	✓	✓	✓	✓
Nitrogen fixation	✓	✓	–	–
Nematode diseases	–	✓	–	–
Nutritional quality	✓	✓	✓	–

Breeding for resistance to foliar diseases

Rust, late leaf spot, and early leaf spots are the major foliar diseases in groundnut. Screening methods for foliar disease resistance have been developed. Resistant sources (*Arachis hypogaea*) have been identified, including 169 lines for resistance to rust, 69 lines to late leaf spot (LLS), 42 lines to rust and late leaf spot, and 32 for early leaf spot. Resistant interspecific derivatives are developed. Through hybridization and selection, scientists have been able to transfer rust and late leaf spot resistance to commercially acceptable and agronomically superior genetic backgrounds. A few resistant cultivars (ICG(FDRS)10, ICGV 86590, ICG(FDRS) 4, Girnar 1, ALR 2) were released but remain largely unadopted due to long-duration, unattractive pod shape and low shelling percentage.

Breeding for resistance to insect pests (foliar and soil)

Achievements in this area include the development of screening methods and the identification of sources of resistance in 15 lines for thrips, 133 lines for jassids, nine lines for termites, four lines for aphids, 14 lines for leaf miner and one for *Spodoptera*. Higher levels of resistance were found in wild *Arachis* species. Elite germplasm (ICGVs 86031, 86252, 86393, 86455, 86462) was developed. Work on pest resistance has been suspended since 1997.

Breeding for resistance to Aspergillus flavus

Aflatoxin contamination is a complex problem and it can occur at preharvest, harvest or postharvest stages in the field as well as during storage at the processor/consumer level. Genetic improvement in the resistance level is considered as one of several approaches to resolving this problem. Genetic resistance, together with better crop management practices and optimal storage conditions, can significantly reduce contamination.

Screening methods and detection tools have been developed. Resistant sources have been identified in 21 lines for preharvest seed infection, 37 lines for *in vitro* seed colonization, and two lines for aflatoxin production. Elite germplasm (ICGVs 88145, 89104, 91278, 91279, 91284) have been developed. Germplasms having resistance to *A. flavus* have been used in the breeding programme. The resulting derivatives have been tested for level and stability of resistance to *A. flavus* and for yield potential in multilocational trials. Scientists have been able to transfer stable resistance into different genetic backgrounds and some of these lines outyield local control varieties at certain trial locations.

Breeding for resistance to viral diseases

Breeding efforts include developing resistant/tolerant varieties to groundnut rosette disease, peanut bud necrosis disease (PBND) and peanut mottle virus (PMV). Scientists have developed diagnostic tools, virus characterization and screening methods, and have identified sources of resistance or tolerance to groundnut rosette disease and peanut bud necrosis disease in 116 and 32 lines, respectively. Bud necrosis disease, caused by tomato spotted wilt virus and transmitted by thrips, occurs in serious proportions in India and is becoming increasingly important in many other countries. By breeding for vector resistance, scientists were able to reduce the incidence of PBND considerably. Both vector-resistant and virus-resistant lines were used to improve the level of PBND resistance. Some of the developed lines having tolerance to bud necrosis, such as ICGV 86032, ICGV 86030, ICGV 86031 and ICGV 86032, have agronomically desirable properties with higher yield potential. Resistance breeding for groundnut rosette disease is in progress in Africa, while resistance breeding for PBND in Asia stopped after 1996.

Breeding for drought tolerance

Achievements include better understanding of the physiology of drought, traits associated with drought resistance and their surrogates, field screening techniques, resistance sources (both to mid-season and end-of-season drought) and resistant varieties. Research activities in this area are in progress. Future work needs to be on trait-based selection procedures for greater efficiency in breeding, knowledge of inheritance of traits associated with drought resistance, and high yielding drought resistant cultivars adapted to different agroecoregions.

Breeding for adaptation to specific environments and requirements

The targets of adaptation breeding were to develop material for varying requirements, from no-stress to multiple-stress situations. After succeeding in developing improved varieties for relatively low-stress conditions by the mid-1980s, breeders then moved to focus on developing breeding lines with resistance/tolerance to multiple-stress factors in different maturity classes – early, medium and late. Achievements include character association and inheritance studies, studies on response to photoperiod in groundnut, enrichment of NARS through improved breeding materials and release of improved varieties jointly by NARS and ICRISAT.

Research Products

Intermediate products

Successful development of new groundnut varieties depends to a large extent on the availability of source germplasm with desirable traits such as high yield, greater oil content, high shelling ratio, disease and insect resistance, drought resistance, improved grain quality for confectionery use, etc. Identification of germplasm with desirable traits and the incorporation of desirable traits to a wide range of germplasm is important for expanding the gene pool. ICRISAT has conducted a massive screening programme to identify germplasm with desirable traits. The team effort of scientists in groundnut breeding, entomology and pathology in screening materials for resistance to various biotic and abiotic stress factors has resulted in the development of elite germplasms. A total of 62 elite lines has been developed. Of these, some display high yield with resistance to one or more stress factors (12 lines); others are noted for high yield (five lines), large seed size (four lines), early maturity (four lines), resistance to foliar diseases (17 lines), resistance to *A. flavus* seed infection (two lines), resistance to insect pests (six lines), seed dormancy (five lines), non-nodulating lines (five lines), and puckered leaves (one line).

Varietal production

A total of 67 improved varieties were released in 22 countries through ICRISAT-NARS partnerships (Table 14.4). Out of the 67 releases, 41 were in nine countries of Asia; 20 varieties were released in ten countries of Africa; and six varieties were released in two other countries. Four released varieties were developed by NARS from ICRISAT germplasm, and seven were released through ICRISAT networks, while all other varieties were bred at ICRISAT and released after adaptive trials by the NARS of the respective countries. Another 40 improved varieties are likely to be released in coming years: 32 varieties in nine African countries, and eight varieties in four Asian countries.

Adoption and Impacts of Improved Groundnut Varieties

Extent of adoption

Table 14.5 reports the level of adoption of improved groundnut varieties in different countries. The level of adoption of improved varieties in South Africa, Mozambique and Swaziland is about 75%. Botswana, Zimbabwe and Zambia have adoption levels of 70, 50 and 20%, respectively. About 10% of the groundnut area of Malawi and Uganda is under improved varieties. Adoption levels in other sub-Saharan African countries are low. Farmers mostly cultivate traditional varieties or some local varieties recommended by researchers. For example, Chalimbana (a local variety recommended for cultivation) is grown in about 80% of the groundnut area in Malawi. In Uganda, three recommended varieties (Red Beauty, Bukene and Roxo) of regional origin have been in cultivation since the 1960s, and they occupied about 80% of the country's groundnut area in 1999.

Technology adoption in some countries is very high – especially in Argentina, Brazil and China (Freeman *et al.*, 1999). Latin America and China have had remarkable success in promoting improved groundnut technology – improved varieties, fertilizer, crop rotation, and chemical control of weeds, pests and diseases. Almost the entire groundnut area in Argentina and Brazil is sown to improved varieties. In China, improved varieties cover more than 90% of the groundnut area.

More than 17% of the groundnut area in Vietnam is under improved varieties. Improved groundnut varieties grown by farmers are: VD-1, VD-2, VD-3, VD-4, VD-5, VD-6, VD-7, VD-8, VD-9, VD-10, HL25, ICGV 87883 and ICGV 90068 (Phan Lieu *et al.*, 1999). On-farm survey results showed that the improved variety VD-1 has started to replace local varieties since its release in 1995. Two other improved varieties belonging to the VD series, VD-3 and VD-4, have also slowly been taken up by farmers.

Table 14.4. Number of ICRISAT-NARS released groundnut varieties in different countries, 1981–1999.

Country	< 1980	1981–1985	1986–1990	1991–1995	1996–1999	Total released	Likely to be released
Africa		(1)	5(1)	7(1)	4(1)	20	32
Burkina Faso	–	–	–	–	–	0	4
Congo	–	–	1 (1)	–	–	2	–
Ethiopia	–	–	1	1	–	2	–
Gambia	–	–	–	–	–	0	3
Ghana	–	–	1	–	–	1	–
Guinea	–	–	–	1	–	1	3
Lesotho	–	–	–	–	–	0	3
Malawi	–	–	1	1 (1)	–	3	3
Mali	–	–	–	–	–	0	7
Namibia	–	–	–	–	–	0	3
Sierra Leone	–	–	–	3	–	3	–
Swaziland	–	–	–	–	–	0	3
Tanzania	–	(1)	–	–	–	1	–
Uganda	–	–	–	–	2	2	–
Zambia	–	–	1	1	(1)	3	–
Zimbabwe	–	–	–	–	2	2	1
Asia		2(1)	13+4*	13(1)	6(1)	41	8
Bangladesh	–	–	–	–	–	0	2
China	–	–	–	–	–	0	1*
Cyprus	–	–	–	3	–	3	–
India	–	1	9 + 4*	4	–	18	–
Indonesia	–	–	–	–	3	3	–
Korea (South)	–	–	1	–	1	2	–

Continued

Table 14.4. *Continued*

Country	< 1980	1981–1985	1986–1990	1991–1995	1996–1999	Total released	Likely to be released
Malaysia	–	–	–	–	–	0	2
Myanmar	–	1 (1)	–	1	1	4	–
Nepal	–	–	–	–	1 (1)	2	3
Pakistan	–	–	1	2	–	3	–
Philippines	–	–	–	(1)	–	1	–
Sri Lanka	–	–	–	2	–	2	–
Vietnam	–	–	2	1	–	3	–
Caribbean	–	–	2	2	2	6	–
Jamaica	–	–	1	–	–	1	–
Mauritius	–	–	1	2	2	5	–
TOTAL	–	2 (2)	20 (1)+4*	22 (2)	12 (2)	67	40

Note: Figures in parentheses indicate number of releases through ICRISAT network while asterisk indicates number of NARS releases from ICRISAT parent. All other releases are ICRISAT-bred varieties released in respective countries after adaptation trial conducted by NARS.

Table 14.5. Level of adoption of improved groundnut varieties in Asia and Africa, 1999.

		Percentage of area planted to improved varieties				
Country/region	Year (season)	ICRISAT bred	ICRISAT parent	ICRISAT network	Other improved	All improved
Africa						
Botswana	1999	–	–	–	70.00	70.00
Malawi	1999	10.00	–	–	–	10.00
Mozambique	1999	0.20	–	–	75.00	75.20
Namibia	1999	–	–	–	50.00	50.00
South Africa	1999	–	–	–	75.00	75.00
Swaziland	1999	5.00	–	–	70.00	75.00
Uganda	1999	–	–	–	10.00	10.00
Zambia	1999	15.00	–	–	5.00	20.00
Zimbabwe	1999	2.00	–	–	50.00	52.00
Asia						
China	1990s					>90.00
India/Maharashtra						
Nasik, Dhule, Kolhapur	1997 (*Kharif*)	–	–	–	25.56	25.56
Nanded, Parbhani, Satara	1997 (*Kharif*)	–	–	–	94.28	94.28
Nanded, Parbhani, Satara	1997 (*Rabi*)	31.71	–	–	48.78	80.49
Nasik, Dhule, Kolhapur	1997 (Summer)	–	–	–	4.49	4.49
Nanded, Parbhani, Satara	1997 (Summer)	18.23	–	–	67.39	85.62
India/Andhra Pradesh						
Guntur, West Godavari	1997 (*Kharif*)	98.00	–	–	–	98.00
Guntur, West Godavari	1997 (*Rabi*)	58.00			–	58.00
Guntur, West Godavari	1997 (Summer)	31.74	–	–	–	31.74
Anantapur, Chitoor, Prakasam	1997 (*Kharif*)	–	–	–	37.00	37.00
Anantapur, Chitoor, Prakasam	1997 (*Rabi*)	39.00	–	–	1.00	40.00
Vietnam/South	1997	0.25	–	–	17.21	17.46

Source: Impact Monitoring Survey 1999–2000. For India, computed from Bantilan *et al.* (1999); and for Vietnam Phan Lieu *et al.* (1999); and for China, Freeman *et al.* (1999).

Uptake of other improved varieties such as HL 25 and ICGV 87391 remained insignificant. HL 25 is suitable only in elevated areas due to its susceptibility to leaf diseases associated with excess moisture. Reasons for low adoption of ICGV 87391 are related to undesirable traits (thick shell, pod reticulation and constriction), high input requirements and its lower market price compared with local varieties. Although five out of six respondents were aware of improved varieties, most did not grow improved varieties. The major reasons cited for not growing improved varieties were: lack of seed availability, lack of elevated land for seed multiplication in the rainy season, some of the improved varieties were not suitable for local conditions, low price (due to low quality) of unsuitable improved varieties, improved varieties with large seed size have varying qualities across seasons (Phan Lieu *et al.*, 1999).

Bantilan *et al.* (1999) reported that in 1997, farmers of Andhra Pradesh (India) grew several improved groundnut varieties (JL 24, Kadiri and ICGS 44), while farmers of Maharashtra (India) adopted JL 24, TAG 24, UF-70-103, TG 26 and Karad 4-11. ICRISAT varieties are popular in the Guntur and West Godavari districts of Andhra Pradesh and in the Nanded, Parbhani and Satara districts of Maharashtra. Two older varieties, TMV2 and SB11, are widely cultivated in Andhra Pradesh and Maharashtra, respectively. The Government of India recommended these two varieties in the early 1940s. Reasons for the wide cultivation of TMV2 and SB11 are seed availability, drought resistance and yield stability. ICGS 11, ICGS 44, ICGS 21 and ICGS 49 were observed on farmers' field in locations where technology was disseminated and seeds were made available. The main reasons given for low adoption of ICRISAT varieties in Maharashtra are non-availability of seed and longer duration. The most preferred traits for rainy season groundnut varieties in Maharashtra are medium duration, high pod yield, greater oil content and higher shelling percentage. On the other hand, Andhra Pradesh farmers want varieties with high pod yield, with pest and disease resistance.

Impacts of improved varieties

The goals of groundnut breeders were to enhance yield and incorporate resistance to biotic and abiotic stresses. ICRISAT-developed varieties were first released for cultivation in farmers' fields in the early 1980s. Therefore, we have computed yield gains in groundnut in 1997–1999 compared with 1979–1981. We have also estimated percentage annual compound rate of growth in yield for the period 1979–1999. If a country has a large area under groundnut cultivation or ICRISAT varieties have been released (or are likely to be released), then we have included that country in our analysis. Results of this exercise are presented in Table 14.6.

Table 14.6. Yield growth for groundnut in different countries.

Country	Growth rates, 1979–99	Country	Growth rates, 1979–99
Africa		Asia	
Benin	1.6	Bangladesh	0.2
Botswana	2.6	China	3.5
Burkina Faso	2.7	India	1.4
Chad	2.7	Indonesia	–0.5
Congo	0.7	Malaysia	0.5
Egypt	2.3	Myanmar	1.2
Gambia	–1.0	Pakistan	–0.7
Ghana	–1.5	Philippines	0.5
Guinea	3.2	Sri Lanka	0.5
Kenya	0.3	Syria	0.9
Malawi	–0.5	Thailand	1.2
Mali	0.5	Vietnam	2.5
Morocco	1.2	Argentina	1.6
Mozambique	1.3	Brazil	1.2
Niger	–1.0	Haiti	0.4
Nigeria	0.8	Jamaica	0.2
Senegal	1.4	Mexico	0.9
Sierra Leone	0.8	Paraguay	0.4
Sudan	–0.1		
Swaziland	7.3	Developed countries	
Tanzania	0.4	Australia	1.5
Uganda	–0.2	Greece	1.6
Zambia	–1.9	Israel	2.1
Zimbabwe	0.6	Japan	1.3
		South Africa	5.1
		USA	0.3
		World	1.9

Note: Growth rates are percentage per annum.
Source: Estimated from FAO data.

Growth in yield

Groundnut yields increased worldwide by 1.9% per year between 1979 and 1999. Yield has increased in most countries, especially in Asia, Latin America and in the developed countries. Yield has increased at the rate of 3.5% per year in China. The adoption of improved varieties in China is more than 90%, and Chinese farmers have also adopted improved management practices such as organic and inorganic fertilizer, crop rotations, plastic film mulch, and pest and disease control. Between 1979–1981 and 1997–1999, groundnut yields in China increased from 1.5 to 2.8 t ha^{-1}. In India, growth in yield was 1.4% per year and average yields increased from 0.8 t ha^{-1} in 1979–1981 to 1 t ha^{-1} in 1997–1999. However, yields in India vary widely depending on the production system. On rainfed groundnut, which occupies about 80% of groundnut area, yields are roughly 0.9 t ha^{-1}, while the irrigated crop yields about 1.6 t ha^{-1}. Yields in Vietnam have increased at the rate of 2.5% per annum. Though the adoption rate of improved groundnut varieties is only about 17%, Vietnamese farmers have widely adopted improved groundnut production technology including alternative coconut ash (ACA). Traditionally Vietnamese farmers used to grow groundnut using coconut ash, but non-availability of natural coconut ash created problems in groundnut cultivation in Vietnam. Then scientists developed ACA, which is a chemical formulation of all the nutrients (chemicals) available in coconut ash. ACA is very popular in Vietnam. Improved management practice, accompanied by improved varieties to a significant extent, has led to the increase of groundnut yield in Vietnam to 1.4 t ha^{-1} in 1997–1999 from 0.9 t ha^{-1} in 1979–1981.

In some countries in sub-Saharan Africa, growth in groundnut yields has been low, but the countries that have high rates of modern variety adoption have experienced high yield growth. For example, Swaziland had 7.3% annual growth in yield and had 75% area under improved varieties. South Africa had 75% adoption and experienced 5.1% annual compound growth rate. Botswana had 70% adoption and enjoyed 2.6% annual growth rate in yield for the period 1979–1999. Mozambique also experienced 1.3% annual growth in yield. Senegal also has high adoption and experienced 1.4% annual growth rate.

By contrast, farmers cultivating under semi-subsistence systems in Africa (Gambia, Ghana, Malawi, Uganda, Zambia) generally grow low-yielding, late-maturing varieties in marginal land without irrigation. They have experienced negative growth in yield, implying that yield declined over time in those countries. Developing countries in Latin America (especially Argentina and Brazil) have experienced positive growth. Argentina and Brazil have also adopted improved varieties and management practices to a large extent.

Between 1979–1981 and 1997–1999, yields in the USA increased from 2.6 to 2.9 t ha^{-1}. This relatively slow growth was due to several reasons. First, technology adoption (e.g. the introduction of runner varieties in the 1970s) and large-scale commercialization had already taken place earlier. Second, weather variability, including drought in some parts, generally reduced yields and increased fluctuations in yield. Third, public concern over the effects of high levels of fertilizer and pesticide use on environmental and human health led to reductions in the use of agro-chemicals during the 1980s, thus slowing down yield growth (Freeman *et al.*, 1999).[1]

[1] It is also true that in the USA, groundnuts tend to be grown on relatively poor farmland, where high input use does not always pay.

Table 14.7 reports data on the adoption and impacts of popular improved groundnut varieties on yield, unit cost of production and per hectare return in Africa and Asia. Adoption levels of some individual varieties are very high. For example, Sellie – a variety from South Africa – covers 70% of the groundnut area in Swaziland, 60% of area in Botswana, 40% of area in Mozambique, 40% of area in South Africa and 10% of area in Zimbabwe. Some are popular in a specific region. For example, ICGS 44 is cultivated in 98% of the rainy season groundnut area of the Guntur and West Godavari districts of Andhra Pradesh in India. CG7 covers 10% of the groundnut area in Malawi. Improved groundnut varieties provided 5–25% higher grain yield in southern and eastern African countries, compared with the best performing local varieties. Yield gain from improved varieties in India was 13–108% in Maharashtra and 27–53% in Andhra Pradesh in the year 1997. Per tonne cost of production was 15–44% lower in Maharashtra except for TMV 10, which had higher per tonne production costs compared with the best performing local variety (SB11). In Andhra Pradesh, per tonne production costs of improved varieties were 11–37% lower, except for ICGS 44, which had slightly higher per tonne production costs. All improved varieties provided a higher net return on a per hectare basis. Compared with the best performing local variety, per hectare net return was 50–594% higher in Maharashtra and 36–191% higher in Andhra Pradesh.

Spillover impacts

An important objective of international agricultural research institutions is to determine the extent to which research undertaken in one location may impact on other regions of interest. This is because research activities are most often planned to target mandate crops and agro-ecological areas found in many parts of the world. ICRISAT has, as a policy, distributed a wide range of groundnut germplasm and elite materials to breeding programmes in NARS throughout the semi-arid tropics. This has contributed to faster and more cost-effective development of useful final products by the receiving parties and thus has generated technology spillover. For example, ICG 221 was developed for India and was also released in Swaziland (Table 14.8). ICGM 286, ICGV-SM 86066, ICGV-SM 85038 and ICGV-SM 86080 were developed for Malawi and are now grown in Rwanda. RMP 12 was originally developed for Burkina Faso but was also released in Uganda and Mozambique.

These examples indicate that the genetic material used in the development of these cultivars has wide adaptation, thus resulting in spillovers from groundnut genetic enhancement research. This indicates the advantages in targeting wide adaptation of the improved varieties.

Table 14.7. Adoption and impacts of popular improved groundnut varieties.

				Impacts of improved varieties on		
Country (region)	Variety	Year	Adoption (% area)	% yield gain	% reduction in cost per tonne	% increase in net return per ha
Africa						
Botswana	Sellie	1999	60.00	10	–	–
Botswana	55-437	1999	10.00	5	–	–
Malawi	CG7	1999	10.00	50	–	–
Mozambique	Natal Common	1999	30.00	5	–	–
Mozambique	Sellie	1999	40.00	10	–	–
Mozambique	RMP 12	1999	5.00	15	–	–
Namibia	Sellie	1999	50.00	10	–	–
South Africa	Sellie	1999	40.00	10	–	–
South Africa	Anel	1999	15.00	10	–	–
South Africa	Akwa	1999	20.00	20	–	–
Swaziland	Sellie	1999	70.00	10	–	–
Swaziland	ICG 221	1999	5.00	15	–	–
Uganda	Igola-1	1999	10.00	25	–	–
Zambia	MGV4	1999	10.00	25	–	–
Zambia	Luena (JL24)	1999	5.00	15	–	–
Zimbabwe	Falcon	1999	30.00	10	–	–
Zimbabwe	Flamingo	1999	10.00	10	–	–
Zimbabwe	Sellie	1999	10.00	10	–	–
Asia						
India/Maharashtra						
Nasik, Dhule, Kolhapur	JL 24	1997 (*Kharif*)	11.24	31	15	104
Nasik, Dhule, Kolhapur	TMV10	1997 (*Kharif*)	9.08	13	–27	50

Nasik, Dhule, Kolhapur	K2	1997 (*Kharif*)	4.87	66	31	168
Nanded, Parbhani, Satara	JL 24	1997 (*Kharif*)	39.05	24	24	100
Nanded, Parbhani, Satara	Karad4-11	1997 (*Kharif*)	5.71	33	44	254
Nanded, Parbhani, Satara	TAG 24	1997 (*Kharif*)	49.52	–	–	–
Nanded, Parbhani, Satara	ICGS 21	1997 (*Rabi*)	31.71	37	–	–
Nanded, Parbhani, Satara	TAG 24	1997 (*Rabi*)	48.78	86	23	109
Nasik, Dhule, Kolhapur	JL 24	1997 (summer)	4.49	95	27	251
Nanded, Parbhani, Satara	UF-70-103	1997 (summer)	9.94	44	37	133
Nanded, Parbhani, Satara	TAG 24	1997 (summer)	56.35	108	37	445
Nanded, Parbhani, Satara	ICGS 11	1997 (summer)	3.31	19	25	119
Nanded, Parbhani, Satara	ICGS 49	1997 (summer)	14.92	86	24	594
India/Andhra Pradesh						
Guntur, West Godavari	ICGS 44	1997 (*Kharif*)	98.00	50	–	–
Anantapur, Chitoor, Prakasam	JL 24	1997 (*Kharif*)	30.00	57	14	36
Anantapur, Chitoor, Prakasam	Kadiri	1997 (*Kharif*)	7.00	40	37	191
Guntur, West Godavari	ICGS 44	1997 (*Rabi*)	58.00	27	–4	71
Anantapur, Chitoor, Prakasam	JL 24	1997 (*Rabi*)	24.00	53	0	62
Anantapur, Chitoor, Prakasam	Kadiri	1997 (*Rabi*)	15.00	–	–	–
Guntur, West Godavari	ICGS 44	1997 (summer)	31.74	91	11	47
Vietnam	VD-1	1997	12.73	–	–	–

Note: – indicates data not available.
Source: Impact Monitoring Survey 1999–2000. For India, Deb *et al.* (2000) and for Vietnam, Phan Lieu *et al.* (1999).

Table 14.8. Spillover impacts of improved groundnut varieties.

Variety	Production system* and country where originally selected	Spillover into
Chipego	21 Malawi	19, 20, 21 Zambia
ICG 221	9 India	20 Swaziland
ICGM 286	21 Malawi	21, 23 Rwanda
ICGMS 42	21 Malawi	21 Zambia
ICGV-SM 86066	21 Malawi	21, 23 Rwanda
ICGV-SM 85038	21 Malawi	21, 23 Rwanda
ICGV-SM 86080	21 Malawi	21, 23 Rwanda
Johari	9 India	21, 22 Tanzania
RMP 12	15 Burkina Faso	21 Uganda
		21, 22 Mozambique
Roba	8 India	23 Ethiopia
Stella	21 Malawi	22 Mauritius
Veronica	21 Malawi	22 Mauritius

Note:
*Production system 8 (PS 8) is tropical, low rainfall, primarily rainfed, post-rainy season, crops are sorghum/oilseed. Western Deccan Plateau of India is the location included in PS 8. Production system 9 is tropical, intermediate-length rainy season, sorghum/oilseed/pigeonpea interspersed with locally irrigated rice, located in Peninsular India. Production system 15 is intermediate season (125–150 days), rainfed, mixed, sorghum-based, located in Southern Sudanian Zone. Production system 19 (PS 19) covers lowland, rainfed, short-season (less than 100 days) and suitable for sorghum/millet/rangeland, located in Sahelian eastern Africa, and margins of the Kalahari Desert. Production system 20 (PS 20) covers semi-arid, intermediate season (100–125 days) and suitable for sorghum/maize/rangeland, located in eastern Africa and parts of southern Africa. Production system 21 is intermediate season (125–150 days), sorghum/maize/finger millet/legumes, located in eastern and southern Africa. Production system 22 is lowland, sub-humid, mixed, rice/maize/groundnut/pigeonpea/sorghum, located in coastal areas of eastern and southern Africa. Production system 23 is highland, rainfed, long-season (150–180 days), sorghum/maize/teff, located in highland zones of northeastern and eastern Africa. Agro-ecological details of each production system are given in the ICRISAT *Annual Report, 1993*.
Source: *ICRISAT Southern and Eastern Africa Highlights, 1996* (p. 30).

Conclusions

This chapter documents the evolution of genetic enhancement research in groundnut at ICRISAT and provides an inventory of research products. It also reports the level of adoption and impacts of improved groundnut varieties in Africa and Asia. Only public sector institutions are involved in developing improved groundnut varieties, with the international system playing a central role. There is a notable absence of organized private sector effort in developing and marketing groundnut seed. The groundnut breeding programme at ICRISAT, in partnership with NARS, has released 67 varieties in 22 countries of Africa and

Asia, with as many as 40 more varieties 'in the pipeline' at present in 13 countries of Africa and Asia. Varying rates of adoption are observed in sub-Saharan Africa and Asia. Key factors influencing adoption are availability of seed with high yield potential, more oil content, high shelling ratios, and resistance to insects and pests.

Acknowledgements

The authors are grateful to Drs P. Subrahmanyam, P.J.A. vander Merwe, A.J. Chiyembekeza, N. Kameswara Rao, L.J. Reddy, B.R. Ntare, A.B. Obilana, F.T. Bantilan, Dr Joseph and SAARI scientists for information and help in completing the Impact Monitoring Survey in Asia and Africa; K.V. Anupama, V.K. Chopde, Y. Mohan Rao, Irshad Ahmed, Md. Moinuddin, Vijetha and Archana are thanked for their help in completing this study. However, the views expressed in this chapter are those of the authors.

References

Bantilan, M.C.S., Nigam, S.N., Rao, Y.M., Chopde, V.K., Rao, G.D.N. and Deb, U.K. (1999) Adoption of improved groundnut varieties in India. In: *SEPP Annual Report*, Vol. 2. Mimeo.

Deb, U.K., Bantilan, M.C.S. and Nigam, S.N. (2000) *Impact of genetic improvement in groundnut in India.* Paper presented at the NCAP-ICRISAT Workshop on Impacts of Agricultural Research in India 10–11 February 2000. ICRISAT Patancheru, India.

ICRISAT (1996) *Southern and Eastern Africa Highlights, 1996*. ICRISAT, Patancheru, India.

FAO (2000) *FAO Statistical Database* (http://www.fao.org).

Freeman, H.A., Nigam, S.N., Kelly, T.G., Ntare, B.R., Subrahmanyam, P. and Boughton, D. (1999) *The World Groundnut Economy: Facts, Trends and Outlook.* ICRISAT, Patancheru, India.

Kameswara Rao, N. (2000) *ICRISAT genetic resources.* Presentation made during Plant Breeding Review Meeting.

Phan Lieu, Bantilan, C., Sechrest, L., Nguyen Thi Lien Hoa, Ngo Thi Lam Giang, Bui Viet Nu and Duong Hoa Xo (1999) *Adoption of New Technologies for Groundnut in South Vietnam: Assessment using Focus Group and On-Farm Surveys.* ICRISAT, Patancheru, India.

Potato Genetic Improvement in Developing Countries and CIP's Role in Varietal Change 15

T.S. Walker, Y.P. Bi, J.H. Li, P.C. Gaur and E. Grande

Shortly after the founding of the International Potato Centre (CIP) in Peru in 1971, a planning conference was held to establish priorities and recommend specific uses for the wealth of genetic resources becoming available to the institute (CIP, 1974). Many of these recommendations were pursued with vigour. A few high-priority recommendations, such as improving the content and quality of protein, were subsequently dropped and some low-priority recommendations, i.e. the use of sexual propagation via true potato seed (TPS) instead of vegetative propagation from seed tubers, were later emphasized. Some research areas, such as adaptation to very warm growing regions and resistance to bacterial wilt, have proven to be technically difficult, but others, such as late blight resistance, have been highly successful. Over time, CIP's investment in potato breeding has centred on varietal resistance, mainly to late blight and to seed-borne viruses (CIP, 1998).

With hindsight, the opportunity for generating international public goods via potato crop improvement was a risky investment in 1971. Potato production was concentrated in developed countries, and it was not yet apparent that demand would emerge in countries where potato was not a staple food crop.

Another source of uncertainty that was probably not fully appreciated in 1971 was the empirical fact that varietal change was much slower in potato than in other major field crops (Walker, 1994; Huffman and Evenson, 1995). In developed country agriculture, yield growth in potatoes was at least as high as in other annual field crops, but the sources of growth were mainly non-genetic in the form of increased application of inorganic fertilizer, improved irrigation management and healthier planting material. With some notable exceptions, effective demand for varietal disease resistance was limited in North America and Western Europe; on such a high value crop, it paid to control diseases and pests with chemicals and other alternatives. Institutionally efficient seed programmes meant that the demand for resistance to seed-borne diseases was low. Moreover, yield potential in increasingly specialized production environments was so high in potatoes that gains in yield potential did not loom large as a source of growth. Potatoes are also characterized by a very high harvest index, so modifying this productivity parameter was not a viable option.

None the less, the handful of developing countries that had invested in potato breeding prior to CIP's creation had reaped handsome returns on those investments. In Colombia and Peru, *andigena* × *tuberosum* hybrids (from the two major subspecies of the crop) represented a source of productivity growth. The positive experience of the Rockefeller Foundation and the Mexican national programme in transferring late-blight-resistant varieties to Central America and sub-Saharan Africa was another bright spot in the returns picture. Hence, demand for disease resistance looked bright, and so did the prospects for varietal change, provided that demand for the crop was forthcoming.

In this chapter, we take stock of CIP's contribution to varietal change in potatoes in developing countries. The evaluation is based on information collected in NARS surveys which are described in the next section and which cover three topics: scientific staffing patterns, varietal release and varietal adoption. We review information on expected productivity gains from potato breeding. Finally, we assess in an illustrative manner the rate of return to potato breeding at CIP.

Surveys and Participating Countries

The questionnaires

In 1993–1994, potato crop improvement programmes in 20 developing countries were surveyed with a long questionnaire which elicited information on potato-producing zones; released varieties; escapes; seed production; varietal adoption; scientific staffing; and the demand and prospects for CIP programmes. In 1998–1999, a shorter version of this questionnaire was canvassed for 30 countries. Particular attention was given to 14 provincial programmes in China and several large Latin American potato-producing countries where information in the 1993–1994 survey was incomplete.

The 30-country sample

The 30-country sample is evenly distributed across Asia, sub-Saharan Africa and Latin America (Table 15.1). The geographical distribution of potato production in developing countries is very similar to that of wheat. Many countries grow both crops, but production is heavily concentrated in China and India, with a combined 65% share of production. Potato area and yield have expanded rapidly since the early 1960s. Production increased from about 30 million tonnes in the early 1960s to about 110 million tonnes in the late 1990s. During the same period, area has more than doubled from about 3.5 to 7.5 million ha.

More than 80 developing countries produced potatoes in the late 1990s (FAO, 1998). But many of these countries have not invested in potato breeding. A negligible or limited commitment to breeding is present in two situations conditioned by either the presence of alternative suppliers or by the small size of production.

The case of alternative sources of supply pertains to the Middle East and North Africa (MENA). Several countries in this region export potatoes to Europe in the winter season, and they import tuber seed from Europe. Local consumer preferences are shaped by the seasonal export market. Western European varieties, particularly Dutch varieties, do well in this setting, and they are aggressively promoted by seed companies. Ten countries in this region are large enough to warrant a fully fledged potato breeding programme, but none have invested on a sustained basis. Breeding is largely restricted to testing of introduced materials, and the national potato improvement programmes concentrate on other areas such as IPM, seed programmes, agronomy and storage. Research in several of these areas has been successful (Horton *et al.*, 1990; Fuglie, 1995); therefore, the neglect of potato breeding does not imply that returns to crop improvement are low. But prospects are bleak that internationally assisted conventional breeding will be productive in this region, which accounts for about 10% of developing country area and 13% of production.

For very small producing countries with a few hundred or thousand hectares, investing in potato breeding also does not make economic

Table 15.1. Distribution of total and sample potato-producing developing countries by region.

Countries	Middle East and North Africa	Latin America	Sub-Saharan Africa	Asia
Total	20	21	26	18
Sample	0	10	9	11

sense. An annual production level of 25,000 t is the size threshold required to justify an investment in a testing programme that evaluates elite introduced materials (Walker and Fuglie, 1999). Twenty-nine countries in Asia, Latin America and sub-Saharan Africa do not meet this criterion, and we purposely did not sample many of them (Table 15.2).

Likewise, we have sampled only a small number of somewhat larger countries that produce 25,000–250,000 t, a range of production that justifies a testing but not a fully fledged breeding programme where crosses are made and progenitors are evaluated. In contrast, we include 24 of the 26 larger countries with more than 250,000 t of production (Table 15.2). Only North Korea and Cuba were not included.

The sample countries account for about 85% of developing country potato area and production. In a few of the study countries, such as Sudan and South Africa, interactions with CIP have been quite limited. In contrast, about 15 countries not in this sample have released CIP-related material, but they tend to have small potato-growing areas. Their exclusion will not significantly affect an estimate of the extent of CIP-related materials in farmers' fields in developing countries in the late 1990s.

Table 15.2. Distribution of total and sample potato producing developing countries by size of production.

	Size of production		
Countries	Less than 25,000 t	Between 25,000 and 250,000 t	More than 250,000 t
Total	29	20	26
Sample	2	4	24

Staff Capacity in Potato Crop Improvement in Developing Countries

In the 30 countries, about 950 full-time equivalent scientists with a BSc or higher were working in potato improvement in the late 1990s (Table 15.3). Most of these scientists were employed in public sector NARS. A few were in universities. The weighted average research intensity was about 11 scientists per million tonnes of production. The number of full-time equivalent scientists per national programme ranged from about three or four in the smallest national programmes to about 400 in China. The vast majority of potato programmes are highly centralized at the national level. Only China and Brazil have markedly decentralized programmes at the provincial or state level.

Table 15.3. The disciplinary composition of scientist staffing in potato crop improvement programmes in developing countries.

	Scientists	
Area of specialization	No.	%
Agronomy and physiology	212	22.2
Breeding and germplasm	218	20.8
Seed production	143	14.9
Tissue culture	104	10.9
Pathology	146	15.3
Entomology and nematology	51	5.4
Molecular biology and transgenics	37	3.9
Postharvest	23	2.4
Social sciences	15	1.5
Others	7	0.7
Total	955	100.0

Technicians below the BSc level also discharge scientific responsibilities in some programmes, most notably in China, where 250 pre-professionals are engaged in potato crop improvement research. Many of these are individuals who did not have the opportunity to obtain a full university degree during periods of political instability.

Similar to the situation in the USA, potato crop improvement programmes in many countries feature a diversified portfolio of disciplines. Breeding and germplasm maintenance only account for about 20% of the total number of scientists (Table 15.3). Because of the importance of vegetative propagation, scientists working in tissue culture and seed production have a large role to play in potato improvement programmes.

Table 15.3 also contains one or two surprises. Given the importance of insect pests, a 5% share for entomology is lower than expected. Moreover, a 14% share for pathology, including virology, bacteriology and mycology, is less than we would have expected, given the widespread importance of damaging diseases. Lastly, the share of postharvest and other scientists is likely to be understated because details about these areas of disciplinary specialization were not explicitly requested in China.

Research intensity and the size of national production

An inverse relationship between research intensity and the size of national or provincial production is a common finding in studies such as this one. Small countries, states or provinces usually invest more per unit of production than large countries, states or provinces. This empirical reality holds for our survey data on provincial potato breeding programmes in China. Similar observations can be made for states in the USA and for countries in the 30-country data set. Average research intensities are 10–15 times higher in the smallest producing countries than in the largest producing countries.

Comparing scientific strength in wheat and potato breeding programmes

Using a narrow definition of a breeding programme, we find that on average the research intensity is about the same for potatoes as for wheat – five scientists per million tonnes (Table 15.4). The numbers in Table 15.4 also suggest that for both crops, there are a number of small countries with very high research intensities. The location of high research intensity countries also differs. In wheat, these countries are concentrated in sub-Saharan Africa; in potatoes, almost all the high-research intensity small countries are in South and Southeast Asia, where the crop is growing in importance as a vegetable.

Strength of national programmes over time

Baseline data are not available for comparing the strength of potato genetic improvement in developing countries over time. Such comparisons are also increasingly difficult in some of the larger countries

Table 15.4. Research intensities in potato and wheat breeding programmes in developing countries by size of production.

	Potato			Wheat[a]		
Production	No. of NARS	No. of scientists per country	No. of scientists per million t of potato	No. of NARS	No. of scientists per country	No. of scientists per million t of wheat
<0.1	7	4	75	11	8	150
0.1–0.5	8	4	14	8	6	30
0.5–1.0	7	10	12	4	9	11
1.0–5.0	6	16	8	9	26	9
>5.0	2	132	4	6	139	4
All developing countries	30	16	5.5	38	32	5

[a]From Bohn *et al.* (1998), p. 26.

because, with the broadening definition of NARS, part-time researchers may be increasingly numerous. For example, responses to the survey indicate that 68 scientists work on potato crop improvement in Brazil in 19 institutes located in federal, state and university programmes. Many of these scientists allocate only a portion of their time to potatoes.

In spite of the absence of reliable information for a comparative evaluation, it is safe to say that public research capacity probably peaked in the 1980s. Privatization of agricultural research has exacted a heavy toll on public-sector research in Latin America and offsetting private-sector research has not been forthcoming. In several countries, such as China, the average research output in public-sector institutions may also have declined because increasingly staff are engaged in non-research production activities to help defray the costs of their institutes.

Released Varieties

Caveats

Data on varietal release are usually the most available information on the performance of a plant-breeding programme. However, release is indicative of, but does not guarantee, success.

Release does not imply adoption nor is it a necessary condition for success. For example, in North America, the odds are only one in seven that a new potato variety will account for more than 1% of growing area in any year following its release (Walker, 1994). Escapes may result in more practical impact than released varieties. In China, Kexin No. 1 has been one of the most widely grown varieties since the 1960s. Kexin No. 1 was popularized by a field worker of the breeder who made the cross in the 1950s, even though the breeder subsequently discarded this selection (Song BoFu, personal communication). In recognition of its popularity, Kexin No. 1 was finally released in 1984.

Evaluating the impact of IARC germplasm solely from release data may underestimate or overestimate the role of IARC material in the varieties cultivated by farmers. Escapes from IARC material would lead to underestimation. In Kenya, Nyayo, a good chipping clone, has been the leading potato cultivar in the 1990s. The origin of this recently introduced and yet-to-be-released clone is unknown, but it could have been derived from CIP-related material (P. Ewell, personal communication).

For a few countries, such as Argentina, release lists do not include exotic varieties imported from developed countries, even through these varieties make a large contribution to national production. In India, a few large private-sector producers of tuber seed have informally made selections of national programme material. Although the source of this material is the national potato programme, these private sector varieties are not listed, and it is unlikely that these large seed merchants know the pedigree of the material that they name.

The structure of agricultural research and release policy also influences the number of releases. Countries with decentralized emphases, such as Brazil and China, are characterized by substantially more releases than countries, such as India, where national releases are given more prominence.

The above caveats notwithstanding, data on varietal release are still informative about the progress of a plant breeding programme. For potatoes, release is (weakly) associated with adoption, and the absence of released varieties over several years is a sign of inactivity or lack of progress in a plant breeding programme. Escapes are a relatively rare occurrence. For instance in the USA there are a few state programmes with more than five releases and negligible adoption, but there are no programmes without five or more releases and significant adoption.

Information on release is most useful if it is as comprehensive as possible and is not restricted to source of material or institution of release. We tried to assemble the most complete information possible so that CIP's contribution could be placed in a perspective of space and time.

The pace of release

The oldest variety listed in our sample is Kerr's Pink, released in Kenya in 1927. From 1927 to 1998 about 500 varieties were released. The first CIP-related variety was released by a national programme in 1979, 7 years after CIP's founding in 1972. The pace of varietal release has accelerated since the 1950s when, on average, two varieties were released per year, levelling off to about 16 or 17 per annum during the 1980s and 1990s. This trend reflects the pattern of investment in public sector NARS (Alston *et al.*, 1998). In many public sector NARS, growth in the early 1960s to the early 1980s was followed by a period of stagnation and even decline in funding agricultural research.

Events in China have also had a bearing on the temporal pattern of release in the 30-country database. Before 1960 only exotic clones from Europe and North America were released. Between 1960 and 1978, a few clones were released, largely from small county research farms. Political reforms beginning in 1978 gave rise to an open release policy, and many varieties, bred and selected in the 1950s and 1960s, were released in the early 1980s (D.P. Zhang, personal communication).

The increase in varietal release over time has slowed down in Asia and Latin America and has turned negative in sub-Saharan Africa. The annual rate of varietal release peaked at about 3.7 per year in sub-

Saharan Africa in the 1980s. Several potato crop improvement programmes in Central Africa were casualties of civil war and political unrest in the 1990s.

Sources of released varieties

The institutional source of the germplasm on which released varieties are based can be divided as follows:

1. Developing country NARS
   - NARS-bred varieties with germplasm unrelated to CIP
   - NARS-selected varieties from crosses unrelated to CIP
   - NARS-released native variety
   - NARS borrowing from other developing country NARS
2. CIP
   - Developing-country-released varieties distributed by CIP to other developing countries
   - NARS selections from CIP crosses
   - NARS crosses from CIP progenitors
3. Developed country NARS
   - Varieties borrowed from developed country NARS
4. Other
   - Sports (somatic mutations), farmer varieties, etc.

NARS-bred material (unrelated to CIP) has been the dominant source of releases (Table 15.5). Developed countries also supply a large proportion of potato varieties released in developing countries. About 23% of the varieties from the 30-country sample are related to CIP.

The importance of these institutional categories has been changing over time. Shifts in importance are described for roughly three equal time periods that correspond to pre-CIP, early CIP and mature CIP (Table 15.5). CIP's role has also changed over time. Prior to 1978, CIP's contribution was negligible. In the 1980s, CIP emphasized the South–South sharing of NARS-bred material. The salient example of this borrowing is Achirana-INTA, which was bred in Argentina, identified as promising, pathogen-tested and distributed by CIP, and was subsequently evaluated and released by several developing countries ranging from Bhutan to China to Madagascar. (Interestingly, this variety never attained commercial importance in Argentina.)

The role of CIP as a facilitator of varietal spillovers is justified by several considerations. First and foremost, potato is a vegetatively propagated crop which is susceptible to viruses and diseases which are, in turn, transmitted by tuber seed. Pathogen testing and virus elimination is costly and potentially imposes a heavy burden on the oper-

Table 15.5. Institutional composition (percentage) of NARS varietal releases over time.

Institutional source	Time period Before 1978	1979–89	1990–98	Total
NARS alone (without CIP involvement)	61.2	46.2	38.1	48.2
NARS-bred, no CIP role	52.7	42.1	32.4	42.2
NARS-selected, no CIP role	4.2	0.6	1.1	2.0
NARS sharing, no CIP role	2.4	1.2	1.7	1.8
Released native varieties	1.8	2.3	2.8	2.2
Developed country clone, NARS-released	36.4	26.3	17.6	26.8
CIP-NARS partnership	2.4	25.1	41.5	23.4
CIP-distributed, NARS-released	0.6	17.5	7.4	8.6
CIP-cross, NARS-selected	1.8	6.4	30.7	13.3
NARS-cross, CIP progenitor	–	1.2	3.4	1.6
Others	–	2.4	2.8	1.8
Sport, no breeding or CIP involved	–	1.2	2.8	1.4
Farmer or private sector variety	–	1.2	–	0.4
Total %	100	100	100	100
Number of releases	165	171	176	512

ating budgets of national programmes. Such costs probably acted as a deterrent to the sharing of NARS elite germplasm. Evidence for NARS sharing without CIP support is scanty in Table 15.5 and is restricted largely to seed flows from strong NARS, such as India and Colombia, to neighbouring countries. Secondly, a track-record of success was already established by the export of elite late-blight resistant varieties from the Rockefeller and national programmes in Mexico to other countries, particularly Uganda and those in Central America. Indeed, many of the elite clones distributed by CIP and subsequently named and released as varieties in developing countries originated in the Mexican national programme.

Lastly, small national programmes had limited access to disease-resistant elite material. Before 1980, such countries were considerably more dependent on elite clones from developed temperate countries. Elite *tuberosum* clones have outstanding market quality characteristics, but they are often very susceptible to disease, particularly late blight which exacts a heavy toll on potato production in the tropics and sub-tropics. With the exception of China, strong NARS did not in general release CIP-distributed elite material released by other NARS. These varieties found most favour in very small national programmes, especially those with levels of production that did not warrant a sustained investment in potato breeding or even screening of large amounts of introduced material.

Emphasis on the distribution of finished varieties developed by national programmes in the South was needed to buy time and deliver some short-term impact until CIP's potato breeding programme matured. In the 1990s, more released varieties have been derived from CIP-bred material which has been selected by national programmes (Table 15.5). Again, selection from CIP-bred material has been particularly important for smaller national programmes that do not have enough potato production to justify a full-scale breeding effort. CIP-bred material is also being used as parental material to make crosses by larger NARS, such as India, but so far only a few varieties have been released from CIP parents.

CIP-related material is more frequently encountered in released varieties from sub-Saharan Africa, where it figures in more than 60% of the releases, than in Latin America or Asia. Several of the strong NARS in Latin America and Asia have not yet released any varieties related to CIP, although parental material from CIP is usually present in the early stages of breeding activity by these programmes.

Private-sector releases are conspicuously absent in Tables 15.5 and 15.6. This largely reflects the lack of private-sector investment in potato plant breeding in developing countries. Even in North America only a handful of varieties have been released by private companies through conventional plant breeding, which has usually featured a joint venture

Table 15.6. Institutional composition (percentage) of NARS varietal releases by region.

	Region		
Institutional source	Africa	Asia	Latin America
NARS alone (without CIP involvement)	6.3	60.9	50.9
NARS-bred, no CIP role	1.3	55.9	43.4
NARS-selected, no CIP role	–	2.3	2.4
NARS-sharing, no CIP role	3.8	2.7	–
Released native varieties	1.3	–	5.2
Developed country clone, NARS released	30.0	20.0	32.1
CIP-NARS partnership	63.8	16.8	15.1
CIP-distributed, NARS-released	30.0	6.4	2.8
CIP-cross, NARS-selected	31.3	7.7	12.3
NARS-cross, CIP progenitor	2.5	2.7	–
Others	–	2.3	1.9
Sport, no breeding or CIP involved	–	2.3	0.9
Farmer or private sector variety	–	1.2	0.9
Total %	100	100	100
Number of releases	80	220	212

between the public sector and food processing companies. Prospects for transgenic varietal change has markedly increased private-sector investment in potato breeding in the USA and Canada, but such releases still account for only a small percentage of certified seed. In the release data, only one transgenic clone is listed: a virus-resistant version of a leading European variety, Alpha, in Mexico.

Pedigrees of released varieties

Pedigree information on released varieties was also elicited. Although we have not systematically analysed that information, some generalizations can be made. The use of parental material is diverse in NARS-bred varieties. Few, if any, varieties are from the same cross. A wide array of clones were used as parents. Kathadin, the first variety released by USDA in the early 1930s and a prolific producer of pollen, was the most common progenitor in the database, but it accounted for less than 5% of the 215 crosses listed in the NARS-bred category in Table 15.5. Most breeding programmes follow a logical evolution in the utilization of parental material. In the early stages of the programme, elite clones from North America and Europe are frequently used as parents and crossed with locally adapted material. These elite clones are then often replaced by successful cultivars released from the same programme. Clones that were or still are widely cultivated, such as Baronesa in Brazil and Hunikul MAG in Argentina, feature prominently as parents of later releases.

The pedigree information on released varieties also suggests that wide crosses between wild and cultivated species or even between *andigena* and *tuberosum* subspecies, on the one hand, and minor cultivated species, on the other, are not that common. Most breeders appear to find sufficient variability in *andigena* and *tuberosum* materials without having to tap into the resistance offered by wild or minor cultivated species. Alternatively, the cost of poor adaptation may not be worth the benefit of enhanced resistance.

Varietal Adoption

The data on varietal adoption show a very credible performance by NARS-bred material (Table 15.7). About 50% of the released varieties belonged to this category, and they accounted for 60% of area planted to potatoes in the 30-country sample in 1997.

Table 15.7. Regional adoption (percentage) of potato varieties in the 1990s by institutional source.

Institutional source	Region			Total
	Africa	Asia	Latin America	
NARS alone (without CIP involvement)	10.4	65.4	43.4	58.8
NARS-bred, no CIP role	5.3	56.7	37.5	50.6
NARS-selected, no CIP role	–	7.1	–	5.6
NARS-sharing, no CIP role	4.6	1.6	0.8	1.7
Released native varieties	0.4	–	6.0	0.9
Developed country clone, NARS-released	12.5	24.4	33.3	25.0
CIP-NARS partnership	40.6	3.0	7.4	5.9
CIP-distributed, NARS-released	31.2	1.6	0.9	3.2
CIP-cross, NARS-selected	7.8	0.1	6.5	1.6
NARS-cross, CIP progenitor	1.6	1.3	–	1.1
Local varieties	20.8	0.6	13.1	3.7
Native varieties	9.8	–	13.1	2.5
Old, introduced, degenerated material	11.0	0.6	–	1.2
Others	15.7	6.6	2.8	6.5
Sport, no breeding or CIP involved	–	–	–	–
Farmer or private sector variety	–	–	0.8	0.4
Others	15.7	6.6	2.0	6.4
Total %	100	100	100	100
Total area (ha)	407,769	5,472,310	1,052,536	6,932,615

Shares of released varieties and area coverage were also congruent for developed country clones with about one released variety in four and 1 ha planted in 4. (If we make the reasonable assumption that all 800,000 ha of potatoes in the Middle East and in North Africa are planted to Dutch varieties, then about 1 ha in 3 of total developing-country potato area is planted in these varieties.) This correspondence attests to the continuing popularity of these exotic materials.

In contrast to these first two categories, CIP's 23% share in varietal releases did not translate into a comparable rate of area covered. CIP-related materials were planted on about 400,000 ha, equivalent to about 6% of area. About 50% of this area was planted in CIP-distributed NARS-released varieties, and 50% was planted to selections from CIP crosses and varieties with CIP progenitors.

Why is varietal adoption of CIP-related materials substantially lower than CIP's contribution to varietal release? Again, part of the explanation resides in the fact that CIP-related materials have had wider acceptance in smaller NARS with less area planted to potatoes. Equally important, varietal change is slower in potatoes than for other major field crops. For example, the leading varieties in area in China and India in the 1990s were both released before CIP was started.

As with varietal releases, the relative importance of CIP-related material in farmer fields varies markedly by region. CIP's relative contribution was highest in sub-Saharan Africa, where about 40% of the potato-growing area was planted to CIP-related materials. CIP's role was noticed least in Asia where NARS-bred material (without CIP involvement) and developed country clones account for over 90% of growing area.

The meagre private sector contribution in Table 15.7 pertains to the varieties Chola and Superchola in Ecuador where German Bastidas, a hobby breeder, has selected material from crosses he made in the Carchi region. The contribution of a few large seed merchants in making clonal selections of NARS material in India is not reflected in the data. We assume that the genetic make-up of these materials is so similar to NARS-released varieties that they really belong in the NARS-bred category.

One of the surprising aspects of Table 15.7 is the low share of local varieties in cultivated area. Native varieties of potatoes command a large area only in Bolivia and Peru and to a much lesser extent in Ecuador and Colombia. One can also still find degenerated material of very old introduced varieties in several settings. In northern Bangladesh, these low-yielding varieties are widely cultivated. They produce many small-sized tubers per plant that are highly prized in the market. But, in general, old landrace materials comprise only a small share of cultivated area outside the Andes.

In closing, we return to the very impressive performance by NARS in potato breeding that was alluded to at the outset of this section. The adoption data in Table 15.7 clearly indicate an attractive rate of return on investment in potato breeding in developing countries. However, there has been considerable variation in performance across countries and over time within the same country.

Prospects for adoption of CIP-related material

Does past adoption performance indicate future positive consequences? One way to address this question is to group the 30-country sample by release and adoption of CIP-related material. About half (16) of the study countries show both release and adoption of

CIP-related material. For these countries the prospects are bright that CIP-related material will increasingly be cultivated by potato producers. These countries include China, Peru, Bolivia and several small-producing nations in sub-Saharan Africa. Adoption performance will improve further in these countries by focusing even further on large opportunities such as replacing the old German cultivar, Mira, in southwest China. Greater emphasis on farmer participatory selection and more structured international testing also seem to be steps in the right direction. More selection pressure earlier on storage and market-related traits could also lead to greater impact. In the past, scientists may have been too conservative in seeking resistance to secondary diseases that farmers can often control via agronomic improvement.

Three countries have released CIP-related varieties but with no adoption. Breeders in Ecuador have selected and recently released several varieties from one of CIP's late-blight populations; impact is imminent. In Bangladesh, seed availability is still a major constraint. Sri Lanka is the third country in this grouping, and here it is questionable whether production problems should be addressed via breeding, or whether lower cost potatoes can be imported from other countries in the region.

The remaining 11 countries score negatively on both counts. Three countries belong to the Southern Cone where market quality and specialization in production are increasingly paramount. As in the earlier example of Achirana-INTA, the challenge for CIP is to mobilize strong NARS in this region to generate spillovers elsewhere. Several of the remaining eight countries, such as South Korea, South Africa, Taiwan and Sudan, have not been priorities for CIP because of the nature of potato production in the past or because of the relative lack of importance of the commodity. Likewise, Pakistan is not a cause for immense concern because well-adapted varieties from India should be suited to almost all potato production environments in the subcontinent.

Several countries in this group are surprises. India and Mexico are strong NARS; the challenge for CIP is to generate parental material that ultimately translates into varietal release via national breeding efforts. In contrast, in Colombia and in Indonesia, CIP has not been able to generate varietal options to accommodate and reinforce a rapidly expanding demand for the crop. Given the institutional fragility of potato breeding and the existence of a dominant clone in the market, CIP-assisted transgenic varietal change may eventually have better prospects than non-transgenic conventional breeding in countries such as Indonesia.

Productivity Gains from Cultivar Improvement

As described in the introduction to this chapter, productivity gains from varietal adoption have been difficult to document in the genetic improvement of potato in North America and Western Europe. Based on trials of varieties of different vintages, the absence of productivity differences is described as yield stasis (Douches *et al.*, 1996). Nineteenth-century clones selected by hobby breeders are as high-yielding as those bred and released by professional plant breeders in the 20th century. Progress has been made in maturity and in tuber appearance and is presumably reflected in higher market value. A few outstanding bred varieties are referred to as setting new standards or as contributing heavily to the health of the potato industry in a specialized growing area. Such advantages are undoubtedly real, but they are considerably harder to quantify than differences in yield.

Breeding progress has been end-use specific. In North America, the most measurable progress has been made in chipping potatoes, which account for about 10% of production. Since 1960, statistically significant trends in lowering sugars, in improving chip colour, and in lessening chip defects have been documented (Love *et al.*, 1998). Not surprisingly, chipping varieties turn over more rapidly than cultivars primarily destined for table consumption or for processing for frozen chips, the other economically important end-uses.

In the tropics and subtropics, yield differences are more likely to fuel varietal change than in temperate developed regions. Heavy disease incidence during the rainy season often results in lost productivity, as chemical control is usually ineffective in protecting yield potential. Virus resistance may also confer yield benefits over time because efficient seed systems cannot be realized until countries are more institutionally developed. More rustic varieties that better tolerate abiotic stresses are also often characterized by yield gains over farmers' material.

Attributing on-farm productivity to genetic improvement is also more problematic in a vegetatively propagated crop like potatoes than in sexually propagated cereals. The absence of effective seed systems means that a productivity effect from a variety is confounded with the effect of cleaner or physiologically more correct seed. To isolate the variety effect, tuber seed of checks has to meet the same health and physiological standards of prospective clones. Usually, tuber seed needs to be of the same age for all entries including checks.

The on-farm results in Rwanda (Table 15.8) are representative of the size of productivity effects of different components of a successful potato crop improvement programme (Haverkort, 1986). The first row entry shows the effect of cleaning up the seed of two of the best local cultivars. An additional 17% increase in yield was gained with the

Table 15.8. Results of on-farm trials with national programme seed in Rwanda.

Type of seed and variety	Years	Number of on-farm trials	Average yield increase relative to farmer average
Improved local cultivars (Muhabura, Bufumbira)	1980–81	8	23
Improved introduced cultivars (Montsama, Sangema)	1981–82	72	40
Locally selected and named cultivars (Gahinga, Kinigi)	1982–84	106	112

introduction of late-blight-resistant clones bred by the Mexican national programme. Locally adapted improved late-blight-resistant clones selected from CIP-bred populations more than doubled farmers' yield, but some of that increase was attributed to cleaner seed and to earlier varietal change.

The rate of degeneration of improved cultivars implies that these yield differences are unlikely to stay the same over time. Ideally, productivity differences would be estimated over time according to average seed replacement rates. In practice, such trials are costly and are almost never carried out.

Underestimation of yields in national and FAO statistics make crop cuts in farmer fields absolutely necessary in order to sort out varietal productivity effects. Yield estimates based on large samples usually show that on-farm potato productivity is 50–100% higher than levels conveyed in official estimates (Pakistan-Swiss Development Project, 1991; Terrazas *et al.*, 1998). Moreover, because of the unreliability of official government data on root and tuber crop production, even dramatic technical change in the form of widespread high-yielding cultivar adoption with healthier seed may not budge stagnating productivity trends based on secondary data. This problem is especially pervasive in small potato-growing countries in sub-Saharan Africa.

In summary, the productivity effects of varietal change in potatoes in developing country agriculture are likely to be quite large. Yield gains of several tonnes per hectare – large by the standards of cereal agriculture – are clearly feasible.

Productivity gains are also contextual. In case studies of CIP-related varietal change, on-farm 'pure' cultivar yield differences of 2.8 t ha^{-1} resulted in rapid diffusion of high-yielding late-blight-resistant varieties in the 1980s in the Central and East African highlands along the slopes of the Zaire–Nile Divide (Rueda *et al.*, 1996). In the diffusion of a CIP-related drought- and virus-tolerant variety in the interior of North China, the varietal productivity effect was conservatively estimated at

3 t ha^{-1} (Song Bofu *et al.*, 1996). In Vietnam, experimental on-farm data were not available to disentangle the joint effect (6 t ha^{-1}) of first-generation varietal change combined with a new highly cost-effective seed system in a small highland potato-producing region. Second-generation varietal change was conservatively estimated to generate a yield difference of 1.5 t ha^{-1} (Nguyen Van Uyen *et al.*, 1996). In Peru, earlier maturity and better market quality reflected in price premia drove adoption of a high-yielding late-blight-resistant variety in the main growing season. In the shorter secondary growing season, a productivity advantage of about 2.5 t ha^{-1} was the primary source of benefit in planting the improved variety (Fonseca *et al.*, 1996).

Returns on CIP's Potato Breeding

An adoption level of about 6% following a quarter-century of work is not suggestive of a high rate of return on investment. If the question were posed to CIP research management or participants at the first plant breeding planning conference in 1974 on the extent of CIP-related material in farmers' fields in 25 years, it is likely that expectations would have been considerably higher than 6%. Superficially, this relatively low rate of uptake of CIP-related material would appear to translate into a negative or very low rate of return on investment. However, that does not appear to be the case.

Our 'best bet' back-of-the-envelope calculation shows a positive, albeit modest, rate of return on investment (Table 15.9). We assume the following: (i) a project duration of 50 years; (ii) a cost share of 55% for potato breeding as a proportion of total CIP expenditures on potato crop

Table 15.9. Back-of-the-envelope calculation on returns to CIP's investment in potato breeding period: 50 years, 1972–2021.

CIP costs		Real; 55% of expenditure on potato
Area		1997, no trend assumed
Prices		Constant (1993)
Source of benefit		Yield increase of 2.5 t ha^{-1} = \$220 ha^{-1}
Seed, extension and all other research costs		\$110 ha^{-1}
Net benefit		\$110 ha^{-1}
Adoption ceiling in 2021	IRR (%)	NPV (US$ millions)
5.8	15	39
10.0	16	51
15.0	17	71

Notes: IRR, internal rate of return; NPV, net present value.

improvement[1]; (iii) a logistic pattern of diffusion; (iv) a yield increase of 2.5 t ha^{-1} as a source of benefit; (v) a constant price of $110 per tonne; (vi) additional costs of NARS research, seed, and extension equivalent to 50% of net benefits; (vii) 1997 potato growing area; and (viii) variable levels of adoption by 2021 of 5.8%, 10% and 15%.

These assumptions warrant some comment and explanation. First, they are chosen to establish a reasonable lower bound or conservative estimate on the rate of return to investment. Potato breeding includes all costs related to germplasm conservation. The price of $110 per tonne is low. The increasing trend in potato-growing area is not considered, and a reduction of 50% of net benefit to cover all other research, seed multiplication and extension costs is undoubtedly on the high side. As discussed in the previous section, a 2.5 t ha^{-1} increase in yield (net of increased cost) would be high for cereals but is not large for a varietal replacement effect in potatoes. The adoption levels at 2021 are arbitrarily chosen and reflect the current situation of 5.8% and a doubling and tripling of the present levels. Given materials now in the pipeline and the momentum of varietal change, it is likely that adoption of CIP-related materials would increase even with no further breeding. Therefore, doubling or tripling the present level would seem to be very attainable with a continuation of past investment.

[1] This estimate is based on CIP's project portfolio in 1995 when the institute had 36 research and training projects; 19 related to potato crop improvement. Of these, we estimate that 8 were primarily genetic and 11 non-genetic although there are overlaps in many of the projects. In 1997, the genetic projects had an expenditure share of 55% of all expenses on potato crop improvement; we have assumed that that share has remained constant since CIP's founding in 1971–1972.

Projects which were included in Genetic Crop Improvement were titled: Adaptation and Integration of Potato Production Technologies, Potato Collection and Characterization, Potato Germplasm Enhancement and Application of Molecular Technologies, Control of Potato Late Blight, Combining Resistances to Potato Viruses, Potatoes with Resistance to Major Insects and Mite Pests, Sexual Potato Propagation, and Maintenance and International Distribution and Monitoring Performance of Advanced Potato Germplasm. Potato projects which were not viewed as primarily 'genetic' were: Evaluation of the Impact and Sustainability of Potato Production Technologies, Characterization of Constraints and Opportunities for Potato Production, Integrated Control of Potato Bacterial Wilt, Detection and Control of Potato Viruses, Molecular Approaches for Detection and Control of Pathogens, Integrated Methods for the Control of Andean Potato Weevils, Integrated Methods for the Control of Potato Cyst Nematodes and False Root Knot Nematodes, Propagation of Healthy Clonal Potato Planting Materials in Diverse Agricultural Systems, Abiotic Stresses and Potato Crops Management, and Expanding Utilization of Potato in Developing Countries.

Our illustrative calculation shows that the internal rate of return (IRR) on investment is clearly positive (ranging from 15 to 17%) and is quite insensitive to assumptions about ceiling levels of adoption. This result is to be expected because the investment in potato breeding is pretty much 'a done deed' in terms of return from capital. In contrast, the net present value of the investment evaluated at a 10% discount rate is highly sensitive to the level of adoption at the end of the project.

Conclusions

The adoption of CIP-related potato material in developing countries has been modest compared to the IARC performance in wheat, rice and maize. None the less, a back-of-the-envelope calculation shows a solid rate of return on investment of about 15% in CIP's potato breeding activities. Over time and space a high-value vegetatively propagated crop like potato also offers opportunities for technological change from other avenues of research. Historically, integrated pest and disease management, improved seed quality, and better storage practices have figured among these.

The prospects are bright that CIP-related materials will increasingly find a home in farmers' fields. In Bolivia, the national programme has recently released several late-blight-resistant varieties. Mira, the dominant blight-susceptible clone in southwest China, looks ripe for replacement by provincial NARS selections from CIP populations of disease-resistant cultivars. However, the pace of varietal change is quite slow in potatoes, and 25% of potato-growing area in CIP-related materials in developing countries by 2020 would be an impressive performance.

Proportionally, and even absolutely, CIP's impact has been greatest in small NARS. There is no evidence to suggest that CIP's activities have crowded out programmes in large NARS.

Acknowledgements

Many potato scientists contributed to this project. We are particularly grateful to Oscar Hidalgo and Jim Bryan for their support in starting and in contributing to later stages of this work. Charles Crissman was responsible for coordinating the first questionnaire, and he also helped design the project. The work could not have been completed without the commitment and active participation of NARS plant breeders. CIP regional representatives and regional staff gave unselfishly of their time in coordinating survey responses. We are indebted to the following individuals:

Primo Accatino, Francis Alacho, Héctor Andrade, A.A. Ashandi, Song BoFu, Enrique Carrasco, Alwyn Chilver, Enrique Chujoy, Hilario Da Silva Miranda Filho, P.S. Dahiya, Rolando Egúsquiza, Ramzy El-Bedewy, Nelson Estrada, Peter Ewell, Fernando Ezeta, Francisco Flores, Khaled Forooq, Keith Fuglie, Wang Guilin, J.J. Hakiza, Dieudonne Harahagazwe, Marcelo Huarte, Dao Huy Chien, Jackson Kabira, Rogers Kakuhenzire, Julio Kalazich, Bekele Kassa, T.B. Katwal, Peter Kinyae, V.S. Khatana, Aziz Lagnaoui, Verónica Larenas, Berga Lemaga, C.M. Mayona, Humberto Mendoza, Il-Gin Mok, Hyun Mook Cho, Laura Niño, Pedro Oyarzum, Noel Pallais, Dai Peters, Ramón Pineda, M. Rakotondramanana, Ana Ramos, M.M. Rashid, Jorge Recharte, José Santos Rojas, S.K. Tharmarajah, Graham Thiele, Rudy Torres, Jocelyn Tsao, Xuan Tung, Mahesh Upadhya, Troung Van Ho, Peter Vander Zaag, Percy Vilca, Yi Wang, Greta Watson, Gebremedhin Woldegiorgis, and Dapeng Zhang. We are grateful to the Impact Assessment and Evaluation Group (IAEG) of the CGIAR for partially funding this work.

References

Alston, J.M., Pardey, P.G. and Smith, V.H. (1998) Financing agricultural R&D in rich countries: what's happening and why. *Australian Journal of Agricultural and Resource Economics* 42, 51–82.

Bohn, A., Byerlee, D. and Maredia, M.K. (1998) Investment in wheat improvement in developing countries. In: Maredia, M.K. and Byerlee, D. (eds) *Research Efficiency in the Presence of Technology Spillovers: The Case of National and International Wheat Improvement Research.* CIMMYT Research Report. CIMMYT, El Batan, Mexico, pp. 16–29.

CIP (1974) *Strategy for Utilization. Report of the International Potato Center's Planning Conference on Utilization of Genetic Resources.* CIP, Lima, Peru.

CIP (1998) *Working Document on CIP Plant Breeding Strategies for Potato and Sweetpotato.* Internally commissioned external review (ICER). CIP, Lima, Peru.

Douches, D., Maas, D., Jastrzebski, K. and Chase, R.W. (1996) Assessment of potato breeding progress in the USA over the last century. *Crop Science* 36, 1544–1552.

FAOSTAT (1998) November. www.fao.org

Fonseca, C., Labarta, R., Mendoza, A., Landeo, J. and Walker, T.S. (1996) Economic impact of the high-yielding, late-blight-resistant variety Canchán-INIAA in Peru, 1996. In: Walker, T. and Crissman, C. (eds) *Case Studies of the Economic Impact of CIP-Related Technologies.* CIP, Lima, Peru, pp. 51–64.

Fuglie, K.O. (1995) Measuring welfare benefits from improvements in storage technology with an application to Tunisian potatoes. *American Journal of Agricultural Economics* 77, 162–173.

Haverkort, A.J. (1986) Forecasting national production improvement with the aide of a simulation model after the introduction of a seed potato production system in Central Africa. *Potato Research* 29, 119–131.

Horton, D., Cortbaoui, R., Hattab, H. and Monares, A. (1986) Impact of agricultural research: a seed potato project in Tunisia. In: Walker, T. and Crissman, C. (eds) *Case Studies of the Economic Impact of CIP-Related Technologies.* CIP, Lima, Peru, pp. 67–81.

Love, S.L., Pavek, J.J., Thompson-Johns, A. and Bohl, W. (1998). Breeding progress for potato chip quality in North American cultivars. *American Journal of Potato Research* 75, 27–36.

Maredia, M.K. and Byerlee, D. (1998) Efficiency of wheat improvement research investments in spill-ins. In: Maredia, M.K. and Byerlee, D. (eds) *Research Efficiency in the Presence of Technology Spillovers: The Case of National and International Wheat Improvement Research.* CIMMYT Research Report. CIMMYT, El Batan, Mexico, pp. 99–118.

Maredia, M.K., Ward, R. and Byerlee, D. (1998) Assessment of the potential international transferability of wheat varieties across countries. In: Maredia, M.K. and Byerlee, D. (eds) *Research Efficiency in the Presence of Technology Spillovers: The Case of National and International Wheat Improvement Research.* CIMMYT Research Report. CIMMYT, El Batan, Mexico, pp. 45–64.

Nguyen Van Uyen, Truong Van Ho, Pham Xuan Tung, Vander Zaag, P. and Walker, P.S. (1996) Economic impact of the rapid multiplication of high-yielding, late-blight-resistant varieties in Dalat, Vietnam. 1996. In: Walker, T. and Crissman, C. (eds) *Case Studies of the Economic Impact of CIP-Related Technologies.* CIP, Lima, Peru, pp. 127–138.

Pakistan-Swiss Potato Development Project (1991) Potato yield estimation survey of autumn crop of Punjab – January, 1991: a study carried out jointly by Directorate of Agriculture, Crop Reporting Services, Punjab, Ministry of Food, Agriculture, and Cooperatives, Government of Pakistan, and the Pakistan-Swiss Potato Development Project, PARC, Islamabad.

Rueda, J.L., Ewell, P.T., Walker, T.S., Soto, M., Bicamumpaka, M. and Berríos, D. (1996) Economic impact of high-yielding, late-blight-resistant varieties in the Eastern and Central African Highlands. In: Walker, T. and Crissman, C. (eds) *Case Studies of the Economic Impact of CIP-Related Technologies.* CIP, Lima, Peru, pp. 15–30.

Song Bofu, Tian Weiming, Wang Jimin, Wang Chulin, Yan Zhengui, Wang Shengwu and Huarte, M. (1996) Economic impact of CIP-24 in China. In: Walker, T. and Crissman, C. (eds) *Case Studies of the Economic Impact of CIP-Related Technologies.* CIP, Lima, Peru, pp. 31–50.

Terrazas, F., Suárez, V., Gardner, G., Thiele, G., Devaux, A. and Walker, T. (1998) *Diagnosing Potato Productivity in Farmers' Fields in Bolivia.* Social Science Department Working Paper No. 1998-5. CIP, Lima, Peru.

Walker, T.S. (1994) *Patterns and Implications of Varietal Change in Potatoes.* Social Science Department Working Paper, No. 1994-3. CIP, Lima, Peru.

Walker, T.S. and Fuglie, K. (1999) The economics of generating international public goods from investing in potato plant breeding. In: *CIP's 1997–98 Program Report: Impact on a Changing World.* CIP, Lima, Peru, pp. 229–237.

The Impact of IARC Genetic Improvement Programmes on Cassava

16

N.L. Johnson, V.M. Manyong, A.G.O. Dixon and D. Pachico

Global Perspective on Cassava

Cassava (*Manihot esculenta*) is a root crop that is widely grown throughout tropical and subtropical Africa, Asia and Latin America. Together these regions consistently account for over 99% of global cassava production (FAO). Cassava is grown in many different environments and for many purposes. However, it is almost always grown by the poor as a food or cash crop.

Africa is the largest producer of cassava. In 1998, Africa produced 90 million tonnes, over half of total world production (Table 16.1). Five African countries are among the top ten producing countries (Table 16.2). During the 1980s, production nearly doubled in Africa, while it declined slightly in the rest of the world due to a reduction in area planted (FAO).

Table 16.1. World production and yield of cassava.

	1980			1998		
Region	Area planted (Mha)	Production (Mt)	Yield (t ha^{-1})	Area planted (Mha)	Production (Mt)	Yield (t ha^{-1})
Africa	7.1	48.3	6.8	10.8	90	8.3
Asia	3.9	45.9	11.8	3.3	44.4	13.6
Latin America and Caribbean	2.6	29.7	11.2	2.4	27.5	11.4

Table 16.2. Top ten cassava producing countries, 1998.

Country	Production (t)
Nigeria	30,409,250
Brazil	19,808,690
Congo	16,500,000
Thailand	15,958,500
Indonesia	14,728,290
Ghana	7,171,500
Tanzania	6,192,516
India	5,868,300
Mozambique	5,639,000
China	3,600,744

Yields have also grown faster in Africa than in any other part of the world; however, at 8.3 t ha^{-1}, they remain the lowest in the world (Table 16.1).

Cassava is the most important staple food crop in many countries in sub-Saharan Africa. According to the FAO, 95% of cassava in Africa is for human consumption, either boiled fresh or lightly processed. According to the Collaborative Study for Cassava in Africa (COSCA), cassava serves multiple roles as a food staple, famine reserve crop or cash crop (CICRTCR, 1999). The remaining 5% of cassava produced in Africa is used as animal feed. The use of cassava for industrial purposes is minimal.

The situation in Africa contrasts sharply with that of Asia, the world's second largest cassava producing region (Table 16.1). In Asia, relatively little cassava is used for direct human consumption. Most goes to industry, either for chipping and drying, or for starch production. Much of this is for export. In 1994, cassava starch was Thailand's second biggest export (Hershey, 1994). Average yields in Asia are the highest in the world, 13.6 t ha^{-1} in 1998 (FAO). In addition to root yields, the starch content per root has also grown over time in Asia in response to the industrial demand.

The situation for Latin America and the Caribbean (LAC) is somewhere in between. Brazil is the major producer in the region, and the second largest in the world (Table 16.2). In southern Brazil, cassava is used for industrial purposes, while in the north and northeast, it is a staple food crop. Recently a regional public–private research consortium, CLAYUCA,[1] was established with the support of poultry and livestock producers, who are interested in developing and promoting cassava varieties as an alternative to maize for animal feed.

[1] Consorsio Latinamericao para la Yuca (Latin American Cassava Consortium).

Cassava Genetic Improvement

International agricultural research centres (IARCs)

Cassava improvement work in the CGIAR is done at two centres, the International Centre for Tropical Agriculture (CIAT) in Cali, Colombia, and the International Institute of Tropical Agriculture (IITA) in Ibadan, Nigeria. CIAT, which began working on cassava in 1969, has a global mandate for the crop. While Latin America is not the major producing region, it is the centre of origin for cassava, which means that genetic diversity in cassava is higher in Latin America than in any other part of the world. IITA has a regional mandate for Africa and began working on roots and tubers in 1971. In 1982, CIAT established a regional cassava breeding programme in Asia.

Unlike other crops, such as rice, maize or even beans, little systematic work on cassava improvement had been done prior to the establishment of the IARCs. IARC researchers had to start from scratch, acquiring basic knowledge about the crop and assembling collections of genetic materials from which to begin characterization, selection and breeding activities. The first 5 years of CIAT's cassava programme were dedicated to developing a broad-based germplasm collection. The first breeder was appointed in 1974. In 1976, CIAT began sending germplasm accessions and seed populations to Asia and Africa.

Table 16.3 shows CIAT total spending on cassava research as a measure of IARC investment in cassava genetic improvement over time, as well as the number of principal scientists and breeders in the cassava programme.[2] The data show an increase in overall investment in cassava research between 1975 and 1990, followed by a moderate decline.

Table 16.3. CIAT financial and human investment in cassava research.

	1970	1975	1980	1985	1990	1997/8
Investment in cassava research (thousands of US 1990 dollars)	886	2280	4069	5148	5603	3942
CIAT principal staff in cassava improvement	1	6.5	13	12	15	12.5
Total number of breeders	0.2	2	2	2	4	2
Breeders as % of total staff	20	31	15	17	27	16
Number of breeders in Asia				1	1	1
% of breeders in Asia		0	0	50	25	50

Source: CIAT annual reports and from expert opinion.

[2] Similar data are not available for IITA.

Investment in cassava breeding, as measured by the number of breeders, remained essentially constant over the period, growing in 1990 but then declining again by 1998. Breeders make up a relatively low proportion of total cassava scientists, compared with other crops. One reason is that the cassava programme has engaged in a relatively wider range of activities, requiring a more diverse disciplinary background for its scientists. For example, during the 1980s, the cassava programme carried out research on postharvest technology, with the goal of increasing the participation of small farmers in cassava processing.

In the late 1990s, the low proportion of cassava breeders may have been due to the fact that CIAT was making a significant investment in cassava biotechnology. Geneticists and others in biotechnology are working on varietal improvement along with conventional breeders. To date, no genetically engineered varieties have been released or even tested in the field.

National agricultural research systems (NARS)

In spite of its importance, cassava did not receive much attention from either public or private research institutions until relatively recently. There were a few exceptions: Brazil, Indonesia and Madagascar. In 1985, Cock calculated that research investment on cassava, measured as scientists per value of production, was roughly one-tenth of research on sorghum, maize or potatoes.

Within a few years of the creation of CIAT and IITA, national programmes were established in many LAC, Asian and African countries (Hershey, 1994). Many of the scientists in these programmes were trained at CIAT and IITA. Training was, and continues to be, an important part of IARC cassava work. Tables 16.4 and 16.5 show the number of scientists trained at the IARCS since their establishment. At IITA, 1381 professionals have received training on cassava research for sub-Saharan Africa since 1970. CIAT trained 1855 professionals in Latin America and the Caribbean (LAC), Asia, Africa, North America and Europe.

Table 16.6 presents data on the number of breeders working in national agricultural research systems (NARS) for selected years. The NARS include national and state level institutions, as well as universities and private companies. The private sector continues to play a small role in cassava breeding. Indonesia is the only country reported to have a private-sector breeder in 1998 (R. Howeler, 2000, personal communication), although the private sector funds research in Latin America via the consortium CLAYUCA.

Table 16.4. Number of scientists trained at IITA cassava research in sub-Saharan Africa, 1970–1998.

	1970–79	1980–89	1990–98	Total
PhD level				23
Men	3	6	13	22
Women	–	–	1	1
MSc level				28
Men	5	18	4	27
Women	–	1	–	1
Research training associates				42
Men	17	21	2	40
Women	–	–	2	2
Visiting student research scholars				10
Men	3	4		7
Women	–	1	2	3
Group trainees				1278
Men	73	676	375	1124
Women	4	66	84	154
Total				1381
Men	101	725	396	1220
Women	4	68	89	161

Source: Manyong *et al.* (1999).

Table 16.5. Professionals trained at CIAT-sponsored courses, 1972–1994.

Specialization	LAC	Asia	Africa	North America and Europe
General production/research and project management	239	94	1	2
Physiology	12	4	1	7
Breeding/genetics/germplasm management	35	23	2	3
Entomology	58	4		14
Phytopathology	45	7	3	6
Agronomy/soils/seed systems	110	15	2	9
Socio-economics	12	2		11
Utilization	70	5	2	8
Tissue culture/biotechnology	46	1	2	
Total	627	155	13	60

Source: Hershey (1994), p. 15.

At the regional level, available data for Asia and LAC show that the number of breeders has remained relatively constant in Asia over time, increasing in the early 1990s and then declining again in recent years. LAC shows the same pattern, but the increase around 1990 is more dramatic.

Table 16.6. Cassava breeders in national programmes, universities and the private sector.

Country	1980	1990	1999
China	1	4	3
India	6	6	6
Indonesia	4	3	2
Malaysia	4	1	0
Philippines	4	4	3
Thailand	6	8	5
Vietnam	0	1	4
Asia total	25	27	23
Country	1988	1994	1998
Brazil	8	7	8
Mexico	4	4	0
Cuba		10	4
Ecuador	4	4	4
LAC total	16	25	16
Africa total			49

Sources: Manyong *et al.* (1999), expert opinion and Iglesias (1994).

Table 16.7 shows measures of breeding intensity in national programmes. Breeding intensity is defined as the number of breeders per million tonnes of cassava production. Breeding intensity appears to have been remarkably steady over both time and space, on average just over half a scientist per million tonnes of production.[3]

Table 16.7. Cassava breeders/million tonnes of cassava production, by region.

Breeders/million tonnes of production	1980	1990	1999
Asia	0.544	0.542	0.516
Africa			0.570
	1988	1994	1999
LAC	0.547	0.803	0.573

[3] One caveat on these data is that they come largely from expert opinion, and may vary in terms of how a breeder is defined. An attempt was made to include only breeders in 'senior' positions; however, the definitions could vary between regions.

Germplasm Exchange and the Production of Improved Cassava Varieties

Goals, constraints and strategies

Cassava is a crop largely grown by poor farmers who use few purchased inputs. In addition, it is a crop with a long production cycle (9–12 months) which means that inputs would have to be used over a long period of time and for a wide variety of pests and diseases (C. Hershey, 2000, personal communication). As a result, breeding efforts have focused on substituting biological adaptation for purchased inputs, especially pesticides and fungicides. In general, attempts to find durable sources of resistance have been successful in cassava. 'Cassava improvement has maintained the broad-based and stable resistance found in landrace varieties' (Hershey, 1994, p. 34). This was accomplished largely by exchanging germplasm within regions, and by introducing new germplasm from LAC to Asia and Africa.

In Asia, germplasm was supplied directly to national programmes. It is thought that more genetic diversity has been introduced into Asian cassava varieties in the past two decades than in all previous history (Hershey, 1994). It could be argued that breeders in Asia have had an easier time than their counterparts in Africa and Latin America. Compared to Africa and LAC, Asia has fewer cassava pests and diseases, which allowed breeders to focus more on yield and on starch content of roots, a priority for industrial use. There are also relatively few edaphoclimatic zones in Asia, meaning that promising varieties could be widely disseminated. In 1998, one variety, Kasetsart 50, was estimated to have been planted on approximately 150,000 ha in Thailand, 30,000 ha in South Vietnam, and in some areas of Sumatra and the Philippines (R. Howeler, 2000, personal communication).

In Latin America and Africa, breeders faced more constraints, not only from ecological conditions but also because they had to incorporate human consumption characteristics into their selection criteria. The breeding strategy in these regions focused on resistance more than on yield potential. To meet the challenge in Africa, IITA introduced selected cassava lines from Asia and South America, bringing in germplasm from CIAT and from the Brazilian national programme. Well-adapted materials from all of Nigeria's major cassava-producing regions and germplasm from other African countries were also used extensively in the breeding programme.

Varietal releases

Table 16.8 shows the number of improved cassava varieties with IARC input that have been released by national programmes. According to the data, a total of 29 varieties have been released in LAC countries, 33 in Asia and 130 in Africa. Africa has a significantly higher release rate, even when adjusted for research investment or the number of countries releasing varieties.[4]

Trends in number of varieties released over time

Figure 16.1 shows the average number of varieties released per year with IARC content. Data are presented in terms of 5-year averages. Again, these data show that Africa has had consistently higher release rates than either Asia or LAC. They also show that releases have been relatively constant over time.

Composition of released varieties

Not all new varieties released by national programmes contain material from IARCs. Figures 16.2–16.4 show the number of varieties released by NARS, and the number which are IARC-related. The difference between the two lines is the number of varieties that were produced entirely by national programmes alone without input from either IITA or CIAT.

IARC-related varieties are the vast majority of total varieties released, reflecting the important role that CIAT and IITA play in cassava genetic improvement. In all three regions, the total number of varieties and number of IARC-related varieties tend to move together. There is no evidence for the idea that IARCs have crowded out the NARS in terms of varietal production. In fact, the relationship appears to be complementary. The complementarity of NARS and IARCS is also supported by data from Africa that show that total releases jumped significantly in the early years after IITA was established. During the 1970s, before IITA-related varieties began to become widely available, the average number of releases per year was 1.67. During the 1980s and 1990s, the number of releases by NARs with no IARC content increased to 2.28, an increase of 36% (Manyong *et al.*, 1999).

[4] It is important to control for the number of countries that have released varieties, since the same variety is often released in different countries, often with a different name. On a per country basis, LAC countries released an average of 3.6 IARC-related varieties, Asian countries released an average of 4.7 and Africa released an average of 7.6.

Table 16.8. Cassava varietal releases with CIAT and/or IITA material.

Country	Number of cassava varieties
LAC	
Brazil	11
Colombia	7
Cuba	2
Dominican Republic	2
Ecuador	2
Haiti	2
Mexico	2
Panama	1
Total	29
Asia	
China	6
Indonesia	3
Malaysia	2
Philippines	8
Thailand	7
Vietnam	7
Total	33
Africa	
Togo	8
Nigeria	10
Tanzania	8
Gabon	14
Côte d'Ivoire	2
Zimbabwe	8
Congo	9
Rwanda	4
Cameroon	13
Sierra Leone	6
Chad	15
Uganda	9
Benin	5
Guinea	12
Angola	2
Kenya	2
Ghana	3
Total	130

Source: Johnson (1999) and Manyong *et al.* (1999).

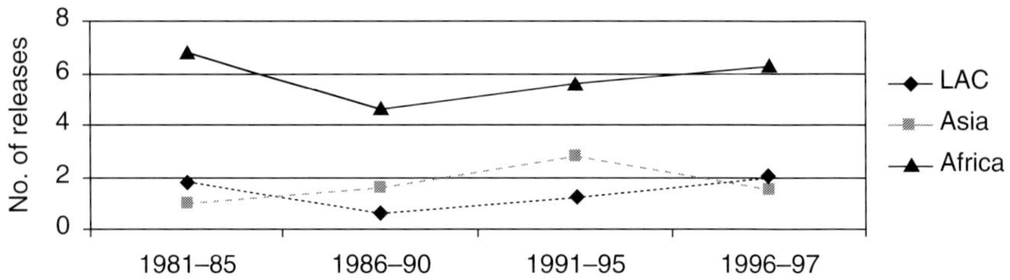

Fig. 16.1. Average number of cassava varieties released per year.

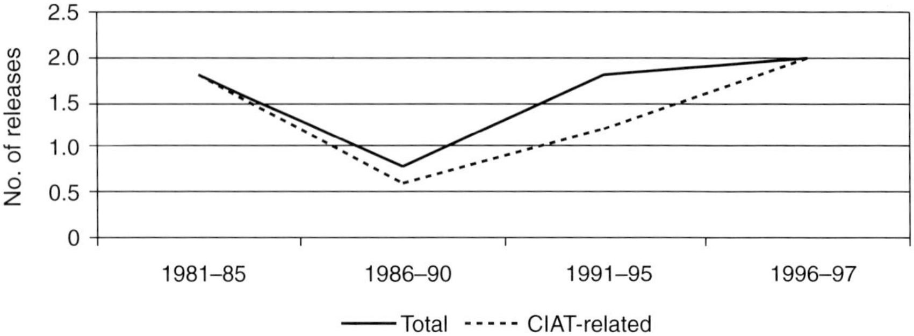

Fig. 16.2. Average annual cassava variety releases (LAC).

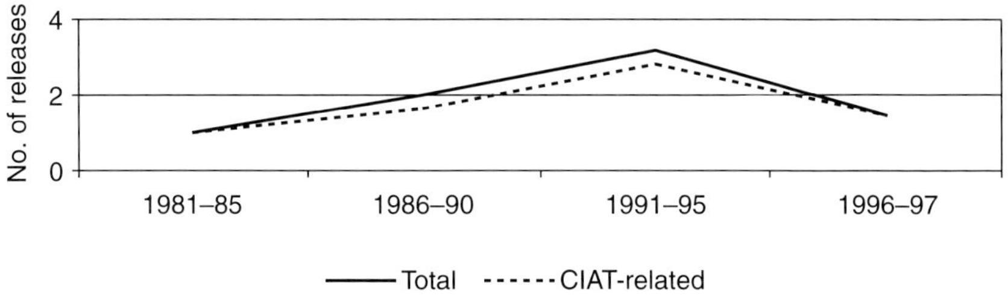

Fig. 16.3. Average annual cassava variety releases (Asia).

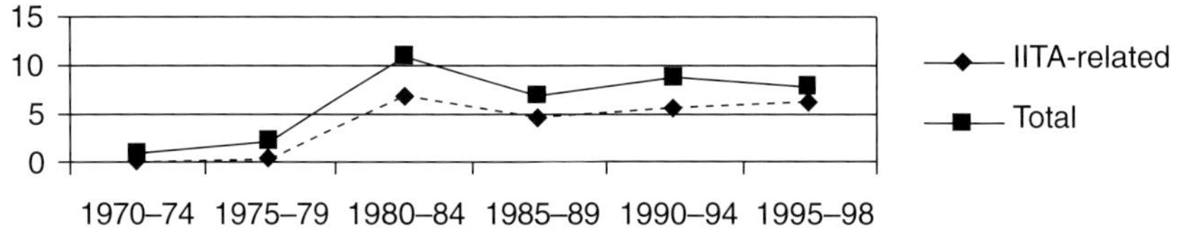

Fig. 16.4. Average number of varieties released per year.

Cassava varietal release data for LAC and Asia identify three different ways that an IARC can contribute to an improved variety. The variety can be: (i) a landrace from an IARC germplasm bank; (ii) a cross made in an IARC; or (iii) a NARS variety with an IARC parent. In the case of Africa, the data do not distinguish varieties as landraces and crosses, but rather categorize them by how much local selection or improvement was done by NARS upon receipt of the material from IITA.

The changes in composition of releases reflect a general evolution of breeding programmes. In the LAC and Asian cases (Figs 16.5 and 16.6), production gains were initially realized by selecting and disseminating promising landraces. These varieties are represented in the data by the GRUs, which were a significant portion of CIAT-related cassava varieties in the 1980s. Later production gains were achieved by replacing selected landraces with crosses, and this process is clearly visible in the data. The proportion of GRU releases dropped significantly between the 1980s and the 1990s as crosses increased.

Another trend suggested by the data is the increase in local selection and adaptation being done by NARS. Between the 1980s and the 1990s in LAC and Asia, the proportion of crosses made in CIAT

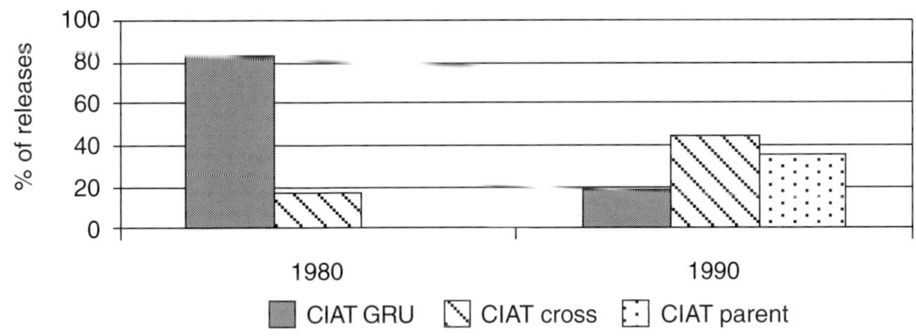

Fig. 16.5. Composition of CIAT-related cultivars in LAC.

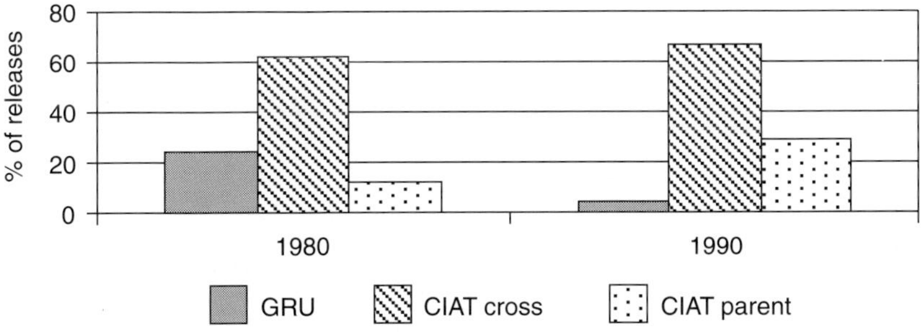

Fig. 16.6. Composition of CIAT-related cultivars in Asia.

declined and the proportion of crosses made in NARS using CIAT ancestors increased (Figs 16.5 and 16.6).

In the case of Africa, we have data on how IITA germplasm has been used by NARS (Table 16.9). Germplasm is divided into four categories: (1) released directly as a variety, (2) released as a variety with some additional selection, (3) released after additional improvement and selection, and (4) used as basic germplasm for NARS improvement and selection activities. The first two Africa categories would roughly correspond to the category 'IARC-variety' in the LAC and Asia data. The second and third Africa categories would roughly correspond to the category 'NARS variety with IARC parent material'.

Over the entire period (1970s–1990s) no clear trends emerge from the African data. During the 1980s and 1990s, however, the same pattern that emerged in LAC and Asia is also visible in Africa. The share of germplasm that is either used directly as a variety or is released with little additional input from NARS drops from 89% to 78%, as stronger NARS took on more responsibility for selection and breeding and were perhaps able to conduct their own breeding.

Adoption and Impact of IARC-related Varieties

In order for new varieties to have an impact on agricultural production and human welfare, they must be adopted by farmers, and must outperform local varieties. Tables 16.10–16.12 present data on the adoption and impact of IARC-related varieties. In the absence of empirical surveys, the data come from expert opinion from a variety of sources in LAC, Asia and Africa (see Manyong *et al.* (1999) and Johnson (1999) for more details on sources).

Table 16.9. Patterns of use of cassava germplasm from IITA in the cultivars released by NARS in sub-Saharan Africa (% of cultivars).

Category	1970–79 (*n* = 11)	1980–89 (*n* = 56)	1990–98 (*n* = 78)	All (*n* = 145)
1	11	3	18	10
2	56	8	5	12
3	11	59	45	48
4	22	30	33	30
Total	100	100	100	100

n, number of cultivars; 1, basic germplasm (substantial improvement and selection done after being received from IITA); 2, IITA germplasm, with some improvement and selection for local adaptation; 3, IITA germplasm, with some selection for local adaptation; 4 = direct use of IITA material, no additional improvement and selection done except planting material multiplication.

Table 16.10. Estimated economic benefits from the use of improved cassava cultivars per country, 1998.

	Area planted in 1998		Average yield			
Country	Total ha	% Improved	Local (t ha^{-1})	Advantage (%)	Incremental production (t)	Gross benefit (US$)
Brazil	1,607,760	0.02	12.43	24	96,000	2,880,000
Colombia	184,718	0.10	9.68	31	57,000	1,710,000
Cuba	65,000	0.31	2.92	103	60,000	1,800,000
Dominican Republic	19,500	0.51	4.72	64	30,000	900,000
Ecuador	19,760	0.10	6.69	45	6,000	180,000
Haiti	85,000	0.71	2.18	137	180,000	5,400,000
Panama	5,400	0.02	5.56	54	300	9,000

Table 16.11. Estimated economic benefits from the use of improved cassava cultivars per country, 1998.

	Area planted in 1998		Average yield			
Country	Total ('000 ha)	% improved variety	Local variety (t ha^{-1})	Advantage (%)	Incremental production ('000 t)	Total benefit (US$)
Benin	158	8	7	71.43	65	1,079,235
Côte d'Ivoire	270	16	7.5	20	66	3,985,200
Nigeria	2,950	19	13.41	44.97	3,355	78,948,903
Sierra Leone	48	19	7	71.43	45	2,977,942
Togo	111.7	12	9	44.44	52	5,809,936
Malawi	76	8	8	62.5	31	2,704,688
Swaziland	14.6	34	8	50	20	1,634,100
Tanzania	580	31	3.14	78.34	439	33,833,804
Zimbabwe	38	8	8	80	20	1,312,148
Rwanda	40	2	10	40	4	633,007
Uganda	450	30	7	20	187	12,062,093
Cameroon	80	31	17.3	27.17	116	9,635,820
Gabon	43	16	5	60	21	2,207,276
Congo	2,455.1	24	8	50	2,374	221,698,648
Chad	45	15	5.5	81.82	31	732,281
Zambia	165	0	5	89.5	2	48,437
Guinea	140	17	5.8	18.43	26	3,987,073
Angola	576	14	6	20	96	2,890,598
Kenya	98	16	9	1	16	610,736
Ghana	630	25	6.5	125	1,259	6,292,188
Total	9018.4	18	7.7	48.83	8,226	335,729,600

Table 16.12. Percentage of total crop area planted to varieties with CIAT content for selected countries in Asia, 1998.

	Area planted in 1998		Average yield			Starch yield				
Country	Total (ha)	% Variety	Local (t ha^{-1})	Advantage (%)	Incremental production (t)	Local variety (%)	Imp (%)	Benefit root (US$)	Benefit starch (US$)	Gross benefit (US$)
China	230,060	0.01	15.65	10	3,159	25.6	29.7	94,770	219,585	314,355
Indonesia	1,233,550	0.01	12.22	39	31,872	18.1	23.4	956,160	1,312,373	2,268,533
Philippines	240,000	0.04	8.51	43	30,600	19.0	19.6	918,000	107,908	1,025,908
Thailand	1,200,000	0.57	14.26	17	1,507,380	16.1	18.6	45,221,400	44,446,672	89,668,072
Vietnam	238,700	0.11	7.70	72	134,768	20.0	23.0	4,043,025	2,149,089	6,192,114

In LAC, IARC-related cultivars were estimated to have been planted on 143,250 ha, or 7% of total cassava area. Within the region, there is a great deal of variation between countries. Haiti and the Dominican Republic have the highest adoption levels, 71 and 51% respectively, while Brazil had only 2% adoption.[5] Given the size of Brazil's cassava area, Brazil still ranks second after Haiti in terms of total number of hectares planted to CIAT-related varieties. The gross value of increased production associated with the new varieties in LAC was approximately 430,000 t, with a value of nearly US$13 million in 1998 alone.

At the regional level, these adoption rates would not be expected to have a notable effect on national-level average yields, with the exception of a few cases. According to FAO, LAC yields increased 2% between 1980 and 1998, while Central American and the Caribbean yields increased 9%. Haiti and the Dominican Republic report yield increases of 9% and 25%, respectively, which is less than would have been expected given the estimated yield advantage of the improved varieties.

In Africa, estimates of area planted to IARC-related varieties in 1998 were not available; however, data are available on total area planted to improved varieties. Using these data, area planted to IARC-related varieties was approximated using the proportion of the total African varietal pool that is IARC-related. During the 1990s, 82% of released varieties were from IARCs, so we assume that 82% of area planted to improved varieties is planted to varieties from IITA or CIAT.

Based on these estimates, it appears that in Africa over 9 million ha or 18% of total area was estimated to have been planted to IARC-related varieties in 1998, ranging from less than 1% in Zambia to over 30% in Swaziland, Uganda and Tanzania (Table 16.11). The total value of the incremental production due to improved cassava cultivars was estimated at US$327.5 million in 1998. Benefits were highest in the DR Congo, followed by Nigeria.[6]

[5] This may significantly underestimate the CIAT contribution because several important varieties released in Brazil in the early 1990s (Mae Joana, Zolhundinha and EMBRAPA 8) were developed collaboratively between CIAT and the Brazilian national programme. For the sake of consistency with the other studies in the IAEG project, these varieties were classified as NARS varieties. If they are included as CIAT-related varieties, area in these varieties would be around 10%.

[6] For the purpose of comparison, it should be pointed out that for LAC and Asia, a world price of US$30 per tonne for fresh roots was used. In the case of Africa, local prices were used. These prices varied quite a bit between countries, with a regional average price of US$40 per tonne.

The 49% yield increase is not inconsistent with aggregate FAO data which report average cassava yields increases of 22% for Africa between 1980 and 1998. Over the same period, area planted increased by 50% (FAO). Nigeria and the DR Congo, which together account for more than half of the benefits associated with improved varieties, had yield increases of 12% and 13%, respectively (FAO).

In Asia, an estimated 722,500 ha or 23% of area was planted to CIAT-related cultivars.[7] Thailand is by far the biggest adopter of CIAT-related varieties, with 57% of area planted to CIAT-related varieties in 1998. Vietnam was estimated to have 11% of cassava area planted to CIAT-related varieties in 1998, which would imply a very rapid rate of diffusion given that the first variety was not released in Vietnam until 1993.

In Asia, the benefits of improved cassava varieties are measured and valued in terms of both increased root yield and increased starch yield per root (Table 16.12). The estimated increase in root yield associated with the new varieties varied from 10% in China to 72% in Vietnam, which has the lowest average national yield (FAO). In Thailand the yield increase associated with CIAT-related varieties was estimated to have been 17%.

At the regional level, the associated increase in production among the five countries was 1.7 million t, worth over US$51.2 million at the standard root price. The increased starch content per root, for which farmers received a price premium, was worth an additional US$48.2 million, bringing the total value associated with the new varieties to US$99.5 million in 1998.

While national-level data are not available on starch yields, we can compare these results with FAO root yield data. Average cassava yields in these five countries have risen about 4% since the mid-1980s, when CIAT-related varieties first began to become widely available. In Thailand aggregate yields have increased 7% over the period. Area planted to cassava in Thailand rose rapidly from the early 1980s until around 1990, at which point it began to decline to previous levels (FAO). Production followed a similar pattern.

[7] As was the case in Brazil, this may significantly underestimate the CIAT contribution because several important varieties in Indonesia (e.g. Adira 4) were developed collaboratively and classified as having had CIAT technical assistance. In order to be consistent with other IAEG studies, these varieties were considered NARS varieties for the purposes of this analysis.

Return on Investment in Cassava Improvement

To calculate the return on cassava improvement research, we need to compare the stream of benefits associated with improved cassava varieties with the investments made in breeding over the years. The adoption and impact data can be used as a rough estimate of the total benefits associated with the research. These figures overstate the benefits to the extent that production increases were accompanied by higher costs, e.g. labour, processing and other inputs.

In terms of research costs, to accurately estimate the internal rate of return (IRR) of improved varieties, we would need to include both IARC and NARS costs. Since cost data are not available for NARS, we will make an arbitrary assumption that NARS investment is equal to IARC investment. This is the same as attributing half the benefits of improved varieties to IARC research and half to NARS research.

There are two options for calculating IARC costs. The first is to use data on total cassava research investment. These data were presented in Table 16.3 for LAC and Asia. The problem with using these data is that they include all cassava research, not just breeding research, and might therefore result in an underestimation of the rate of return. Work on improved agronomic and management practices, training, and other activities not directly associated with cassava genetic improvement are all included in this cost figure, yet their benefits are not included in the estimate of incremental production increases associated with improved varieties. Using this measure of total investment in research, the net benefits to cassava research became positive in 1991, and in 1998 the internal rate of return to cassava research was 9%.

The other alternative is to use only the proportion of research costs devoted to breeding. While CIAT cost data are not broken down this way, one way to estimate the portion of resources devoted to breeding research is to assume that it is the same as the proportion of total scientists in the cassava programme who are breeders. These percentages are also shown in Table 16.3. This method is likely to underestimate the total amount of resources devoted to crop improvement because in many cases scientists such as geneticists, pathologists or entomologists work in support of the genetic improvement programmes. Using this estimate of total investment in breeding research, net benefits to cassava breeding became positive in 1984, and as of 1998 the internal rate of return was 22%.

Given the shortcomings of each method, it is likely that the true rate of return to cassava genetic improvement lies somewhere within the 9–22% range. The rate of return may also differ according to where the research is done. There is some evidence that research returns were particularly high in Asia, but it is difficult to estimate this precisely because of lack of data and conflating factors.

Conclusions

There has been significant progress on cassava genetic improvement since the establishment of cassava research programmes at CIAT and IITA. As of 1999, a total of 192 IARC-related improved cultivars had been released by national programmes, with 29 in LAC, 33 in Asia and 130 in Africa. These varieties are estimated to be planted on 23% of cassava area in Asia, on 18% in Africa, and on 7% in LAC. The incremental root production associated with improved varieties is estimated to be 10.3 million t in 1998. The value of increased root and starch production from the new varieties was worth nearly US$440 million. About 74% of the benefits were realized in Africa, 23% in Asia and 3% in LAC.

The results of the study also suggest that there is a complementary relationship between IARC cassava research and cassava research in the national programmes. The founding of the IARCs appears to have contributed to the establishment of national programmes in many countries, and they have played an important role in training scientists in those programmes over the years. The varietal release data suggest that the NARS are playing an increasingly important role in the production of improved varieties.

Acknowledgements

To the CIAT cassava programme and CLAYUCA, especially Clair Hershey, Bernardo Ospina, Reinhard Howeler and Rupert Best.

References

CICRTCR (1999) *Roots and Tubers in the Global Food System: a Vision Statement to the Year 2000.* Report to TAC of CGIAR prepared by the Committee on Inter-center Root and Tier Crops Research, 21 September.

Cock, J.H. (1985) *Cassava: New Potential for a Neglected Crop.* Westview Press, Boulder, Colorado

FAO, Agricultural Database, various years.

Hershey, C. (1994) *Research for development: the CIAT cassava programme.* CIAT working paper, CIAT, Cali, Colombia.

Iglesia, C. (1994) (ed.) Interfase entre los programas de mejoramiento, los campos de los agricultores y los mercados de la yuca en America latina, memorias de la tercera reunion Panamericana de fitomejoradores de yuca. Documento de trabajo 138, CIAT, Cali, Colombia, 279 pp.

Johnson, N. (1999) *Global impacts of CIAT germplasm improvement programs, 1967–1997: progress report on the IAEG study.* Paper presented at 1999 Annual Review. CIAT, Cali, Colombia.

Manyong, V. *et al.* (1999) *Contribution of IITA germplasm to cassava production and food security in sub-Saharan Africa.* Paper presented at the Workshop on Economic Impacts and Genetic Resources Enhancement by IARCS, 8–9 December. ICRISAT, Bulawayo, Zimbabwe.

17 Objectives and Methodology for Country Studies

R.E. Evenson

When farmers evaluate improved crop varieties and decide to adopt those varieties, they move from one equilibrium to another. They change the quantity of crops that they produce. They usually change the quantities of variable factors of production, such as fertilizer use, labour use and machine use. The technology embodied in the varieties is also likely to change the economic returns to investment in fixed factors of production, particularly in irrigation infrastructure. A number of studies of the adoption of improved crop varieties of rice and wheat in Asia, for example, have shown that the new varieties raised the productivity of irrigation investments and caused farmers to undertake investments in private tube well irrigation. Multiple cropping patterns and seasonal timing of planting also changed as a consequence.

In order to evaluate the contribution that crop genetic improvement makes to farm production, the evaluator faces two sets of problems. The first set is inherent in the comparison between the economic equilibrium before adopting crop genetic improvement and the economic equilibrium after adopting crop genetic improvement. The second set of problems is associated with achieving consistency between 'micro' evidence for plots or farms and 'macro' evidence for aggregations of farms at an aggregate level (e.g. a district in India).

The first set of problems requires explicitly modelling the 'endogenous' consequences of the availability of new technology to farmers. These consequences include the adoption of improved crop varieties, as well as changes in input use, investments in fixed factors of production and in production. Changes in multiple cropping and in timing of crop production are also endogenous consequences of the availability of new technology.

The second set of problems arises because, even though considerable evidence on crop yield comparisons between modern variety and traditional variety (or other modern variety) production is available from experiments, either on research station fields or on farmers' fields, this micro-evidence is not easily made consistent with macro-evidence for an aggregate of farms. This is because the micro-evidence is not collected using a sampling design to achieve micro–macro consistency. Several of the crop chapters in this volume review micro-evidence comparing production for different classes of varieties. While this evidence is useful, it is subject to the micro–macro consistency problem.

In recognition of this consistency problem, three country studies for China (Chapter 18), India (Chapter 19) and Brazil (Chapter 20) were included in the study design. These three countries represent major proportions of cropped area in their regions. China has 25% of the cropped area of Asia. India has 35% of the cropped area of Asia. Brazil has 50% of the cropped area of South America. In each of these studies, macro-data at the Province, District and State level are analysed. This ensures macro-consistency. In each study several crops are subject to analysis in a common framework. This enables some degree of cross-commodity consistency.

The studies differ in one very important respect. For the China and Brazil studies, the key MV adoption variable is a 'turnover' variable based on seed sales data. This variable measures the replacement of existing varieties by new varieties. Typically, the existing replaced varieties are actually modern varieties, not traditional varieties. For the India study, the variable is 'percentage area planted to modern varieties' and it measures the replacement of traditional varieties by modern varieties. The two variables are related. The data for China and Brazil suggest that, as modern varieties replace traditional varieties, i.e. as the percentage area planted to modern varieties goes from zero to 90%, the turnover variable is roughly 300%.

The estimation strategy employed in these studies is illustrated in the India study. For five crops in India, district data were available to allow estimation of a four-equation structural model. The estimation technique was three-stage least-squares. The model structure is depicted in Fig. 17.1.

The four endogenous variables were:

HYV: the percentage of cropped area planted to modern varieties.
IRR: the proportion of cropped area that is irrigated.
Area: the share of cropped area in the district planted to the crop.
Yield: production per hectare in kilograms.

Fig. 17.1. Model structure.

Endogenous dependent variables	Endogenous independent variables	Exogenous independent variables
PCT HYV	PCT IRR	EXT, STRES, PR, MARKET, E, C, PRRES
PCT IRR	PCT HYV	W, E, C
AREA	PCT HYV, PCT IRR	RESSH, E, C
YIELD	PCT HYV, PCT IRR, AREA	EXT, STRES, PRRES, WAGEFERT, C, W, E

The variables treated as exogenous to farmers, i.e. not under their control, include:

STRES:	State Research stocks
RESSH:	Relative Crop Research Shares
PRRES:	Private Sector R&D
PR:	Relative Crop Prices
WAGEFERT:	Relative Input Prices
MARKET:	Market Development
EXT:	Agricultural Extension
E:	Soil Indicators
C:	Climate Indicators
W:	Weather Variables

The key parameters measuring the impact of *HYV* on yield include the direct parameter estimated in the *Yield* equation and the indirect parameters associated with *HYV* impacts on irrigation, and area.

One of the key features of the three country studies is the identification strategy employed to estimate a 'structural' model. If all endogenous variables are 'jointly' determined by the full set of exogenous variables, the study authors are forced to resort to a 'reduced form' estimation strategy where each endogenous variable is regressed on the full set of exogenous variables. This means that one cannot obtain estimates of the *HYV* (or MV/TV) variable on crop yields or area.

With identifying restrictions, however, one can obtain structural estimates of *HYV* impacts on other endogenous variables. The India chapter discusses identification more fully. One of the major identifying variables is current weather. Current weather affects yields and may affect irrigated area, but decisions regarding area planted and varieties planted are made before the weather is known.

The China and Brazil studies are modifications of the India study. In these studies, a varietal turnover variable replaces the *HYV* variable from the India study. In the China study, sufficient data are available to compute a crop-specific TFP index. This essentially combines the *Yield*, *Area* and *Irrigation* measures into a single TFP index. Thus, the China study is based on two endogenous variables, varietal turnover and crop-specific TFP. The Brazil study did not have sufficient data to compute a TFP index. Instead, it used a yield index as a proxy for a TFP index. Price variables as exogenous variables were used to control for input supply effects (as in the India study). The Brazil study did address a problem associated with the varietal turnover measure. (The China study made corrections for this problem.) This measure is based on seed sales, but farmers often save their own seed, depending on their interest in switching to new varieties. The proportion of farmer-saved seed was treated as an endogenous variable in the Brazil study.

Estimation of varietal impacts on production utilizing macro-data has its limitations. It is unusual to obtain data on crop-specific inputs as in the China study, so the analyst must resort to a quasi-supply function specification, where prices correct for the intensity of production, i.e. for fertilizer, labour and other variable factor use. These prices do not always handle the problem. In some cases, prices themselves may be endogenous, although typically this is not a serious problem.

Public investments in research, extension and farmer schooling programmes may also respond to farm conditions and this raises endogeneity issues. These investments, however, are long-term investments, and variables indexing their 'services' are usually based on capital stock concepts. Thus the research services relevant in time are based on past investments, not current investments. This provides a time lag element that helps avail the most serious endogeneity problems.

The methods utilized in the three country studies reported here do represent evolvement over past studies.

The three country studies were expected to achieve two forms of comparison consistency not always possible in 'micro' studies of crop genetic improvement impacts. The first form of consistency was across crops. By applying a common statistical model to data for a single country, where measurement standards are similar, a degree of cross-crop consistency is attained. The second form of consistency is with actual aggregate production data for the country. Many micro-studies of experimental fields or farm surveys do not achieve this consistency.

18 The Impact of Investments in Agricultural Research on Total Factor Productivity in China

S. ROZELLE, S. JIN, J. HUANG AND R. HU

A number of recent studies have documented the importance of agricultural research and extension in promoting the expansion of crop production in the world over the past 30 years. Rosegrant and Evenson (1992) have shown the effect of new varieties and extension effort on Indian productivity. Pingali *et al.* (1997) identified the contributions made by the Green Revolution in South and Southeast Asia. Several studies have also measured the impact of agricultural research investments on China's agricultural output, e.g. Rozelle *et al.* (1996), Fan and Pardey (1997), and Lin (1991). However, for the case of China, no previous research has systematically analysed the determinants of growth in total factor productivity (TFP). Understanding the sources of technological improvements for food production in the world's most populous country is important, particularly since changes in agricultural TFP have historically been the main force driving growth in agricultural output and farm income in countries that have modernized their economies (Huang and Rozelle, 1996).

Past analyses of productivity change have suffered from two common shortcomings, both of which have limited their ability to characterize the relationship between technology change and output growth. First, researchers have typically focused on supply responses, yield responses or production function analysis and have not examined changes in TFP. Moreover, with the exception of Rosegrant and Evenson (1992), the analysis has been highly aggregated, across states or provinces and especially across crops.

Second, methodological limitations and data availability have made it difficult for researchers to link technological improvements to investments in research and extension. Most researchers use only rough proxies for research and extension inputs, and many studies ignore the complexity of the research production, extension, and adoption processes. This makes it difficult to identify and assess the impact of research carried out in a national programme or with its international partners.

The overall goal of this chapter is to create a framework for studying the effect of national and international investment on research and extension in China and to measure the impact that such investments have had on creating productivity-increasing technology. Investments will specifically include the flow of germplasm between the international agricultural research centres (IARCs) and China's national agricultural research system (NARS). Our purpose is to provide measures of crop-specific investment in research and of the use of materials from IARCs. Specifically, we use a new measure of varietal technology to track the changes in the quantity and quality of genetic resources in China's major rice-, wheat- and maize-producing provinces from 1981 to 1995 for rice, from 1983 to 1995 for wheat and maize. We also analyse how the technology – and specifically, the research programme and extension system that produce and disseminate the technology – affects provincial-level productivity of rice, wheat and maize over the same period.

We have chosen to limit the scope of our project in several ways due to data requirements. Since information is needed on the names, traits, genealogies and adoption of every major variety in each province for each year – as well as measures of other factors that make up and explain TFP – we limit our attention to major grain crops and to key rice-, wheat- and maize-growing provinces.[1]

[1] The 16 rice-growing provinces are Heilongjiang, Jilin, Liaoning, Hebei, Jiangsu, Anhui, Hubei, Hunan, Jiangxi, Zhejiang, Fujian, Guangdong, Guangxi, Yunnan, Guizhou and Sichuan. Together the 16 rice-growing provinces make up more than 90% of China's rice sown area and output in 1995. The 14 wheat-growing provinces are Hebei, Shanxi, Jiangsu, Anhui, Shandong, Henan, Sichuan, Gansu, Guizhou, Heilongjiang, Hubei, Shaanxi, Yunnan and Xingjiang. The 14 wheat-growing provinces account for 92% of China's wheat sown area and 95% of its output in 1995. The 13 maize-growing provinces include Guangxi, Hebei, Heilongjiang, Henan, Jiangsu, Jilin, Liaoning, Shanxi, Shandong, Shaanxi, Sichuan, Xingjiang and Yunnan. The maize-growing provinces account for more than 89% of China's maize sown area and 92% of its output in 1995.

Data and Methodology for Creating TFP Measures

Historically, estimates of China's crop TFP have been controversial, arriving at significantly different conclusions. Poor data and differences in assumptions may account for the disagreements. Researchers gleaned their data from a variety of sources; both input and output series were of erratic quality (Stone and Rozelle, 1995).

In this chapter, we overcome some of the shortcomings of the earlier literature by taking advantage of data that have been collected for the past 20 years by the State Price Bureau (SPB). Using a sampling framework with more than 20,000 households, enumerators have collected data on the costs of production of all of China's major crops. The data set contains information on quantities and total expenditures of all major inputs, as well as expenditures on a large number of miscellaneous items. Each farmer also reports output and the total revenues earned from the crop. Provincial surveys by the State Price Bureau supply unit costs for labour that reflect the opportunity cost of the daily wage forgone by farmers who work in cropping. Over the last few years, these data have been published by the State Development and Planning Commission (*The Compiled Materials of Costs and Profits of Agricultural Products of China*, SPB, 1988–1998). The data have previously been used in analyses on China's agricultural supply and input demand (Huang and Rozelle, 1996; Huang *et al.*, 1996; World Bank, 1997).

The key information that we bring to the analysis from our own data collection efforts is a set of land rental rates. In 1995, we conducted a survey in 215 villages in eight provinces, and obtained estimates of the average per hectare rental rate that farmers were willing to pay for farming. These rates were clearly asked net of all other payments that are often associated with land transfer transactions in China (e.g. taxes). Rental rates from our sample provinces are used to construct rental rates in adjacent provinces.

Our methodological approach is similar to that of Rosegrant and Evenson (1992) and Fan (1997) in that we use standard Divisia index methods to calculate TFP. In essence, TFP measures the difference between aggregate output and aggregate inputs. Conceptually, it can be thought of as the gap between the output and input index lines in Fig. 18.1.

Analysing Productivity in Reform China

During China's reform period, the rapid and monotonic expansion of real output for major food crops ranks as one of the nation's great achievements. Output indices, or price-weighted output data series of rice, wheat and maize, rose sharply between 1982 and 1995 (Fig. 18.1). Rice output increased by 20% over this period, wheat by 80%, and maize by 95%. Not all of this increase is due to technological improvement, however; much is attributable purely to expanded use of inputs. It is true that Divisia indices of aggregated inputs for rice, wheat and maize – including land, labour, fertilizer and other material inputs (see methods and data section for more details) – actually fell for all the crops, but this is mainly due to the decline of labour in the early reform period and to a contraction of

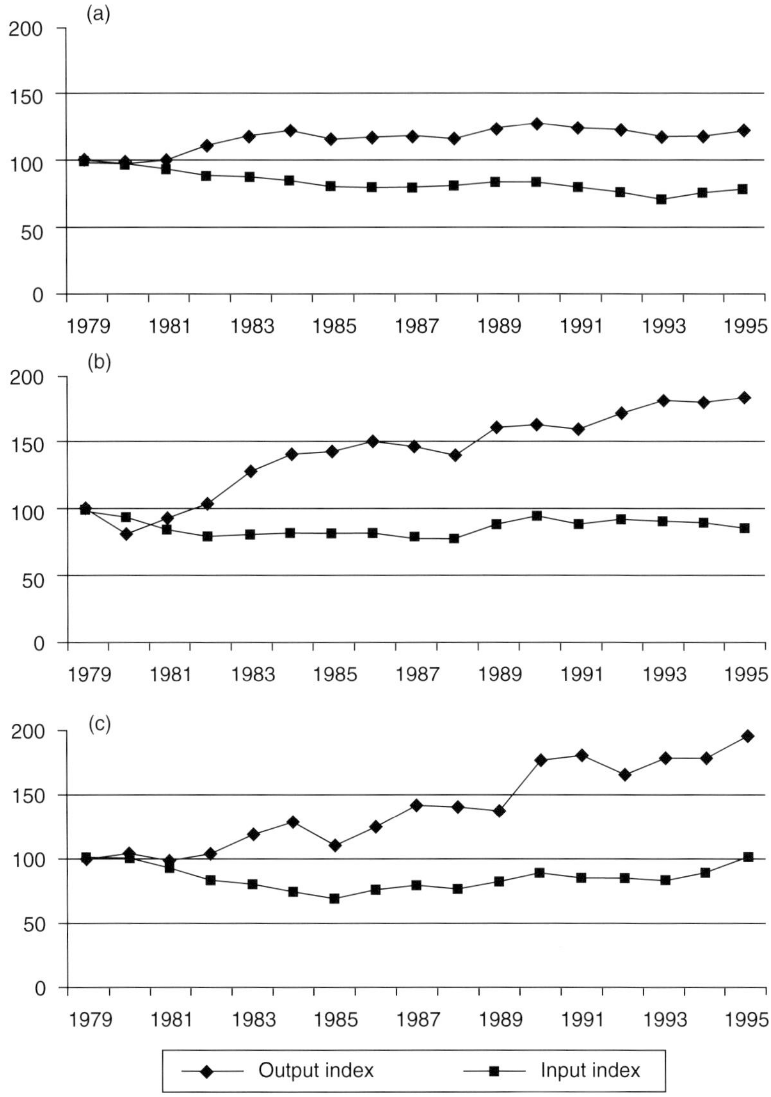

Fig. 18.1. Output and input indices for major rice- (a), wheat- (b) and maize-growing (c) provinces in China, 1979–1995. (Source: Authors' calculations based on Divisia–Tornquist formula.)

sown area later. Material inputs, including fertilizer, pesticide and other factors, rose sharply, increasing at an annual rate of 32% for rice, 26% for wheat and 30% for maize (rates consistent with the overall trends of fertilizer use in China, according to the State Statistical Bureau, 1998).

It is not clear whether the future holds any scope for continued increases in inputs. The rise in fertilizer and pesticide use slowed sharply in the 1990s. As wage rates rise, environmental awareness grows and resource limitations begin to bind, farmers will face pressure to reduce input use. When countries near input plateaus, further growth in output must rely more on technological change. Accordingly, our need to understand the record of past TFP performance and its determinants also increases.

TFP Trends in Reform China

Although we ultimately base our analysis on estimates of provincial TFP by year, it is also instructive to consider national aggregates. These illustrate an upward, but variable, trend in rice, wheat and maize productivity (Fig. 18.2).[2] In general, the TFP of all crops rose rapidly in the early 1980s, the earliest period of China's reforms. Such an unparalleled rise in TFPs, however, could not be sustained. The average TFP of our sample provinces was at about the same level in 1990 as it was in 1985 for all crops. During the 1990s, however, TFP began to rise once again. There are, of course, substantial differences across crops and regions. For example, wheat TFP rose 3–4% annually in Hebei and Shandong Provinces, but less than 1.5% annually in Sichuan and Shanxi Provinces.

Agricultural Technology in China

The nature of technological change in China: quality and quantity of new varieties

By the early 1980s, China's research and development system for agriculture reached its peak. By that time, China had developed one of the strongest research systems in the world. China's agricultural scientists

[2] Pairwise correlation coefficients among our index and three other indices (two used in Wen (1993); and one used in Lin (1990)) all exceed 0.95. The rise in the early 1980s is undoubtedly at least in part caused by the new incentives (Lin, 1992). Huang and Rozelle (1996), however, show that public investment in research and irrigation also contributed at least as much to TFP as increased incentive during the early reform.

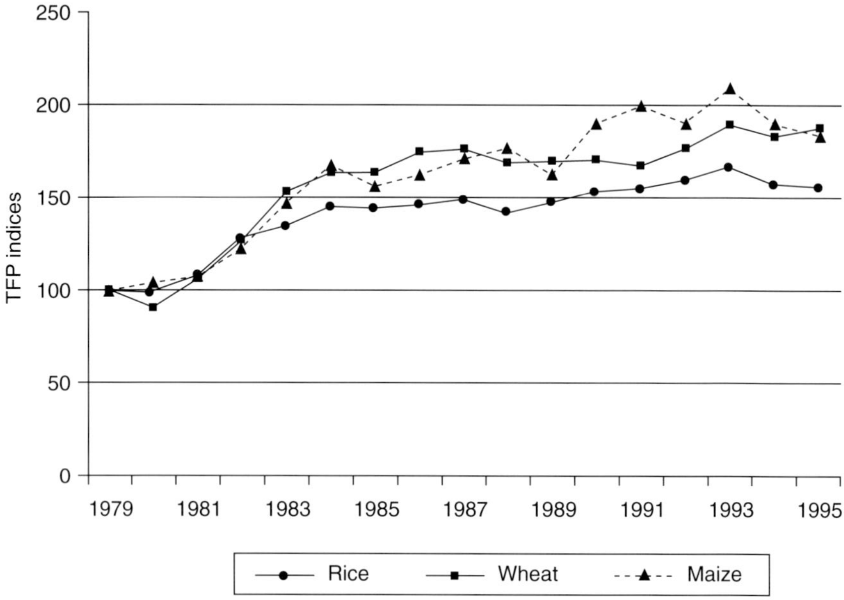

Fig. 18.2. Total factor productivity indices (sown area weighted average) for rice, wheat and maize in China, 1979–1995.

and the extension system have a long record of developing and disseminating technology throughout the People's Republic. Building on past achievements, reform era breeders turned out a constant stream of varieties (Table 18.1).[3]

China's breeding efforts have enhanced the quality of varieties available to farmers. Using experiment station yields of each major variety during the year that the variety was certified, two measures of quality were developed: a 'yield frontier' variable and an 'adopted yield potential' variable.[4] The yield frontier, which is created by using the

[3] A 'major' variety in our sample is any variety that covers at least 10,000 *mu* (or 667 ha) in a province. Since our database is built on this concept, we do not have full coverage. In fact, for the rice, wheat, and most of the maize growing sample provinces, the proportion of area covered by 'major' varieties exceeds 90% in each province.

[4] 'Yield frontier' is defined to be non-decreasing. If a major variety (defined in note 3) used by farmers in the field has the highest yield one year, it is assumed that the yield frontier in that province has reached that yield level and will not fall, even in the rare case that farmers have stopped using that variety and all other varieties have lower certified yields in the following years.

Table 18.1. Total and provincial average of the number of major varieties planted by farmers in China's rice-, wheat- and maize-growing provinces, 1982–1995.

	Rice		Wheat		Maize	
Year	Total	Average per province	Total	Average per province	Total	Average per province
1982	379	24	211	15	130	10
1983	333	21	274	20	130	10
1984	380	24	277	20	130	10
1985	424	27	313	22	156	12
1986	419	26	303	22	156	12
1987	373	23	313	22	156	12
1988	381	24	301	22	130	10
1989	365	23	337	24	143	11
1990	412	26	333	24	156	12
1991	395	25	350	25	156	12
1992	403	25	338	24	156	12
1993	392	25	341	24	182	14
1994	416	26	330	24	182	14
1995	391	24	311	22	208	16

Notes: These are totals for the 16 rice-growing provinces, 14 wheat-growing provinces and 15 maize-growing provinces in our sample. See footnote 1 for list of provinces.
Source: Authors' data gathered from the Ministry of Agriculture.

highest yield of any *one* major variety in the field in each province during a given year, is a measure of the ultimate yield potential of the current technology used by farmers in each province. The other variable, adopted yield potential, is the *average* of the experiment station yields of *all* major varieties that have been adopted by farmers. According to the two measures, China's research system has created a steady stream of quality technology (Table 18.2).

Farmers, however, have not always chosen (or perhaps have not been able to choose) the highest yielding varieties. The annual growth rate of average adopted yield potential of major varieties in the sample area has risen between 1.0% (wheat) and 1.4% (rice) during the reforms (see Table 18.2, rows, 2, 6 and 10). When compared with the farmers' actual yields in 1980 (rows 3, 7 and 11), the differences ranged from 31 to 58% (rows 4, 8 and 12), gaps that are actually not high by the standard of developing countries (Pingali and Rosegrant, 1995; Pingali *et al.*, 1997). In part reflecting the rapid rise in material inputs (see discussion above), the gap fell for all crops, although that for wheat narrowed more than those for rice and maize (ranging from 31% to 14% for rice, from 58 to 31% for wheat and from 51 to 38% for maize).

Table 18.2. Experiment station yields (yield frontiers and adopted yield potential), actual yields and yield gaps in sample provinces in China, 1980–1995.

	1980 (t ha^{-1})	1995 (t ha^{-1})	Annual growth rate (%)[c]
Rice			
Yield frontier[a]	6.6	9.1	2.3
Adopted yield potential[b]	6.1	7.2	1.4
Actual yield	4.2	6.2	2.1
Percentage gap between adopted yield potential and actual yields	31%	14%	
Wheat[c]			
Yield frontier[a]	6.3	7.5	1.3
Adopted yield potential[b]	4.6	5.2	1.0
Actual yield	1.9	3.6	3.2
Percentage gap between adopted yield potential and actual yields	58%	31%	
Maize			
Yield Frontier[a]	7.6	11.0	2.5
Adopted yield potential[b]	6.1	7.9	1.8
Actual yield	3.0	4.9	3.2
Percentage gap between adopted yield potential and actual yields	51%	38%	

[a]Yield frontier is the *highest* experiment station yield of a variety that has been extended to the field. The variable is non-decreasing in the sense that if in some subsequent year the highest yielding variety has a lower yield, the previous period's yield is maintained. In this table, the figure is the average of sample provinces.
[b]Adopted yield potential is the *average* experiment station yields of *all* varieties being adopted by farmers. In this table, the figure is the average of sample provinces.
[c]Annual growth rates are calculated by running a regression of natural log of various yields on a time trend.
Source: Yield frontier and average experiment station yields from authors' data. Actual yield from State Statistical Bureau-ZGTJNJ, 1981, 1983 and 1996.

There are two ways to interpret the yield gaps that currently exist in China. On the one hand, there appears to be a great deal of yield potential left in varieties in the field (the difference between the adopted yield potential and the actual yield), and even more when considering the differences between the yield frontier and the actual yield.[5]

[5] The researchers that argue that the yield gap is 'big' and that there is a lot of potential left in China's current germplasm technology are bolstered by the fact that China's yields may be understated because sown area is probably understated.

On the other hand, it can be argued that, in fact, the relatively low level (between 14 and 38%) and narrowing trend of the percentage difference between actual yields and adopted yield potential mean that China's yield potential is not that large, and the nation will need more breeding breakthroughs if the pace of yield growth is to be maintained on the effort of its domestic research system. The gap between adopted yield potential and actual yield for rice is small compared to wheat and maize, it is even smaller when compared to other rice countries. In 1987, China's gap was only 1.0 t ha^{-1} (or 15%). Similar (although not exactly comparable) gaps ranged from 5 t ha^{-1} (or 65%) in the Philippines and 3.5 t ha^{-1} (or 58%) in India (Pingali *et al.*, 1997). Relatively low yield gaps may imply that the further gains in realized total factor productivity of rice in China may be more difficult, since most of it must come from the creation and adoption of new varieties.

Creating and spreading new varieties in China

One of most impressive accomplishments of China's research system is that it has been able to consistently create and deliver to the field varieties demanded by farmers, inducing them to constantly upgrade their seed stock. Our data show that Chinese farmers adopt new varieties with great regularity (Table 18.3, columns 1, 2 and 3).[6] For example, maize farmers turn their varieties over the fastest, averaging more than 33% per year. Every 3 years on average farmers replace all of the varieties in their fields.

China's domestic research system has produced most of the new technology. The rise of the stock of research in the early reform era mostly reflects the commitment of the leadership during the Mao era (Stone, 1988). Researchers, however, differ in their views about the performance of the government in research and extension in recent years. Adjusting the data to account for the fact that some expenditures support institutes that do not appear to carry out research, as suggested by Rozelle *et al.* (1998), research investment shows a flat pattern from 1985 to early 1990. In the early 1990s, investment levels rose at a slow pace, until 1995 when they moved up sharply. Extension expenditure trends follow a similar pattern. Given the usual research lags, we would expect that slowing investment trends during the 1980s would most likely start to show up as stagnating research stocks in the mid- to late 1990s.

[6] Variety turnover is a measure of how fast major varieties that first appear in China's fields are able to replace the older varieties. Details of the calculations are provided in the data section.

Table 18.3. Trend of sown area weighted averages of varietal turnover (VT_1) and CG contribution in China's rice-, wheat- and maize-growing provinces, 1982–1995.

	Varietal turnover[a]			CG contribution[b]		
	Rice	Wheat	Maize	Rice	Wheat	Maize
1982	0.35	n.a.	0.47	16	1	2
1983	0.22	0.35	0.43	18	2	2
1984	0.20	0.26	0.40	22	2	2
1985	0.19	0.24	0.37	23	3	2
1986	0.28	0.27	0.41	23	3	2
1987	0.28	0.20	0.45	25	3	2
1988	0.26	0.19	0.34	25	3	3
1989	0.17	0.19	0.24	24	4	2
1990	0.24	0.21	0.24	25	4	2
1991	0.13	0.25	0.33	24	4	3
1992	0.29	0.22	0.32	22	3	1
1993	0.19	0.26	0.25	22	3	4
1994	0.25	0.23	0.32	20	3	1
1995	0.22	0.27	0.28	18	3	2

[a]See footnote 6 for definition and computation of varietal turnover.
[b]See footnote 10 for the calculation of CG contribution.

China also has access to genetic materials from international sources for all three crops (Table 18.3, columns 4, 5 and 6). Especially for rice, China has drawn heavily on the international research system for genetic material.[7] For example, material from the International Rice Research Institute (IRRI) comprises a large share of China's rice germplasm. Nationwide, we can trace around 20% of the germplasm to IRRI varieties. The proportion varies over time (from 16 to 25%) and also varies by province, reaching more than 40% in Hunan Province, one of China's largest rice-growing provinces, in the late 1980s. Although the national use of wheat and maize materials from the CG system (mostly from CIMMYT) is lower (columns 6 and 7), there does exist great variability among provinces. In some provinces (especially those in CIMMYT mandate areas, such as Yunnan Province for wheat or Guangxi Province for maize), material from IARCs makes up around half of the germplasm.[8]

[7] It should be remembered that China has also contributed significantly to the world stock of genetic resources for rice and maize.

[8] In large part the reason for the low overall contribution of the CG system to wheat and maize stems from the fact that CIMMYT's mandate area covers tropical and subtropical environments and most of China is in a temperate zone.

A General Framework of Endogenous Technology and Productivity Growth

Determinants of TFP and model specification

TFP indices for rice, wheat and maize in China vary not only across province but also over time. Factors that could account for variations in TFP include changes in technology, institutional reforms, infrastructure development, improvements to human capital, and other factors. Given our data and the research question, a framework for explaining TFP changes over time can be specified as:

$$\text{TFP} = f\,(\textit{Technology, Infrastructure, Institutional Reforms, Z}) \qquad (1)$$

where Z is a vector of control variables with the elements representing weather, agro-climatic zones, and some fixed but unobserved factors that differ across regions. In most countries, technology and infrastructure are thought to be the major factors that drive the long-term TFP growth (Rosegrant and Evenson, 1992). Most of other determinants contribute either to short-term fluctuations or one-time-only fixed shifts in TFP over time.

In this chapter, two measures on technology are specified, where varietal turnover (VT_1 and VT_2), is defined as:

$$VT_t = 1 \text{ for } t = 1, \text{ and}$$
$$VT_t = VT_{t-1} + \Sigma_k\, [V_{kt} = W_{kt} - W_{kt-1} \text{ if } W_{kt} - W_{kt-1} > 0, \text{ otherwise } V_{kt} = 0] \text{ for } t > 1, \qquad (2)$$

where V_k is the area share change for those varieties that have positive sign, and W_k is the area share of the kth variety in the *total sown area* for VT_1, and W_k is the area share of the kth variety in the *sown area of all major varieties* for VT_2.[9] Equation (2) defines technological change as the extent to which newly introduced varieties replace existing varieties and the extent to which existing higher yielding varieties replace existing inferior varieties. Assuming farmers are rational, variety replacement occurs if, and only if, the new variety is of a higher 'quality' than the variety it is replacing, where quality can be cost-reducing, yield-enhancing or include some new taste characteristic. One of the main questions in the chapter is answered by examining the coefficient of the VT variable in an equation explaining TFP.

A potentially serious statistical issue arises, however, with using VT as a measure for testing the effect of technology on TFP, as in equation (1). Since the farmer may be simultaneously making production decisions

[9] Major varieties are defined in footnote 3.

that affect both TFP and technology adoption, an OLS regression of TFP on VT is likely to be problematic because the error term may be correlated with VT. To avoid the endogeneity of VT in the estimation of the TFP equation, we take an instrumental variables approach. Our strategy for identifying the effects of technology on TFP uses the assumption that the technology delivered by the national and international research systems affects technology adoption (and hence VT), but does not affect TFP except through the seeds that farmers adopt. If the assumptions are valid, we can use three variables as instruments: the investments made by the government in crop research (or more precisely the nation's stock of crop research); germplasm that flows into each province from the domestic research system *and* from international agricultural research centres as instruments; and yield frontier, a variable representing the yield-increasing potential of technology generated by the research system.[10]

Based on the discussion above, we use a simultaneous, two-stage least squares (2SLS) estimator to estimate the effect of technology and other variables (infrastructure, institutional change, etc.) on TFP. The empirical specifications are:

$$TFP_{iht} = f_i\,(VT_{iht}, Extension_t, Irrigation_{ht}, D_{90-95}, Weather\ Event\ Index_{ht}, Provincial\ Dummies) + e1_{iht} \quad (3)$$

$$VT_{iht} = g_i\,(Extension_t, Irrigation_{ht}, D_{90-95}, Weather\ Event\ Index_{ht}, Provincial\ Dummies; Research\ Stock_t, CG_{iht}, Yield\ Frontier_{iht}) + e2_{iht} \quad (4)$$

where *i* indexes crops; *h* indexes provinces; *t* indexes years; total factor productivity (TFP) and VT are defined as above; *Extension* is a variable reflecting all expenditures made on the extension system, aggregated to the national level; *Irrigation* is measured as the ratio of irrigated land to

[10] We utilize three variables as instrumental variables to identify the VT. First, crop breeding *research stock* is used as a proxy for public investment in the creation of new varieties. Since most research is either embodied in the seed itself, or requires delivery by the extension system, the effect of which we have already accounted for, this is a conceptually sound instrument. Second, a measure of the *yield frontier*, a variable representing the yield-increasing potential of technology generated by the research system (which is defined as the maximum yield of any variety in the field up to time period *t*), is also a variable that conceptually should explain the adoption of new seed technology, but have no effect of its own on TFP. Finally, we define a variable that represents the proportion of genetic material in China's germplasm for each crop that comes from the CG system (*CG contribution*). This variable is created using pedigree data for all varieties in the field in each period, and assigning geometric weights to parents (0.25/parent), grandparents (0.06/grandparent), and so on. *CG contribution* represents the proportion of germplasm that has parents and grandparents or older generations that are identified as being from an international centre (IRRI for rice; CIMMYT for wheat and maize).

cultivated land; and, D_{90-95} is an indicator variable which equals 1 for the period between 1990 and 1995 and is included to measure the effect of other period-varying factors on TFP during the period of market liberalization that China experienced in the early 1990s. We also include two variables to account for yield fluctuations due to the effect of flood and drought events (*Flood* and *Drought Index*), and provincial dummies to control for unobserved non-time varying fixed effects associated with each province. The three instruments in equation (4), $Research\ Stock_t$, CG_{iht}, $Yield\ Frontier_{iht}$, are defined in note 10.

Data

In addition to the cost of production used in the creation of the TFP indices, we also compiled from numerous sources a nationwide database on China's major rice, wheat and maize varieties. Information on crop-specific varieties and the amount of area sown to each variety in each province are from the Ministry of Agriculture (MOA, Varieties – 1981–1997). The MOA compendium reports on 'major' varieties that cover at least 100,000 *mu* (6667 ha) in a province in any one year. Varietal yield information and pedigree data were mostly collected by the authors through an extensive desk survey that included use of materials in national genealogical databases (published and on-line), information in the national library, and records from the national seed company. After the desk survey, however, information for some crops for some years and some provinces were still missing. Our data collection team made calls and visits to hundreds of provincial and prefectural research institutes, breeding stations, seed companies, individual breeders, and bureaux of agriculture. A table showing the means of the major variables is in Appendix 1.

Results

The determinants of new technology

The varietal turnover equations – which can be viewed as showing determinants of technology change – give good results, both in their own right and for instrumenting the TFP equations. (For VT_1, see Table 18.4, columns 2, 4 and 6.)[11] Using OLS, the R^2 for the technology equations

[11] Since the results for VT_1 version models (Table 18.4) and VT_2 version models (Table 18.5) are very similar, only the results for VT_1 version models are reported in the text.

Table 18.4. Two-stage least-squares estimates of the determinants of total factor productivity for rice, wheat and maize in China.

	Rice		Wheat		Maize	
	TFP	Technology (VT_1)	TFP	Technology (VT_1)	TFP	Technology (VT_1)
Technology variables						
Varietal turnover (VT_1)	15.50 (9.25)***		18.65 (5.84)***		15.75 (6.51)***	
Extension	–0.014 (1.68)*	0.0004 (2.29)**	–0.02 (1.39)	0.0008 (5.49)***	–0.06 (2.88)***	0.0005 (1.47)
Weather, irrigation and period dummy						
Flood index	–8.63 (1.76)*	0.04 (0.37)	–102.29 (5.23)***	0.04 (0.23)	–13.92 (1.94)*	0.02 (0.18)
Drought index	–23.83 (2.56)**	–0.30 (1.42)	–51.81 (3.08)***	–0.11 (0.64)	–38.72 (5.59)***	–0.08 (0.67)
Irrigation index	–100.05 (3.19)***	–0.92 (1.26)	–87.09 (1.19)	–1.24 (1.69)*	–14.45 (3.95)***	–0.09 (1.40)
D90–95 (Index for 1990s)	1.54 (0.40)	–0.29 (3.15)***	6.65 (0.96)	–0.13 (1.94)*	11.60 (1.25)	–0.55 (3.30)***
Instruments						
Research stock		0.02 (19.65)***		0.015 (22.76)***		0.03 (22.73)***
CG contribution		–0.27 (0.76)		0.013 (2.86)***		0.73 (1.05)
Yield frontier		–0.002 (3.03)***		0.003 (5.27)***		–0.003 (4.37)***
No. of observations	240	240	196	196	195	195

Note: All regression equations include provincial dummies to hold constant unobserved fixed effects. For definition of variables, see Table 18.2 and methodological section. *T*-ratios in parentheses. ***, ** and * signify that the coefficients are statistically significant at the 1, 5 and 10% levels.

exceed 0.90 for all three crops. Hausman (1983) tests for exclusion restrictions – designed to test the validity of the instruments – show that our three instruments are statistically valid.[12] At least in a statistical sense,

[12] Since the farmers are simultaneously making production decisions that affect both TFP and technology adoption, the variable measuring technology adoption, *VT*, is likely endogenous. To properly account for the endogeneity, the predicted values from the technology equations can be used as instruments if the variables on the right-hand side of the technology equations affect technology but are uncorrelated with the structural disturbances of the TFP equation. To test if the set of identifying

the instruments do well at predicting varietal turnover, but they do not affect TFP except through their influence on varietal adoption. The results for the model using VT_2 are similar and are reported in Table 18.5.

Substantively, these first-stage equations provide interesting insights into the process of technology creation in China. In all the specifications, and for all crops, the *Research Stock* variable has a positive and highly significant coefficient. This demonstrates the importance of investment in the research system in stimulating farmers' adoption of new varieties. Increases in national research stocks are associated with a more rapid pace of varietal turnover (Table 18.4, columns 2 and 4, row 7). If technology is the engine that will drive China's food supply in the future (Huang and Rozelle, 1996), the results here emphasize the necessity of maintaining the level and growth of public investment in crop research and development. As a cautionary note, however, the negative sign on the time dummy for the market liberalization period (in all but one of the first-stage VT equations) calls for heightened attention to the health of the research system. The factors that have slowed technological change in the 1990s appear to be the source of declines in TFP in 1994 and 1995.

The impact of the yield-increasing technology (created by each province's research system – the *Yield Frontier* variable) is more complicated. Breakthroughs in yields lead to faster diffusion of new varieties and replacement of old varieties for some crops but not others. The positive and significant signs of the *Yield Frontier* variables in the wheat VT equations (Table 18.4, columns 4 and 5, row 9) demonstrate that when higher-yielding wheat varieties appear, farmers turn their varieties over more frequently. In contrast, higher values of *Yield Frontier* variables in the rice and one of the maize equations are associated with slower turnover (Table 18.4, columns 2 and 4, row 9). Such a finding is consistent with our gap

instruments are exogenous, a Lagrange multiplier test can be used (Hausman, 1983). The χ^2 distributed test statistic with 3 degrees of freedom, is NR^2, where N is the number of observations, and R^2 is the measure of goodness of fit of the regression of the residues from the TFP equation on the variables which are exogenous to the system. The test statistics are 0.86 and 0.22 (with VT_1 and VT_2 specifications, respectively), and 0.25 and 0.18 for wheat which indicate that the null hypothesis that there is no correlation between the exogenous instruments and the disturbance term from the TFP equation for rice and wheat can not be rejected. However, the case for maize is less clear. The test statistics are 11 and 5.25, so the hypothesis of no correlation between the exogenous instruments and the disturbance term from TFP equation is rejected for the VT_1 specification. When only two instrument variables, *research stock* and *wcg*, are used in the system, the test statistics are 0.02 and 0.07 which indicates that these two instrumental variables are not correlated with the disturbance term from the TFP equation.

Table 18.5. Two-stage least-squares estimates of the determinants of total factor productivity for rice, wheat and maize in China.

	Rice		Wheat		Maize	
	TFP	Technology (VT_2)	TFP	Technology (VT_2)	TFP	Technology (VT_2)
Technology variables						
Varietal turnover	10.50		26.86		23.01	
(VT_2)	(9.18)***		(6.02)***		(7.42)***	
Extension	–0.01	0.0007	–0.02	0.0005	–0.06	0.0002
	(1.66)*	(3.14)***	(1.42)	(4.45)***	(3.19)***	(0.63)
Weather, irrigation and period dummy						
Flood index	–8.44	0.03	–98.29	–0.12	–14.62	0.04
	(1.70)*	(0.25)	(5.13)***	(0.80)	(2.11)**	(0.33)
Drought index	–21.29	–0.73	–52.54	–0.06	–40.35	0.002
	(2.26)**	(2.91)***	(3.19)***	(0.42)	(6.02)***	(0.01)
Irrigation index	–91.82	–2.32	–118.91	0.18	–11.61	–0.15
	(2.91)***	(2.68)***	(1.63)*	(0.31)	(3.20)***	(2.46)**
D90–95	2.25	–0.46	5.78	–0.08	9.52	–0.35
(index for 1990s)	(0.58)	(4.29)***	(0.85)	(1.34)	(1.06)	(2.15)**
Instruments						
Research stock		0.02		0.01		0.02
		(23.96)***		(20.52)***		(15.37)***
CG contribution		0.68		0.004		0.14
		(1.64)*		(1.01)		(0.20)
Yield frontier		–0.003		0.002		0.0003
		(3.78)***		(4.20)***		(0.44)
No. of observations	240	240	196	196	195	195

Note: All regression equations include provincial dummies to hold constant unobserved fixed effects. For definition of variables, see Table 18.2 and methodological section. *T*-ratios in parentheses. ***, ** and * signify that the coefficients are statistically significant at the 1, 5 and 10% levels.

analysis and may reflect the fact that farmers (especially those cultivating rice) in the mid- to late-reform period prefer adopting higher quality varieties, even though higher yielding varieties are available.

The impact of IARC material

The impact of the materials from international research centres is mainly a story of China's breeders using IRRI and CIMMYT varieties for yield enhancement. If it can be assumed that, when China's breeders incorporate foreign germplasm into their varieties, the material contributes to a rise in productivity, then the test of the direct impact of CG material can be seen in the results of the TFP equation (next sub-section). If technology is important in all the TFP equations, by virtue of the fact that material from IRRI is used more extensively than material from the other IARCs, then IRRI would be making a proportionately large contribution to China's TFP in the reform era.

It is possible, however, that foreign material may be bringing in an extra 'boost' of productivity – beyond its contribution to the varieties themselves – by increasing the rate of turnover of new varieties.[13] Such an effect would show up in the VT equations. If the coefficients of the IARC variables were positive and significant, they would indicate that the presence of material from CG centres makes the varieties more attractive to farmers and contributes to technological change in China in a second way. The evidence of this effect is mixed across crops. In rice, the presence of IRRI material does not appear to be important in increasing the turnover of rice varieties (Table 18.4, row 8, column 2). If farmers are in fact mainly looking for characteristics other than yield (such as grain quality), they may not be finding these characteristics in IRRI materials. This would be consistent with a story in which IRRI contributes a great deal to TFP growth but not much to varietal turnover. A similar interpretation is called for in the case of wheat and maize (Table 18.4, columns 4 and 6, row 8); although the coefficients for IARC content are positive in the VT equations, the standard errors are large relative to the size of the coefficient in all but one case.

This finding does not conflict with the view that IARCs have made large contributions of genetic material to China. For example, CIMMYT wheat and maize germplasm have had large effects on the productivity of some of China's poorest areas. For example, CIMMYT-bred varieties accounted for more than 50% of Yunnan Province wheat varieties and more than 40% of Guangxi Province maize varieties in the late 1980s and early 1990s. Yunnan and Guangxi Provinces are both very poor provinces and some of the poorest populations in China are in the

[13] One alternative way to identify the extra impact of CG material on TFP is to interact it with VT in the TFP equation directly. Since this variable is also simultaneously determined with TFP, we would have to estimate another equation to create an instrument for use in the second-stage equation. We estimate one equation for *VT* and one for *VT*CG* and use the predicted values from these equations in the TFP equations, estimating the three equations as a system. The results are similar to our less formal test; varieties with high content of CG germplasm do not have an extra effect (results not shown for brevity).

mountainous maize-growing areas. Elsewhere (Rozelle *et al.*, 1999), we have shown that the impact of IARC material in poor provinces, in general, is more important than its effect in rich areas. This finding encompasses both the direct contributions of IARC material to productivity and, in some cases, the effects of IARC material on inducing more rapid turnover. In the case of CIMMYT, it appears that the institutions focus on tropical and subtropical wheat and maize varieties has limited its impact on China's productivity as a whole but has none the less played an important role in poor areas that would not otherwise be well served by the Chinese research system (Stone, 1993).

Technology, extension, and productivity

Our TFP results, presented in Table 18.4, also generally perform well. The goodness of fit measures (for OLS versions of the equations) range from 0.80 to 0.85, quite high for determinants of TFP equations. In other work, in India for example, the fit of the specification was only 0.17 (Rosegrant and Evenson, 1992). The signs of most of the coefficients are as expected and many of the standard errors are relatively low.[14] For example, the coefficients of the weather indices are negative and significant in the TFP equations in the rice, wheat and maize specifications (Table 18.4, rows 3 and 4). Flood and drought events, as expected, push down TFP measures, since they often adversely affect output but not inputs. (For many crops, input decisions are made before the onset of bad weather.)

[14] One of the most surprising exceptions is the insignificant or negative sign of the irrigation variable's coefficient. According to our results, the ratio of irrigated to cultivated land does not positively influence wheat productivity and negatively affects that of rice and maize. It certainly may be that for any number of measurement or statistical reasons, we are not measuring the true relationship between marginal increases in irrigation area and TFP. However, it may be, as also found by Rosegrant and Evenson (1992), the value of irrigation is already embodied in the land input variable (since areas with high land values have high levels of irrigation), so its positive impact is already removed. Additionally, the negative value for rice may appear since the area in which most of China's new irrigation projects have occurred are not naturally conducive to rice cultivation. In the south, China's main rice-growing region, irrigated area has expanded little, if any, in most provinces during the reform. In north China, if newly irrigated area does lead to new rice cultivation, it may be that the new land brought into production is inherently less productive than the average rice area already being farmed. Such an explanation is consistent with our results.

Perhaps the most robust and important finding of our analysis is that technology has a large and positive influence on TFP. The finding holds over all crops, and all measures of technology. The positive and highly significant coefficients on both predicted measures of the rate of varietal turnovers (VT_1 and VT_2) show that as new technology is adopted by farmers it increases TFP (Table 18.4, columns 1 and 3, row 1). Following from this, the positive contributions of China's research system and the presence of IARC material both imply that domestic investments in agricultural R&D and ties with the international agricultural research system have contributed (and plausibly will continue to do so) to a healthy agricultural sector.

Further analysis is conducted to attempt to overcome one possible shortcoming of using VT as a measure of technological change. It could be that an omitted variable is obscuring the true relationship between VT and TFP. As varieties age, the yield potential may deteriorate (Pingali *et al.*, 1997). In order to try to isolate the age effect from the new technology effect (given the definition of VT, this may be a problem), we add a variable measuring the average age of the varieties (results not shown for brevity). Although we find no apparent negative age impact on TFP in any of the equations (the coefficient is actually positive in the case of maize), in a number of the regressions the magnitude of the coefficient of the VT variable in the TFP equation actually rises, a finding that reinforces the basic finding of the importance of technology.

Sources of TFP growth

To examine the relative size of the impact of different factors, we perform a decomposition exercise for rice, wheat and maize. Between 1981 and 1995, China's total factor productivity in rice grew at an annual average rate of 2.0%. However, TFP growth was not constant over time. TFP grew faster during the early reform period, from 1981 to 1984 (9.4%) and slowed down in the later period, from 1984 to 1995 (1.11%). In order to understand whether the sources of TFP growth differ between the early reform period (1981–1984) and more recent years (1984–1995), we conduct the decomposition analysis over two subperiods, 1981–1984, and 1984–1995. We report the decomposition results using elasticities calculated from the regressions that included VT_2 (Table 18.5). The results would be substantively similar if we used elasticities from the VT_1 calculations.

The rice decomposition results in Table 18.6 show that technology was by far the most important factor driving the sharp increase in TFP in the early reform period. Improvement in technology (measured by varietal turnover) increased the annual growth rate of TFP by 6.01%,

Table 18.6. Decomposition of the sources of rice TFP growth in China.

		1981–1984			1984–1995		
	TFP	Factor annual	Sources of growth		Factor annual	Sources of growth	
	elasticities[a]	growth rate[b]	Rate[c]	Per cent[d]	growth rate	Rate	Per cent
Varietal turnover (VT_2)	0.28	21.47	6.01	63.61	7.81	2.19	197.01
Extension	–0.02	2.03	–0.04	–0.43	3.96	–0.08	–7.14
Flood index	–0.01	29.02	–0.18	–1.93	9.26	–0.06	–5.19
Drought index	–0.02	–13.17	0.21	–2.26	1.24	–0.02	–1.80
Irrigation index	–0.34	0.70	–0.24	–2.58	1.29	–0.44	–39.50
Residual			1.21	56.62		–0.30	–143.94
Actual growth rates			9.45	100		1.11	100

[a]TFP elasticity with respect to each factor is calculated on the basis of coefficients from rice model in Table 18.5.
[b]TFP and factor growth rates are computed by a least square estimate.
[c]Growth rate contributed by each factor is calculated by multiplying factor growth rate (column 2) by elasticity (column 1).
[d]The percentage of total TFP growth explained by each factor is the corresponding figure in column 3, divided by the total growth rate of TFP (which for the period of 1981–90 was 9.45%).

accounting for 63.61% of the total growth in TFP. Interestingly, expenditure on extension and investments in irrigation did not help the growth of TFP. The breakdown of the extension system that began in the early reform period already appears to have limited the effectiveness of extension's impact on TFP (in fact, there is a slightly negative impact). Likewise, irrigation had little effect on TFP.

In the late reform period, from 1984 to 1995, technology remained the most important source of TFP growth (Table 18.6). In fact, during this period, it was the only factor supporting positive TFP growth during that period. Technology alone would have caused TFP to grow by 4.17% annually in this period. The actual growth of TFP, however, was only 1.11% annually, because of other negative factors. For example, the continuing breakdown of the irrigation system significantly reduced the growth of TFP. Using the elasticities calculated in our VT_2 model, the fall in irrigation investment led to a net reduction in TFP growth of 39.5%.

The decomposition analyses for wheat (Table 18.7) and maize (Table 18.8) find that the results are similar to those for rice (Table 18.6). Technology explains most of the TFP growth in the early reform period (column 3 and 4, in Tables 18.7 and 18.8). In the late period, as with rice, wheat and maize technology would have pushed up TFP growth by 2.72 (for wheat) and 3.20 (for maize) had not other factors constrained it (rows 1, columns 6 and 7).

Conclusions

This chapter, more than anything, establishes a basis for policy makers and donors – both in China and in the international community – to confidently invest in agricultural research. The evidence supports the idea that such investments in technology generation and diffusion have led to TFP gains in the past. The decomposition analysis shows the overwhelming importance of technology relative to other factors. TFP has continued to rise in the reform period primarily due to past contributions of technology.

The picture sketched by our study demonstrates that investments in new technology have benefits in many dimensions. Public investments in breeding and extension pay off in terms of higher TFP; but the characteristics of research outputs matter. It is not sufficient for new varieties to increase yields; they must also prove acceptable to farmers. In the case of rice, for example, although breeders are increasing yield frontiers at a rapid rate, the increases in TFP often appear to come less from yield increases than from other productivity-enhancing traits demanded by farmers. If these traits can be identified and incorporated into higher-yielding varieties, the future of China's rice supply appears sound.

Table 18.7. Decomposition of the sources of wheat TFP growth in China.

		1981–1984			1984–1995		
	TFP	Factor annual	Sources of growth		Factor annual	Sources of growth	
	elasticities[a]	growth rate[b]	Rate[c]	Per cent[d]	growth rate	Rate	Per cent
Varietal turnover (VT_2)	0.33	19.00	6.26	49.29	8.23	2.72	247.27
Extension	0.09	2.33	0.21	1.65	3.96	0.36	32.7
Flood index	–0.03	9.60	–0.29	–2.26	3.36	–0.10	–9.09
Drought index	–0.04	–19.18	0.77	6.03	3.66	–0.15	–13.6
Irrigation index	–0.29	0.17	–0.05	–0.41	1.71	–0.50	–45.50
Residual			1.96	54.30		–0.37	–211.78
Actual growth rates			12.72	100		1.10	100

[a]TFP elasticity with respect to each factor is calculated on the basis of coefficients from wheat model in Table 18.5.
[b]TFP and factor growth rates are computed by a least square estimate.
[c]Growth rate contributed by each factor is calculated by multiplying factor growth rate (column 2) by elasticity (column 1).
[d]The percentage of total TFP growth explained by each factor is the corresponding figure in column 3, divided by the total growth rate of TFP (which for the period of 1981–90 was 9.45%).

Table 18.8. Decomposition of the sources of maize TFP growth in China.

		1981–1984			1984–1995		
	TFP elasticities[a]	Factor annual growth rate[b]	Sources of growth		Factor annual growth rate	Sources of growth	
			Rate[c]	Per cent[d]		Rate	Per cent
Varietal turnover (VT_2)	0.40	17.66	7.06	44.37	8.00	3.20	160.80
Extension	–0.38	2.33	–0.89	–5.56	3.97	–1.51	–75.81
Flood index	–0.03	13.23	–0.40	–2.49	1.99	–0.06	–3.00
Drought index	–0.09	–10.25	0.92	5.79	1.66	–0.15	–7.51
Irrigation index	0	0.56	0	0	1.59	0	0
Residual			4.30	42.11		–1.1	–74.48
Actual growth rates			15.92	100		1.99	100

[a]TFP elasticity with respect to each factor is calculated on the basis of coefficients from maize model in Table 18.5.
[b]TFP and factor growth rates are computed by a least square estimate.
[c]Growth rate contributed by each factor is calculated by multiplying factor growth rate (column 2) by elasticity (column 1).
[d]The percentage of total TFP growth explained by each factor is the corresponding figure in column 3, divided by the total growth rate of TFP (which for the period of 1981–90 was 9.45%).

In this chapter, we have focused primarily on the past and marginal effects of research and extension on TFP. If investments in research are reduced, then productivity – according to these results – will also fall. Because future production gains appear to depend more on productivity increases than ever before, China's ability to meet its food economy goals is going to depend heavily on its ability to recover the high productivity gains of earlier years. The negative and significant sign on the dummy variable for the 1990s in the VT equations (Table 18.4, columns 2, 4 and 6, row 6) is cause for concern.

Our results concerning the impact of the international agricultural research system give reason to be optimistic about the future prospects for yield gains through international collaboration, and suggest that China should continue to maintain and strengthen its ties with the rest of the world. In an era of uncertainty concerning future flows of germplasm across national boundaries, China should do all it can to ensure that it can access stocks of genetic material from abroad. The results suggest that by moving into more temperate materials, CIMMYT might be able to increase its contribution to China, though it is unclear if it would be adding value or substituting for alliances that China already has with other countries.

Acknowledgements

We would like to thank the research staff of CCAP for the literally thousands of hours of time that they have put into collecting and cleaning the data and for assistance in computing. We also thank Julian Alston, John Antle, Cristina David, Xinshen Diao, Robert Evenson, Shenggen Fan, Doug Gollin, Erika Meng, Michael Morris, Prabhu Pingali, Carl Pray and Melinda Smale, who have provided insights on the data and methodology. The financial support from CGIAR, the Rockefeller Foundation's Agricultural Science Foundation, and China National Outstanding Youth Science Foundation is gratefully acknowledged.

References

Fan, S. (1997) Production and productivity growth in Chinese agriculture: new measurement and evidence. *Food Policy* 22, 213–228.

Fan, S. and Pardey, P. (1997) Research, productivity, and output growth in Chinese agriculture. *Journal of Development Economics* 53, 115–137.

Hausman, J. (1983) Specification and estimation of simultaneous equation models. In: Griliches, Z. and Intriligator, M. (eds) *Handbook of Economics*. North Holland Press, Amsterdam.

Huang, J.K. and Rozelle, S. (1996) Technological change: rediscovery of the engine of productivity growth in China's rural economy. *Journal of Development Economics* 49, 337–369.

Lin, J.Y. (1990) *Rural Reforms and Agricultural Productivity Growth in China.* Working Paper, Beijing University, Beijing, China.

Lin, J.Y. (1991) The household responsibility system reform and the adoption of hybrid rice in China. *Journal of Development Economics* 36, 353–373.

Lin, J.Y. (1992) Rural reform and agricultural growth in China. *American Economics Review* 82, 34–51.

Pingali, P., Hossein, M. and Gerpacio, R. (1997) *Asian Rice Bowls.* CAB International, Wallingford, UK.

Rosegrant M.W. and Evenson, R.E. (1992) Agricultural productivity and sources of growth in South Asia. *American Journal of Agricultural Economics* 74, 757–761.

Rozelle, S., Huang, J.K. and Rosegrant, M. (1996) How China Will NOT Starve the World. *Choice*, First Quarter 1996, 12–17.

Rozelle, S., Pray, C. and Huang, J. (1998) Agricultural research policy in China: testing the limits of commercialization-led reform. *Contemporary Economic Policy* (1998), 1–20.

Rozelle, S., Huang, J.K., Jin, S.Q. and Hu, R.F. (1999) *Research Investment, Research Impact in China's Rich and Poor Areas.* Working Paper, DARE, University of California, Davis, California.

SPB (State Price Bureau) (1988–1992) *Quanguo Nongchanpin Chengben Shouyi Ziliao Huibian [National Agricultural Production Cost and Revenue Information Summary].* China Price Bureau Press, Beijing.

State Statistical Bureau-ZGTJNJ (1980–1998) *Zhongguo Tongji Nianjian [China Statistical Yearbook].* China Statistical Press, Beijing, China.

Stone, B. (1988) Development in agricultural technology. *China Quarterly* 116.

Stone, B. (1993) Basic agricultural technology under reform. In: Kueh, Y.-Y. and Ash, R.F. (eds) *Economic Trends in Chinese Agriculture: The Impact of Post-Mao Reforms.* Studies on Contemporary China, Clarendon Press, Oxford, pp. 311–359.

Stone, B. and Rozelle, S. (1995) Foodcrop production variability in China, 1931–1985. In: *The School for Oriental and African Studies, Research and Notes Monograph Series*, Vol. 9. London, August 1995.

Wen, G.Z. (1993) Total factor productivity change in China's farming sector: 1952–1989. *Economic Development and Cultural Change* (1993), 1–41.

World Bank (1997) *At China's Table.* Monograph, World Bank, Washington, DC.

Appendix

Mean values of major variables

	Rice model		Wheat model		Maize model	
Variable name	Mean	Standard deviation	Mean	Standard deviation	Mean	Standard deviation
Varietal Turnover1 (VT_1)	2.89	1.15	2.68	1.05	4.23	1.64
Varietal Turnover2 (VT_2)	3.85	1.59	2.22	0.78	3.31	1.28
Extension (million 1985 RMB)	1163	212.5	1176	214	1163	212
Flood index	0.10	0.19	0.06	0.09	0.06	0.04
Drought index	0.11	0.10	0.12	0.11	0.12	0.10
Irrigation index	0.53	0.23	0.44	0.19	0.46	0.13
Research stock (million 1985 RMB)	2950	62.4	302	58.6	295	62
CG contribution	0.16	0.12	0.64	15.05	0.05	0.01
Yield frontier (t ha^{-1})	560	59.0	592.2	59.0	609.5	70.7

Crop Genetic Improvement Impacts on Indian Agriculture 19

J.W. McKinsey and R.E. Evenson

Indian agricultural development experience offers an excellent opportunity to evaluate the impacts of public policies and public investments on agricultural production and productivity. Because its government structure is a federal–state system, certain common policies, particularly international trade and agricultural price policies, affect all regions of the country in a similar way. However, many public investments, notably investments in rural schools, in agricultural extension programmes and in agricultural research programmes, differ from state to state because they are primarily the responsibility of state governments. And, because India is a large country with diversity in climate and soil conditions, there are opportunities to evaluate the effects of these conditions on productivity in agriculture. (See Evenson *et al.* (1999), Azam *et al.* (1991), Dey and Evenson (1991), Rosegrant and Evenson (1992) and Evenson and McKinsey (1991) for prior studies of agricultural research in India and South Asia.)

Crop genetic improvement (CGI) in India has been of major importance to the welfare of Indian farmers and consumers. After Independence in 1947, India embarked on an ambitious programme of economic development. Most of these programmes were codified and organized in a sequence of 5-year plans. In the earliest 5-year plans, industrial development programmes were paramount. However, by the 1960s, Indian planners recognized the magnitude of the increased demand for food grains brought about by a rapidly increasing population. This population boom was itself the result of successful implementation of public and private health measures leading to reductions in death rates. Population increased rapidly, and this expansion has continued to the end of the century, resulting in an increase in India's population from 350 million in 1950 to 1 billion in 2000.

As Indian policy makers struggled with the problems of supplying food to a burgeoning population, they were presented with exceptional CGI opportunities in both wheat and rice. For wheat, these opportunities were in the form of high-yielding semi-dwarf wheat varieties that originated in Mexico in the breeding programmes led by Norman Borlaug. (This programme was originally a Special Project of the Rockefeller Foundation and then became part of CIMMYT, the International Maize and Wheat Improvement Centre.) These high-yielding varieties presented an opportunity in two ways. Firstly, seed imports allowed farmers to directly plant and multiply the varieties. Secondly, the Indian Council of Agricultural Research (ICAR) and state plant breeding programmes utilized these imported varieties as germplasm (parent materials) in their own breeding programmes.

Similar opportunities for rice were presented in the form of high-yielding semi-dwarf plant types from IRRI, the International Rice Research Institute, in the Philippines. These varieties were also of germplasmic value to Indian rice breeding programmes. India was also a central location for the *indica* × *japonica* breeding programme of the Food and Agricultural Organization of the United Nations (FAO) where early semi-dwarf rice varieties were developed.

The success of both the wheat and rice varieties in India in the late 1960s and early 1970s was popularized as a 'Green Revolution'. However, this popular version of the role of CGI in India is far from complete. During the 'post-Green Revolution' period, from the late 1970s to 2000, the Indian population doubled. This doubling in population was accompanied by a more than doubling of food production, and CGI actually played a larger and more important role after 1975 than it did in the 1968–1975 period.

In this chapter we provide an evaluation of the role of CGI, not just in wheat and rice, but in maize, sorghum and pearl millet as well. We attempt to take advantage of the diversity in public investment in CGI research programmes and in schooling and extension programmes in different states. We evaluate CGI impacts on crop yield and area and on irrigation investments.

In the second part of the chapter we develop the statistical model for our estimates and summarize the variables used in the study. In the third part, we summarize our estimates. The final part provides a policy discussion of the estimates.

Statistical Specification

Variables

Our units of observation are Districts (271 districts in 13 states) for the years 1965/66 to 1994/95. Variables defined for these observations include variables that are determined by farmers and are thus endogenous in our structure; variables determined by state governments (with federal assistance); and variables determined by natural conditions, which do not vary over time.

For each of the five crops studied – rice, wheat, maize, sorghum and pearl millet – we have four endogenous variables that characterize farmers' decisions. Thus, we require a four-equation system (which we estimate using the three-stage least squares (3SLS) method).

Table 19.1 provides a short definition of the variables in the study. Appendix 19.1 provides detail as to sources and more detailed description of the variables. Note that we are treating four variables as endogenous in the model: the share of area planted to the crop, the adoption of modern varieties, irrigation investment, and yields.

Exogenous variables include public policy variables, price variables, weather variables, climate variables, edaphic variables and, for rice, genetic resource content variables. In addition to the variables listed in Table 19.1, the 1965 pre-Green Revolution values of the area share, irrigation and yield variables are included in statistical specifications.

Table 19.2 provides means of variables for beginning and ending periods for important variables.

Statistical specification

Our intention in this exercise is to obtain 'structural' estimates of relationships between endogenous variables. In particular, we seek to estimate the impact of MV (modern variety) adoption on irrigation investment and of irrigation investment on MV adoption. We also seek to investigate the impacts of both irrigation investment and MV adoption on area planted to the crop. Finally, we seek to estimate the impacts of area share, irrigation and MV adoption on yield for each crop.

To estimate this structural model we require variable exclusion restrictions. Table 19.3 summarizes these exclusion restrictions. Appendix 19.2 reports statistical tests of these exclusion restrictions.

Table 19.1. Definition of variables used in the study.

Variable	Definition
Classification of endogenous variables	
AREA SHARE	Measured as the share of cropped area planted to the crop in the District. The share specification was required to make the variable independent of District size and commensurate with other variables in the model.
MV ADOPTION $\ln\left(\frac{mv}{(1-mv)}\right)$	Measured as the percentage of the crop planted to 'modern' high-yielding varieties (i.e. varieties released after 1964). This variable measures the displacement of traditional (pre-1965) varieties by modern varieties. It does not measure varietal turnover, i.e. the displacement of older modern varieties by newer modern varieties.
IRRIGATION	Measured as the ratio of gross irrigated area to gross cropped area in the District for the crop.
YIELD	Measured in kilogrammes harvested per hectare. In Districts where the crop is grown in more than one season, this is the yield for the multiple cropped area.
Classification of exogenous variables	
Government policy variables (P)	
EXTENSION	Index of extension services supplied to the state farmers.
STATE RES	Cumulated research stock – public agricultural research expenditures (see Data Appendix).
MARKET	Number of regulated markets in the State.
PRIVATE RES	Private R&D in the chemical and seed industries in India.
Price variables (P)	
P(W/F)	Daily agricultural wages / Price of chemical fertilizers.
F/GCA	Fertilizer consumption (kg) per hectare of gross cropped area (this variable is not crop-specific).
Weather variables (W)	
JUNERAIN	June rainfall in millimetres.
JUARAIN	July and August rainfall in millimetres.
Climate variables (C)	
DROUGHT	Dummy of District classified as drought-prone (ICAR).
WT	Normal (30-year average) Winter (October–January) Temperature (C).
ST	Normal (30-year average) Summer (April, July) Temperature (C).
WR	Normal (30-year average) Winter (October–January) Rainfall (mm).
SR	Normal (30-year average) Summer (April, July) Rainfall (mm).
Edaphic variables (E)	
STRORIE	Index of organic matter content.
DMS_	Soil type dummy variable 2–19.
DMSLP_	Topsoil depth dummies 1–3.
AGROB_	ICAR agrobiological region dummy variables 1–7.
JULYSTEMP	Normal soil temperature in July.
IARC content variables (rice only)	
AVNOLR	Average number of land acres in adopted varieties in 1984.
IRRIX	Percentage of adopted varieties from an IRRI cross.
IRRIP	Percentage of adopted varieties based on an IRRI-crossed parent.
IRRIA	Percentage of adopted varieties based on an IRRI-crossed non-parent ancestor.

Table 19.2. Variable means: key variables 1965/66 and 1994/95

	Rice		Wheat		Maize		Sorghum		Pearl millets	
	1965	1994	1965	1994	1965	1994	1965	1994	1965	1994
AREA SHARE	0.23	0.24	0.10	0.15	0.04	0.04	0.10	0.05	0.07	0.04
MV ADOPTION	–3.9	0.69	–3.9	2.21	–3.9	–1.09	–3.9	–0.47	–3.9	0.24
IRRIGATION	0.43	0.52	0.63	0.83	0.24	0.28	0.07	0.24	0.05	0.36
YIELD	0.85	1.79	0.84	2.06	0.93	1.62	0.58	0.84	0.49	0.79
EXTENSION	2.14	43								
STATE RES	15.1	246								
MARKET	0	0.039								
PRIVATE RES	6.94	1,367								
P (W/F)	0.0016	0.0037								
F/GCA	2,973	13,000								

Table 19.3. Variable exclusions.

Equation	Excluded endogenous variables	Excluded exogenous variables
AREA SHARE	YIELD	EXTENSION; F/CGA; JUNERAIN; JUARAIN.
MV ADOPTION	YIELD; AREA SHARE.	DMSLP1; DMSLP2, DMSLP3; F/GCA; JUNERAIN; JUARAIN.
IRRIGATION	YIELD; AREA SHARE.	DMSLP1; DMSLP2; DMSLP3; F/GCA; JUNERAIN; JUARAIN; MARKET; EXTENSION.
YIELD		DMSLP1; DMSLP2; DMSLP3; JULYSTEMP; DROUGHT; MARKET

The exclusion restrictions are based on the following arguments. First, weather during the growing season, as measured by JUNERAIN and JUARAIN (see Table 19.3), is realized only after planting decisions are made. Thus, these weather variables should be excluded from the AREA SHARE, MV ADOPTION and IRRIGATION equations. Decisions regarding these variables are made based on normal climate (30-year-mean temperature and rainfall variables) are included in all equations. The weather variables, of course, do belong in the YIELD equations.

We also excluded the F/GCA variable from the first three equations, but not the YIELD equation, on the grounds that the wage/fertilizer price variable (P(W/F)) was effectively controlling for price issues. The chief purpose of the variable in the YIELD equations is to provide a quantitative index of fertilizer use to enable 'growth accounting'. McGuirk and Mundlak (1991) argue that Indian policy effectively made the F/CGA variable an exogenous policy determined variable.

We included the slope dummy variables only in the AREA SHARE equation, on the grounds that this was the most important role played in crop allocation. Since AREA SHARE is in the YIELD equation, the effects of slope are conveyed through that variable.

We excluded EXTENSION from the AREA SHARE and IRRIGATION equations on the grounds that its major impact was conveyed through MV ADOPTION effects. The EXTENSION variable was included in the MV ADOPTION and YIELD equations. We utilized a similar argument for excluding the MARKET variable from the IRRIGATION and YIELD equations.

We acknowledge that there is an element of arbitrariness in the non-weather restrictions and these are discussed further in Appendix 19.2, where tests are reported.

We also need to discuss the rationale for including the 1965 pre-Green Revolution levels of variables in the AREA SHARE, IRRIGATION and YIELD equations. The model as specified is designed to measure the MV impact process. MVs were introduced only after 1965. We could use 'District fixed effects' to control for unobserved factors affecting the relevant varieties. The fixed effects procedure effectively takes out the mean effects of all variables at the District level. This does control for unobserved soil and climate effects, but has the disadvantage that it does not allow us to take full advantage of the 'natural experiment' associated with the Green Revolution. The inclusion of the pre-Green Revolution levels does provide control for the unobserved and unmeasured factors determining these starting levels. Thus, in effect, we are analysing determinants of changes from these levels.

Finally, a note on the MV ADOPTION variable. The logistic 'S'-shaped form is widely used in diffusion studies (see Griliches (1957) for the original work). The logic behind the form is that in the early stages of diffusion (i.e. when MV adoption is low) the effort required (as reflected in policy variables) to achieve a percentage increase in adoption is greater than it is when approximately half of the farmers are adopting. This is because farmers have more neighbours experimenting with MV and observation is easier. Conversely, as MV diffusion approaches a ceiling (in our case we set this at 98% and set the minimum at 2%), adoption increments become more difficult to achieve because the last adopters may have conditions associated with lower gains from adoption.

However, we have another reason for adopting the form. The S-shaped form essentially captures two phenomena. As farmers convert land from traditional varieties (TV) to modern varieties, MV/TV conversion occurs first where the MV/TV advantage is highest. This phenomenon indicates that the MV/TV advantage declines as MV levels rise. The second phenomenon is MV/MV conversion, i.e. the replacement of early MVs by later generations of MVs. MV/MV conversion rises as MV/TV conversion increases. The S-shaped curve implicitly combines these two effects.

We have also included pre-Green Revolution AREA SHARE, IRRIGATION and YIELD variables in the MV ADOPTION equation (see below).

Parameter Estimates

Parameter estimates for policy-related variables are reported in Tables 19.4–19.8. Full estimates of all parameters are available from the authors.

MV adoption: origin and rate analysis

We begin with the MV adoption estimates. We report two sets of analyses for MV adoption. The first is an OLS analysis of the 'origin' or year of first adoption (reported in Table 19.4). The second is the 3SLS MV adoption equation in the system (Table 19.5).

The year of first adoption is defined as the year when the percentage of area planted to MVs in the District exceeded 0.02. Table 19.4 reports a simple OLS set of estimates for this variable.

For three of the five crops, significant negative impacts of the 1965 area share are estimated. This is consistent with the findings of Griliches (1957) regarding hybrid maize. The larger the market, the earlier the development of MVs for that market. The 1965 irrigation shares were not significantly related to the year of first adoption.

For wheat and rice, the two major crops, the 1965 yield level was inversely related to year of first adoption. This is also consistent with the findings of Griliches for hybrid maize. Griliches (1957) argues that the economic gains from hybrid (as opposed to open-pollinated) maize were proportional to the initial yield levels.

The level of public (state research investment) and private sector research investment generally reduces in the year of first adoption, but is only significant for sorghum.

The availability of IRRI germplasm in rice varieties is associated with earlier adoption. This is an indication that IARC crosses and parental germplasm provide earlier access to MVs.

Table 19.5 reports the 3SLS estimate for the MV adoption equation in the system. These estimates are effectively a combination of origin and rate of MV adoption (although they are dominated by the rate of MV adoption). We have included the 1965 levels of area share and yield in this specification.

Table 19.4. Origin or first adoption (MV = 0.02) estimates.

Dependent variable: year in which MV first equal to 0.02 or greater in District

	Rice	Wheat	Maize	Sorghum	Pearl millet
Observations	233	230	196	166	0.39
R^2	0.690	0.354	0.374	0.469	0.460
Independent variables					
AREA SHARE	–9.80	1.16	–33.52	3.04	–22.43
	(3.59)	(0.27)	(5.31)	(0.53)	(2.99)
IRR SHARE	–1.50	0.16	0.668	–2.35	1.45
	(0.88)	(0.64)	(0.35)	(0.87)	(0.29)
YIELD (1965)	–3.61	–1.74	0.71	–0.49	–0.19
	(2.19)	(1.59)	(0.81)	(0.36)	(0.09)
STATE RES	0.011	–0.009	0.002	–0.002	0.005
	(1.22)	(1.24)	(0.18)	(0.29)	(0.41)
PRIVAE RES	–0.0013	0.001	–0.0003	–0.0026	–0.0003
	(1.33)	(0.14)	(0.33)	(2.24)	(0.20)
AVNOLR	10.45				
	(4.75)				
IRRIX	–77.44				
	(2.32)				
IRRIP	–114.84				
	(5.55)				
IRRIA	–226.75				
	(4.29)				

We find mixed evidence that these variables affect the broader MV diffusion process. Higher area shares for rice and pearl millet do contribute to faster MV adoption. Higher initial yields of rice, maize and pearl millet do this as well.

Irrigation investments (note that this is an endogenous variable in the system) clearly enable faster MV adoption in all crops except pearl millet.

Public research programmes have mixed contributions to MV adoption (significantly positive only for pearl millet). Private sector R&D, on the other hand, contributes to faster MV adoption in all crops. Public sector extension programmes significantly increase MV adoption in rice, wheat and pearl millet.

Interestingly, market development facilitates MV adoption for all crops in a highly significant manner.

Prices also drive MV adoption. Higher agricultural wages relative to fertilizer prices produce faster MV adoption. Actually, this price variable requires careful interpretation because most cross-section variation in the variable is from agricultural wages. The price of fertilizer tends to be an administered price that is similar in most Districts.

Table 19.5. MV diffusion specification 3SLS system estimates.

Dependent variable: $\ln\left(\frac{mv}{(1-mv)}\right)$

	Rice	Wheat	Maize	Sorghum	Pearl millets
Observations	4880	4486	2457	2241	2365
R^2 (system)	0.617	0.353	0.562	0.514	0.532
Independent variables					
AREA SHARE (1965)	0.303	−0.839	−0.441	−0.515	3.687
	(2.76)	(1.84)	(0.91)	(1.13)	(6.20)
IRR ratio	2.947	5.350	2.384	−9.60	4.117
	(24.73)	(19.69)	(8.68)	(1.53)	(7.76)
YIELD LAND (1965)	0.163	−0.317	0.895	1.972	−0.568
	(2.34)	(2.68)	(8.10)	(15.39)	(1.86)
STATE RES	0.0003	−0.0007	0.00003	−0.00139	0.00110
	(1.00)	(1.37)	(0.06)	(6.09)	(2.11)
PRIVATE RES	0.00004	0.000008	0.000018	0.000015	0.000019
	(3.68)	(0.93)	(1.62)	(3.55)	(0.68)
MARKET	0.867	11.63	16.08	15.22	10.01
	(1.66)	(7.52)	(8.51)	(11.12)	(4.04)
EXTENSION	0.0323	0.0171	−0.0016	0.0549	−0.0127
	(19.15)	(10.45)	(0.71)	(17.19)	(4.28)
P(W/F)	197.25	204.51	445.58	197.10	580.25
	(7.20)	(6.24)	(11.38)	(7.22)	(13.49)
AVNOLR	−0.0353				
	(4.56)				
IRRIX	2.81				
	(3.18)				
IRRIP	8.87				
	(20.02)				
IRRIA	12.85				
	(14.69)				

Irrigation investment estimates

Table 19.6 reports estimates from the major variables in the 3SLS irrigation system equation.

The dominant factor driving irrigation (most of which was tube well irrigation by farmers and groups of farmers) was MV adoption. The availability of MVs clearly raised the productivity of irrigation investments.

Research programmes (public and private) did not have important effects on irrigation investment (except for public sector research on wheat) over and above these contributions to MV adoption.

Relative wages had positive effects on irrigation investment in wheat, sorghum and pearl millet.

IARC germplasm had negative, direct impacts on irrigation investment (but had strong indirect impacts through MV adoption).

Table 19.6. Irrigation investment specifications 3SLS system estimates.

Dependent variable: irrigation areas/gross cropped area

	Rice	Wheat	Maize	Sorghum	Pearl millet
Observations	4880	4486	2457	2241	2365
R^2	0.789	0.537	0.709	0.633	0.767
Independent variables					
MV ADOPTION	0.091	0.077	0.057	−0.002	0.013
	(19.65)	(24.30)	(13.70)	(1.39)	(6.72)
IRR (GCA:1965)	0.656	0.318	0.691	0.568	1.072
	(42.15)	(26.68)	(32.96)	(15.95)	(54.02)
STATE RES	−0.00012	0.00016	0.00003	−0.00005	0.00003
	(3.77)	(3.82)	(0.80)	(5.72)	(1.62)
PRIVATE RES	−2.78E−6	−2.13E−7	−9.81E−7	9.9E−8	−1.1E−6
	(3.19)	(0.28)	(1.06)	(0.46)	(1.11)
P(W/F)	−15.316	3.558	−17.928	3.626	4.668
	(4.57)	(1.14)	(4.65)	(2.70)	(2.75)
AVNOLR	0.0037				
	(4.23)				
IRRIX	0.0405				
	(0.47)				
IRRIP	−0.6620				
	(9.69)				
IRRIA	−1.049				
	(8.23)				

Area share estimates

Table 19.7 reports estimates for the area share equations in the system.

For rice and wheat, MV adoption produces increased area shares. For other crops, shares are reduced. This is consistent with the general tradeoffs between crops and with international markets. We expect farmers to expand shares in the crops with highest MV advantages (rice and wheat) and to contract shares in crops with lower MV advantages. In addition, demand conditions affect shares through price effects. Non-traded crops (maize, millet and sorghum) have local markets and these markets can be served with reduced areas planted.

Irrigation investments reduce market shares in rice and wheat, reflecting the productivity effects of irrigation.

Public sector research generally reduces area shares. Private sector R&D has mixed impacts. Improved markets produce higher shares except in wheat.

Relative wages have mixed effects on shares.

The impacts of policy variables on area shares are complex because farmers are making area choices over more than one crop and because of the 'locality' of markets. Our chief interest in the share estimates is to assess impacts on yield.

Table 19.7. Area specification 3SLS system estimates.

Dependent variable: share of gross cropped area planted to crop

	Rice	Wheat	Maize	Sorghum	Pearl millet
Observations	4880	4486	2457	2241	2365
R^2 (System)	0.934	0.697	0.903	0.848	0.851
Independent variables					
MV ADOPTION	0.0084	0.0319	−0.0018	−0.0180	−0.0162
	(3.41)	(21.54)	(1.26)	(8.32)	(13.03)
IRRIGATION	−0.022	−0.131	0.0051	0.0139	0.0315
	(2.55)	(13.66)	(1.08)	(0.44)	(3.18)
AREA SHARE (1965)	0.907	0.818	0.867	0.717	0.865
	(128.84)	(52.67)	(107.72)	(50.72)	(64.56)
STATE RES	−0.00007	−0.00002	6.34E-6	−0.00004	−0.00004
	(5.71)	(1.45)	(0.93)	(3.15)	(4.68)
PRIVATE RES	4.01E−7	−3.13E−7	1.89E−7	4.87E−7	−5.98E−7
	(1.35)	(0.13)	(1.20)	(2.05)	(1.10)
MARKET	0.265	−0.612	0.078	0.382	0.138
	(7.91)	(10.53)	(2.11)	(4.28)	(2.65)
P(W/F)	13.439	−5.049	−2.694	−9.587	3.351
	(11.12)	(4.85)	(3.79)	(6.36)	(3.47)
AVNOLR	0.0002				
	(0.52)				
IRRIX	0.319				
	(9.87)				
IRRIP	0.060				
	(2.07)				
IRRIA	−0.256				
	(5.52)				

Yield estimates

Table 19.8 reports estimates for the yield equation. Crop yields are our productivity measure and we are especially interested in the impact of MV adoption and yields.

Our estimates clearly show highly significant impacts of MVs on crop yields for all crops.

Irrigation investments have positive impacts on yields, except for rice.

Area share impacts may be positive or negative, reflecting the relative productivity of lands associated with area expansion and contractions. These are positive for wheat and maize. The direct impacts of public sector research programmes on yields are positive for rice and wheat. Private sector R&D has low direct effects on yields. Direct extension effects are negative (but indirect effects through MV adoption are positive) (see below).

Table 19.8. Yield specification: 3SLS system estimates.

Dependent variable: Yield (kg ha^{-1})

	Rice	Wheat	Maize	Sorghum	Pearl millet
Observations	4880	4486	2457	2241	2365
R^2 (system)	0.683	0.791	0.521	0.483	0.483
Independent variables					
MV ADOPTION	0.208	0.104	0.145	0.191	0.075
	(8.55)	(6.28)	(7.47)	(5.70)	(6.10)
IRRIGATION	−0.1922	0.316	0.287	0.314	0.8452
	(2.66)	(5.33)	(3.75)	(1.53)	(0.94)
AREA SHARE	−0.292	0.143	0.503	−0.372	−0.347
	(5.00)	(8.81)	(3.30)	(3.78)	(3.11)
YIELD (1965)	0.693	0.649	0.170	0.677	0.678
	(22.81)	(16.91)	(4.69)	(7.79)	(15.40)
STATE RES	0.00033	0.0008	−0.00013	−3.34E−6	0.00003
	(3.93)	(8.81)	(0.96)	(0.03)	(0.37)
PRIVATE RES	−6.5E−6	−5.89E−7	1.76E−6	−1.59E−10	2.51E−6
	(2.98)	(0.40)	(0.62)	(0.11)	(0.68)
EXTENSION	−0.0027	−0.0012	0.0022	−0.0045	0.00634
	(3.63)	(2.52)	(2.65)	(2.90)	(9.51)
P(W/F)	82.13	83.45	5.29	−4.99	−11.100
	(9.45)	(13.11)	(0.41)	(0.43)	(1.42)
F/GGA	5.82E−6	0.00003	0.00002	−2.42E−6	−0.00002
	(1.62)	(6.49)	(2.79)	(0.41)	(1.84)
JUNERAIN	0.00022	−0.00009	0.00056	0.00008	0.00031
	(4.05)	(0.92)	(4.81)	(1.08)	(4.30)
JUARAIN	0.000087	0.000001	−0.00019	−0.00063	0.00017
	(3.04)	(0.42)	(4.00)	(2.03)	(6.37)
AVNOLR	0.0008				
	(0.39)				
IRRIX	0.863				
	(3.60)				
IRRIP	−0.654				
	(0.58)				
IRRIA	−0.833				
	(2.13)				

Relative prices affect crops differently. Higher wages and lower fertilizer prices increase rice and wheat yields. These crops are the major users of fertilizer. The F/GCA variable is also designed to pick up fertilizer impacts. It does share impacts for the fertilizer using crops (sorghum and pearl millet are not generally users of fertilizer).

The rainfall variables appear to be controlling for weather effects in the summer crops (wheat is produced in the winter).

Policy Implications

In order to facilitate discussion of the estimates in Tables 19.4–19.8 it is useful to compute impacts of the policy variables on crop productivity. These are reported in Table 19.9 for each crop and for the area-share-weighted total for the five crops. For comparisons, the 1965 and 1994 crop yields means and area shares are reported in Table 19.9.

The computed elasticities are based on both the Direct impacts on crop yields as reported in Table 19.8 and the Indirect impacts through impacts on MV adoption, irrigation investment and area shares. For comparison purposes, the MV-related impacts are reported for the research, extension and markets variables.

The MV impacts on productivity are central to the chapter. These are reported for 98% adoption and for the actual adoption levels achieved from 1965/66 to 1994/95. For the 1994–1965 calculations, productivity impacts are the product of the change in the policy variable times the estimated direct and indirect impacts on yield.

We note first that all MV impacts are estimated to be high. For all crops combined (weighted by area), full adoption (98%) of MVs is estimated to produce a yield increase of 1.27 t ha^{-1}. This is a large percentage of the 1965 yields. The actual 1994–1965 MV impact was 0.68 t ha^{-1}. The sum of the impacts from all sources is 1 t ha^{-1} for all crops combined. This exceeds the actual yield increase (0.85 t ha^{-1}). The MV share of total impacts by crops ranges from 41% to 84% and for all crops the MV impacts share of total impacts is 67%. This is a large contribution to productivity (see Chapter 21 for more comparisons).

Table 19.9. Productivity impacts.

						All crops	
	Rice	Wheat	Maize	Sorghum	Pearl millet	Total	MV related
Yields 1965	0.85	0.84	0.93	0.58	0.49	0.75	
Yields 1994	1.80	2.06	1.62	0.85	0.80	1.60	
Area share 1965	0.23	0.10	0.04	0.10	0.07		
Area share 1994	0.24	0.15	0.04	0.05	0.04		
MV adoption							
Full adoption (98%)	1.31	1.00	1.31	1.57	0.86	1.24	
94–63 adoption	0.68	0.84	0.46	0.74	0.41	0.68	
State research (94–63)	0.092	0.179	−0.025	−0.034	0.030	0.080	0.032
Private research (94–63)	0.003	0.0003	0.005	0.004	0.007	0.003	0.006
Extension (94–63)	0.165	0.023	0.0602	0.168	0.213	0.129	0.93
Markets (94–63)	0.004	0.021	0.092	0.107	0.027	0.032	0.269
Fertilizer	0.073	0.379	0.253	−0.022	−0.022	0.147	

The public (state) research system produced the MVs in collaboration with the relevant international centres (see Chapter 21). State research, however, contributed to productivity over and above the contribution through MVs, particularly in rice and wheat. The private-sector research impacts, by contrast, are relatively small.

The productivity impacts of state research, private research, extension and market development are all positive and, except for private-sector research, quantitatively important. For these impacts, calculations of the indirect impacts through MV adoption accelerations are reported for aggregate crops. This calculation is designed to test the 'Transformation' hypothesis. This hypothesis was first stated by T.W. Schultz in his classic book, *Transforming Traditional Agriculture*. Schultz claimed that farmers using traditional agricultural technology were relatively efficient. Unless new technology, particularly in the form of MVs, was introduced, activities such as extension and market development would not be expected to have large impacts on already efficient production.

The MV-related impacts, i.e. those due to MV acceleration, would be expected to be high if the Schultz hypothesis is correct. The presumption is that extension activities and market development activities are not themselves transforming, while MVs are.

The MV-related impact calculations show that state research programmes may have transforming power independently of MVs (where 40% of the total impact is through MV acceleration), but that for extension and market development the impacts are through MV acceleration. Actually the MV acceleration impacts exceed the total impacts for both of these.

The contribution of fertilizer use to yield improvement was partial for rice, wheat and maize, but non-existent for sorghum and pearl millet. This generally occurs with evidence on fertilizer use. (We have not included the P(W/F) effects because they affect the use of other inputs.)

The general picture thus obtained from these impact calculations is one where CGI programmes have contributed to crop productivity in a major way. Approximately 60–70% of total impacts are estimated to be due to MVs; this is also the case when only direct impacts are considered. It should be noted that this calculation is not the traditional 'total factor productivity' (TFP) decomposition, because we have not considered labour and power inputs.

It is probably the case that labour per unit land and machinery per unit land do account for significant yield impacts as well (these factors probably account for 25% or so of output growth; Evenson and Kislev, 1975). Thus the MV share of TFP growth might be roughly 50–55% when these factors are considered.

None the less, these calculations are of policy interest. The fact that extension effects are large and due heavily to MV adoption acceleration is important. It supports the general proposition that extension is productive when the extension service has new technology to extend. Similarly the improvement of markets is important primarily because this facilitates MV adoption or diffusion.

These findings confirm the proposition that MVs did contribute to Indian agricultural productivity. The MV adoption linkages to extension and markets as well as to irrigation investment also suggest that MVs were 'transforming' events. Had MVs not been delivered to Indian farmers, the MV diffusion process would not have taken place. Thus, without MVs the contributions of extension, markets and irrigation (which of course is costly) would have been modest.

References

Azam, Q.T., Bloom, E.A. and Evenson, R.E. (1991) *Agricultural Research Productivity in Pakistan.* Economic Growth Center Discussion Paper No. 644. Economic Growth Center, Yale University, New Haven, Connecticut.

Dey, M.M. and Evenson, R.E. (1991) *The Economic Impact of Rice Research in Bangladesh.* Bangladesh Rice Research Institute, Gazipur, Bangladesh; International Rice Research Institute, Los Baños, Philippines; and Bangladesh Agricultural Research Council, Dhaka, Bangladesh.

Evenson, R.E. (1968) The Contribution of Agricultural Research and Extension to Agricultural Production. PhD Dissertation, University of Chicago.

Evenson, R.E. (1984) Benefits and obstacles to appropriate agricultural technology. In: Heston, A.W. and Pack, H. (eds) *Technology Transfer: New Issues, New Analysis. The Annals of the American Academy of Political and Social Science* 458, November 1981. [Also Chapter 24 in *Agricultural Development in Third World*, Eicher, C.K. and Staatz, J.M. (eds) (1984). Johns Hopkins University Press, Baltimore.]

Evenson, R.E. and Avila, A.F.D. (1995) Productivity change and technology transfer in the Brazilian grain sector. *Revista de Economia Rural* 34, no. 2.

Evenson, R.E. and McKinsey, J. (1991) Research, extension, infrastructure, and productivity change in Indian agriculture. In: Evenson, R.E. and Pray, C.E. (eds) *Research and Productivity in Asian Agriculture.* Cornell University Press, Ithaca, New York.

Evenson, R.E. and Kislev, Y. (1975) *Agricultural Research and Productivity.* Yale University Press, New Haven, Connecticut.

Evenson, R.E., Pray, C.E. and Rosengrant, M.W. (1999) *Agricultural Research and Productivity Growth in India.* IFPRI, Washington, DC.

Griliches, Z. (1957) Hybrid corn: an exploration in the economics of technological change. *Econometrica* 25(4), 501–522.

McGuirk, A. and Mundlak, Y. (1991) *Incentives and Constraints in the Transformation of Punjab Agriculture.* Research Report 1987. IFPRI, Washington, DC.

McKinsey, J.W. (1999) Climate change and productivity in Indian agriculture. PhD dissertation, Yale University, New Haven, Connecticut.

Rosengrant, M.W. and Evenson, R.E. (1992) Agricultural productivity and sources of growth in South Asia. *American Journal of Agricultural Economics* 74, 757–791.

Appendix 19.1

Introduction

This Appendix describes the variables used in this study. These variables come from two distinct data sets. One, which is sometimes called the 'original' data set, was created between 1980 and 1990, and has been used in numerous studies of production and productivity in Indian agriculture. This first data set was updated in 1996 and again in 1999. This data set has been collected, updated and maintained primarily by researchers at ICRISAT and at Yale University, including the authors of this country study. The second data set, created in 1996 by McKinsey, added edaphic and climatic variables. A complete description of the data set is available in McKinsey (1999).

Geographical coverage

The data set covers nearly all the districts within 13 of the states of India: Andhra Pradesh, Madhya Pradesh, Rajasthan, Bihar, Maharashtra, Tamil Nadu, Gujarat, Orissa, Uttar Pradesh, Haryana, Punjab, West Bengal and Karnataka. The major agricultural states which are absent from the data set are Kerala, at the southern tip of the subcontinent, and the eastern state of Assam. The data have been adjusted to account for numerous changes in district boundaries and definitions over the time period covered. The data set contains observations for each of the variables for the agricultural years 1966/67 through to 1994/95.

Crop coverage

The data set used in this country study contains data pertaining to five crops: rice, wheat, maize, sorghum and pearl millet. For each of these five crops the data set includes information on: area planted, area irrigated, HYV area, and share of total cropped area. The primary sources for these data are various reports from the Government of India and individual states.

For each of these five crops the data set also includes crop-specific research information. In particular, the data set includes district-level measures of research on each crop. The construction of these variables is described below.

For rice, but not for the other crops, the data set also includes variables measuring the origin and content of released modern varieties.

Non-crop-specific inputs

Institutional indicators

The yield equations include the ratio of the annual labour cost to the nitrogen fertilizer price. Agricultural wages are obtained from *Agricultural Wages in India*, published by the Directorate of Economics and Statistics. The fertilizer data source is *Fertilizer Statistics*, published annually by the Fertilizer Association of India.

Agro-climatic inputs:

1. **Soil:** Two sets of edaphic (soil-based) variables appear in the data: a set of soil type dummies, and a set of slope dummies. The soil type dummies take the value of 1 for each of the two soil types most prevalent in the district, as displayed in detailed soil maps published in S.P. Raychaudhuri *et al.*'s, *Soils of India* (1963). Soils are classified according to traditional Indian soil types. The slope dummy variables, obtained from Plates 44–49 of the *National Atlas of India*, denote the predominant slope of land in each district.
2. **Water:** Data relating to irrigation are reported in several forms: area irrigated by source (e.g. by canal or tank or tube well), area irrigated under certain crops, or total areas irrigated. Rainfall is measured daily in most districts in India at so-called 'meteorological observatories' established by the Indian Meteorological Department. Monthly sub-divisional data are published in a number of sources, including *Agricultural Situation in India*. Annual sub-divisional data are reprinted in many sources, most conveniently in *Fertilizer Statistics*. District-level (that is, non-aggregated) data are also published in some States' *Crop and Season Reports*, *Statistical Abstracts*, and in some specialized meteorological publications such as the occasional *Climatological Tables of Observatories in India*; a number of States augment the Indian Meteorological Department's data collection (and publication) with their own data.

The data set contains seven rainfall variables: rainfall in June (at the beginning of the monsoon in most States), and rainfall in July–August (the remainder of the monsoon in most parts of India). The other five rainfall variables measure long-term averages rather than actual annual variation. One denotes those districts designated as 'drought-prone' by the ICAR. The remaining four rainfall variables measure the normal, or 30-year average, rainfall in each district in the months of January, April, July and October. These data for most districts in India are available in the Indian Meteorological Department's *Climatological Tables of Meteorological Observatories in India*. Normal rainfall data for the other

districts was obtained from the Indian Meteorological Department's *Monthly and Annual Rainfall and Number of Rainy Days, 1901–1950.*
3. Temperature: three sets of temperature variables are used in this study: two measure air temperature and the third measures soil temperature. All of the temperature variables are climate variables, representing 30-year averages; the two sets of air temperature variables are measured again in the months of January, April, July and October, while the soil temperature variable is measured only in July. Most of the air temperature data are again obtained from the Indian Meteorological Department's *Climatological Tables of Meteorological Observatories in India*; for some districts the data was obtained from the Indian Meteorological Department's *Agro-climatic Atlas of India*. The data set also includes a measure of the average maximum soil temperature at a depth of 5 cm.
4. Agro-climatic Indices: the data include a 'soil rating index', based on an indexing scheme proposed by Storie in 1959. The data used in this study were obtained from Shome and Raychaudhuri, who modified Storie's scheme slightly. This scheme reflects three soil characteristics: soil profile, topography and climatic suitability (i.e. salinity, stoniness, tendency to erode). In addition, there are seven dummy variables reflecting the agro-climatic regions of India, as proposed by Papadakis.

Public sector inputs

One of the most important inputs into agriculture provided by the public sector is research results. Research activities are undertaken by all of the states as well as by numerous Central schemes and projects, focusing on practically every crop grown in India as well as many inputs and all of the basic agricultural sciences. The research variables are based on three sets of data. First is the indigenous State agricultural research expenditures series, covering the years 1953 through to 1971, which was reported in Mohan *et al.* (1973). Second is a data set which contains the number of articles reporting research results which were abstracted in *Indian Science Abstracts* from 1950 to 1979. This data set provides crop-specific and State-specific data measuring the output of the research activity. Third is recent state budget information regarding research spending, especially at the State Agricultural Universities during the late 1970s and the 1980s. These three were combined to create commodity-specific expenditures data series for each of the states from 1950 to 1983. In addition, for each state, there is a 'general' expenditures data series.

For each State and each commodity a research 'stock' variable was then defined by cumulating past research activity utilizing several patterns of time-shape 'inverted-V' weights as first used in Evenson (1968). Finally, this variable was weighted by the share of the crop in the total value of output, summed across districts, and by the gross cropped area planted to that crop in the state.

The data set also includes a variable measuring private research activity, which has increased in importance markedly during the past two decades. This variable is based on data collected by Prof. Carl Pray (personal commumication) measuring research spending by private firms in the fertilizer and related chemical input industries. From this expenditure data, a research stock variable was constructed using a linear 5-year lag structure with no decay. From the private research stock was then created a variable measuring the *local contribution* (or potential) of this private research knowledge within each district, by multiplying the year's stock of fertilizer research items by the district's input share of fertilizer.

The extension variable is based on three sets of information. The first is data measuring the size of the extension service staff in 1975, 1980, 1983 and 1986 in each state, based on surveys by the World Bank. The second is the number of villages in each state. And the third is data published in various years' annual *Reports* of the Department of Community Development of the Ministry of Agriculture, covering the years 1955–1972. The staffing data were interpolated to obtain estimates for the years 1976–1979, 1981 and 1982. Then the staffing data were divided by the number of villages to obtain a measure of the number of extension workers per 100 villages, interpreted as an indicator of extension presence. This variable was then extended backwards, from 1975 to 1956.

Reference

Mohan, R., Jha, D. and Evenson, R. (1973) The Indian agricultural research system. *Economic and Politcal Weekly* VIII, 31 March.

Appendix 19.2

Identification of the 'structural' estimates reported in this chapter require that exogenous variables be excluded from equations. Table 19.3 summarizes these exclusion restrictions. The text provides a discussion of the rationale for exclusion.

An exclusion test is used to assess these exclusion restrictions. It entails first obtaining the residuals from the estimate equations. Then these residuals are regressed on the excluded variables in each equation. The suggested test (Johnston) is on nR^2 where n is the number of observations. This is distributed as χ^2 with n degrees of freedom.

With a high number of observations (over 4000) the R^2 has to be very low, less than 0.01, to pass this test. The Table below notes that very few of the R^2 values meet this condition. For a modest number of observations (say 200) the critical R^2 can be around 0.03. With this standard we still have a few cases where this standard is not met. With the sorghum estimates being the least viable, it is of interest to note that the YIELD equations perform best in this regard.

Given the uncertainty regarding the test itself and the rationale for exclusion, we believe that the estimates obtained can be given credence.

R^2 values: exclusion tests

	Rice	Wheat	Maize	Sorghum	Pearl millet
AREA SHARE					
(1)	0.0131	0.0385	0.1076	0.1911	0.0333
(2)	0.0077	0.0393	0.1231	0.0462	0.0276
MV ADOPTION					
(1)	0.0089	0.0157	0.0049	0.1609	0.0353
(2)	0.0069	0.0187	0.0038	0.0716	0.1058
IRRIGATION					
(1)	0.0358	0.0292	0.0290	0.2611	0.0547
(2)	0.0158	0.0234	0.0175	0.2007	0.0436
YIELD					
(1)	0.0137	0.0089	0.0157	0.0382	0.0114
(2)	0.0096	0.0062	0.0102	0.0138	0.0001

Notes: (1) includes all excluded variables; (2) includes all excluded variables except pre-1965 variables.

20 Brazil

A.F.D. Avila, R.E. Evenson, S. De Silva and F.A. de Almeida

Brazil's agricultural sector has had a record of dynamism and change in recent decades. Productivity measures show that considerable changes have taken place in both crop and animal production. Regional change has occurred, with the Central-West region (Cerrados) achieving very impressive production increases over recent decades (Avila and Evenson, 1995).[1]

Brazil's agriculture benefits from the research programmes of private firms as well as research conducted in state and federal government programmes. Several major seed companies have research programmes developing new crop varieties. (AgroCeres, Pioneer, Cargill and Novartis all develop hybrid maize varieties.) Cooperative foundations of producers have breeding programmes in some crops (e.g. FUNDACEP and CODETEC in wheat and soybeans). State research programmes in a number of states also manage breeding programmes.

The major plant breeding programmes, particularly where IARC collaboration is achieved, are those of the Brazilian Agricultural Research Corporation (EMBRAPA), which maintains specialized federal crop research programmes in different regions in Brazil. The development of improved crop varieties is the core activity of these research pro-

[1] Brazil has experienced a somewhat erratic record of economic growth over recent decades, with extraordinary growth in some periods and high rates of inflation and poor economic growth in other periods. In general, the agricultural sector in Brazil has been its most consistent sector in terms of economic growth.

grammes. Brazilian scientists in these programmes often work in close collaboration with IARC scientists. Exchange of crop breeding 'germplasm' in the form of landraces, breeding lines and varieties, takes place between IARCs, state and EMBRAPA units.[2]

Varietal releases are governed by release boards and are subject to careful scrutiny. Seed multiplication and sale is undertaken by private companies and by organized seed growers' associations. The proportion of farmer-saved seed varies by crop and has changed over time.

In this chapter, we report the findings of a study of the production impacts of crop genetic improvement in Brazil. The study covers five of the ten crops included in the larger crop genetic improvement (CGI) study: wheat, rice, maize, beans and potatoes, plus soybeans and cotton, which are not included in the SPIA study. This study focuses on the experience of a single country, Brazil, to provide a more comprehensive analysis of the production impacts of crop genetic improvement. We first present a brief discussion of CGI activities for the crops studied. We then present the statistical specification used in the evaluation. Estimates are then reported. The final section summarizes the economic implications of the estimates.

Crop Genetic Improvement Programmes

This study is based on data measured at the Brazilian state level for the period 1978–1998. Table 20.1 provides a summary of the crops, production and states included in this statistical study.

Figure 20.1 depicts yield changes in the Brazil study crops over the period at the national level. In the statistical specification outlined in the next section, we show how we utilize yield data to develop our estimates of crop genetic improvement impacts on crop production. Here we note that crop yields were increasing for all crops except wheat over the period studied.

The study uses a 'varietal turnover' variable reflecting farmer adoption of 'new varieties' and the subsequent disadoption of 'old varieties' (see below for a specific definition). The delivery of varieties to farmers (enabling them to experiment and adopt) is achieved by both private and public enterprises where incentives for private sector appropriation of returns for breeding effort are provided. Where such

[2] The EMBRAPA research units represent the national or federal system and this system is a relatively recent development. Established in the mid-1970s, EMBRAPA is now well established and is providing scientific leadership in many fields of agricultural science.

Table 20.1. Average production and major Brazilian states producers by selected commodities. Average 1978–1998.

Selected commodity	Average production (t)	Major state producers (abbreviations)	Major states participation (%)
Beans	2,604,869	PE, CE, BA, MG, ES, SP, PR, SC, RS, MS, GO	83.85
Maize	29,283,921	CE, PB, PE , BA, MG, ES, SP, PR, SC, RS, MS, MT, GO	95.80
Rice	9,631,576	MA, BA, MG, ES, SP, PR, SC, RS, MS, MT, GO	88.92
Wheat	2,811,183	PR, SC, RS, MS, MG, SP	99.71
Soybeans	23,249,026	BA, MG, SP, PR, SC, RS, MS, MT, GO	99.03
Cotton	1,444,841	CE, RN, PB, BA, MG, SP, PR, MS, MT, GO	96.92
Potato	2,421,031	MG, ES, SP, PR, RS, SC	99.80

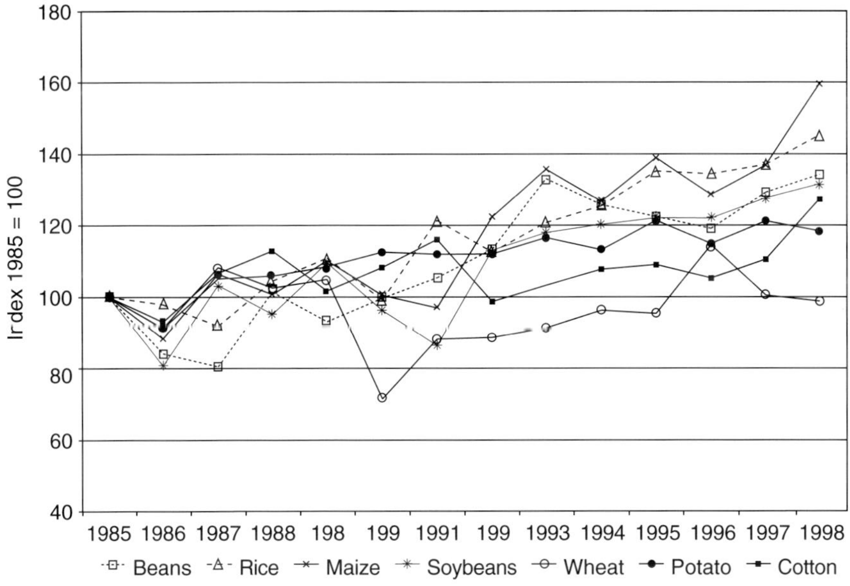

Fig. 20.1. Productivity evolution index for selected commodities in Brazil, 1985–1998.

incentives are not provided (as in the case of beans, rice and potatoes), only public enterprises, including producer cooperatives, are engaged in crop breeding.[3]

[3] Brazil realized MV/TV conversion a number of years ago for most crops. Most of today's varietal change is MV/MV conversion. The turnover variable is designed to reflect the replacement of an earlier modern variety by more recently developed modern varieties.

In the case of 'hybrid' crops (chiefly maize in Brazil), a natural incentive for private-sector plant breeding exists because the 'heterosis' component of yield is not passed on in seed from the crop. This must be produced in new seed each year through a complex process of inbreeding and crossing. Thus, a hybrid seed company can sell its seed for a premium if farmers value the seed sufficiently to pay the premium. Thus, while some 'open-pollinated' maize seed is sold in Brazil (farmers can save this seed), most maize seed is hybrid seed. The large private seed company suppliers include Agroceres (now part of Monsanto), Cargill, BrasKalb, Novartis and Pioneer. Smaller companies, such as FT Semetes, and a producer group, Codetec, also sell hybrid maize seed. EMBRAPA competes in the market as well, with BR 201 and BR 106 as important varieties at present.

Brazil is developing stronger Breeders' Rights laws at present, but these are not yet seen as strong incentives for private firms.[4] As they are developed, it is expected that a number of the private hybrid-maize firms will develop breeding programmes for non-hybrid crops, especially for soybeans and cotton. At present, producer cooperatives are important producers of wheat and soybean seed. For soybeans, Codetec and Fundacep have breeding programmes. Some soybean varieties are still being imported from the USA (Brossier, Cobb, Davis), but these are declining in importance. The chief supplier of soybean varieties, including varieties suited to the Cerrados region, is EMBRAPA.

Wheat varieties are produced by both EMBRAPA and state programmes (IAC, IAPAR, FEPAGRO) and cooperatives (Codetec and Fundacep).

Cotton varieties have traditionally been supplied by the São Paolo state research programme (IAC in Campinas). In recent years, EMBRAPA varieties and those from Minas Gerais state (EPAMIG) have become important.

Brazil produces irrigated rice in the southern states of Rio Grande do Sul and Santa Catarina. The leading producer of irrigated rice varieties is the Rio Grande do Sul programme (IRGA). The Santa Catarina programme (EPAGRI) and EMBRAPA are also important developers of varieties. Brazil is also a major producer of 'upland' rice in other

[4] Breeders Rights are relatively well developed in the USA and Europe (these are the Plant Variety Protection Rights in the USA). They have stimulated considerable private-sector breeding research in some hybrid crops in a number of countries, but are not well developed in developing countries. They are, however, the most likely intellectual property rights to be implemented under the terms of the World Trade Organization rules.

regions. Other parts of the world have made little progress in the development of improved upland rice varieties. However, EMBRAPA has produced a number of improved upland rice varieties in recent years.

The improvement of bean production in Latin America and Africa has similarly been a difficult task. However, the EMBRAPA research programme, with IARC contributions from CIAT, has again achieved significant gains in this crop. This will be apparent in the estimates reported below.

Finally, potato varieties in Brazil are delivered to farmers by the private sector. Most are imported from European countries, especially from Germany and the Netherlands.

Figure 20.2 provides a summary of seed production by origin for each crop for the period 1991–1997.

Statistical Specification

The statistical model applied in this study is specific to each crop. Table 20.2 provides an overview of the three equations in the model. There are three variables that are treated as endogenous in the model: YINDEX, VTC and RUS. The variables are defined below. The model is estimated using three-stage least squares. Units of observation are for Brazilian states for the years 1991–1998.

Endogenous variables

The economic model underlying the statistical specification postulates that farmers in Brazil make choices that affect three variables in the model. Accordingly, these three variables are treated as endogenous variables in the estimation procedure. These three variables are listed below.

Crop yield index (YINDEX)

Farmers make decisions that determine crop yields. In order to account for differences in state soil and climate conditions, we index these yields to a 1985 base period. This takes out state-to-state differences in 1985 yield levels. Thus, we are analysing only changes in each state from this base.

Varietal turnover (VTC)

Farmers are faced with crop varietal decisions each year. They will typically experiment with a new variety, and, if the results justify it, increase planting of the variety in future years. This variable is

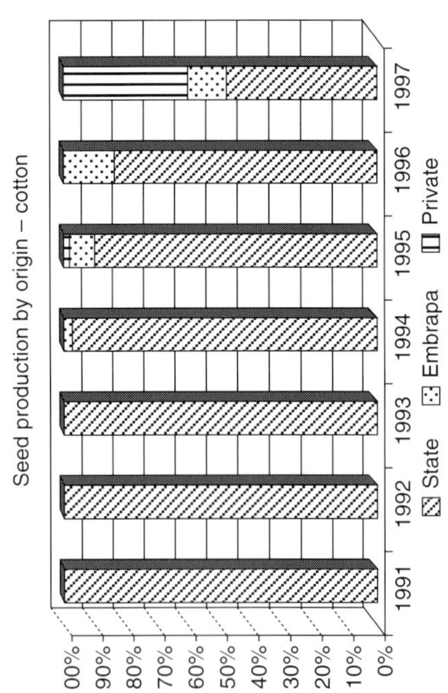
Seed production by origin – cotton
100%
90%
80%
70%
60%
50%
40%
30%
20%
10%
0%
1991
1992
1993
1994
1995
1996
1997
State
Embrapa
Private

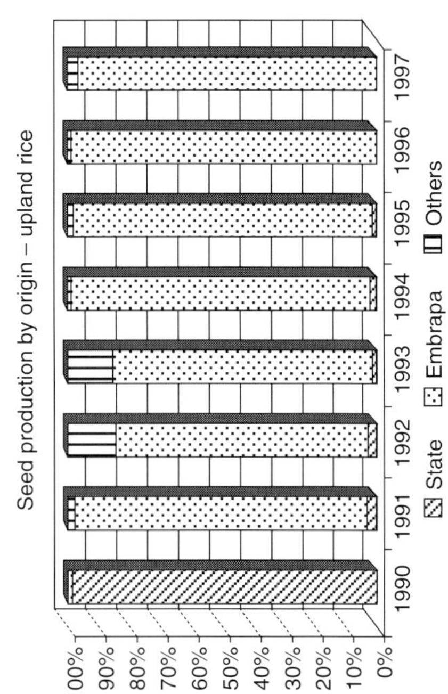
Seed production by origin – upland rice
100%
90%
80%
70%
60%
50%
40%
30%
20%
10%
0%
1990
1991
1992
1993
1994
1995
1996
1997
State
Embrapa
Others

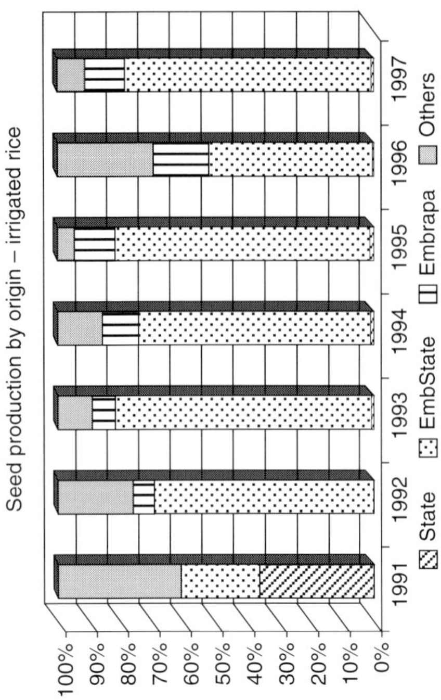
Seed production by origin – irrigated rice
100%
90%
80%
70%
60%
50%
40%
30%
20%
10%
0%
1991
1992
1993
1994
1995
1996
1997
State
EmbState
Embrapa
Others

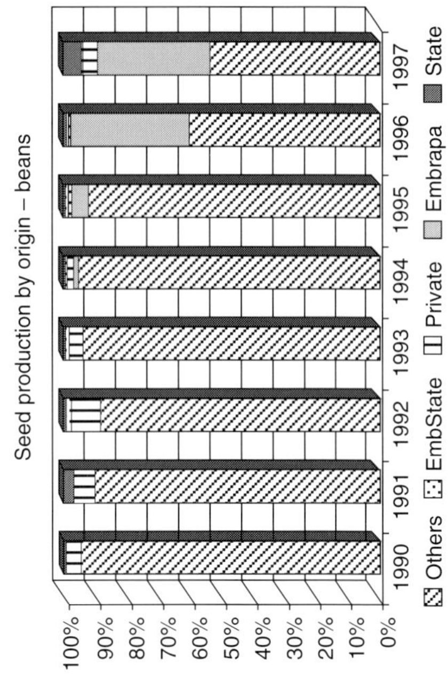
Seed production by origin – beans
100%
90%
80%
70%
60%
50%
40%
30%
20%
10%
0%
1990
1991
1992
1993
1994
1995
1996
1997
Others
EmbState
Private
Embrapa
State

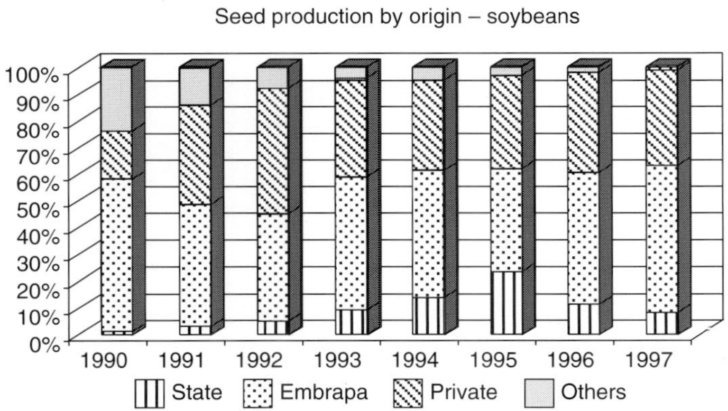

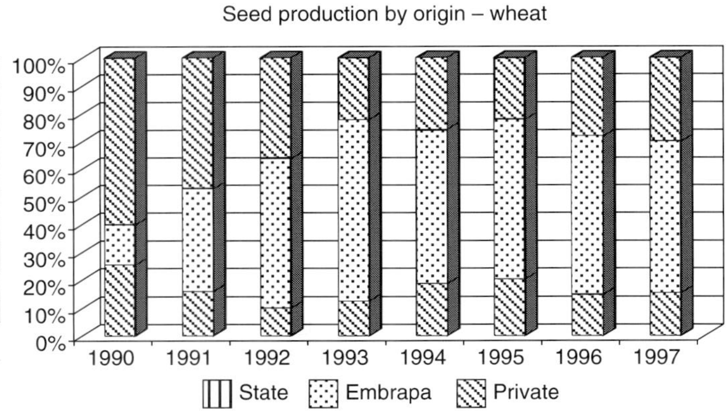

Fig. 20.2. Brazilian seed production by origin: 1990–1997.

Table 20.2. Statistical model structure.

Dependent endogenous	Independent endogenous	Independent exogenous
YINDEX	VTC, RUS	PRPRF, Z
VTC		CG, RES, EXTSTK, SCHOOL, CG, X, Z
RUS	VTC	PRPRF, ARSHARE, Z

measured from seed sales for each crop. In any given year in a state, some variety seed sales will increase while others decrease. The sum of percentage increases measures the turnover in a given year. The variable VTC is the cumulated varietal turnover in the state beginning with the base year. This definition makes the variable commensurate with the yield index.

Rate of seed utilization (RUS)

Farmers also choose how much seed to sow from their own production and how much to purchase from seed suppliers. This variable, defined as the percentage of seed that is purchased, may affect yields.[5] However, since VTC, our varietal turnover variable, is based on seed sales, this variable also serves as a control variable for incomplete information on varietal turnover.

Exogenous variables

The exogenous variables are treated as being beyond the control of farmers.

Agro-ecological zone dummy variables (Z in Table 20.2)

Table 20.3 describes ten Brazilian agro-ecological zones as developed by EMBRAPA. For each state in the data set, a variable for each zone is defined to be the presence or absence of the zone in that state. Thus, ten zone variables (Z) are included in all three equations.[6]

[5] It should be noted that the reliability of the varietal turnover variable (VTC) as a true measure of varietal turnover is affected by RUS, the proportion of seed saved.

[6] These agro-ecological zones have been developed by EMBRAPA to guide research projects and programme focus.

Table 20.3. AEG study – Brazilian agro-ecological zones: characteristics of agro-ecological zones.

Zones	Vegetation	Soil texture	Soil fertility	Soil drainage
Zone87	Tropical forest	Medium to clay	Low to medium	Moderate to well drained
Zone6770	Subtropical forest	Clay to medium	Low	Moderate to well drained
Zone8992	Tropical forest	Clay to medium	Medium to high	Well drained
ZoneCerrados	Savannah	Medium to very clay	Very low	Well drained
Zone7577	Perennial tropical forest	Clay	Medium	Well drained
Zone4853	Tropical forest	Clay	Low to medium	Moderate to well drained
Zone1743	Caatinga	Sandy and medium to clay	Low to high	Well drained
Zone2058	Savannah	Sandy and sandy to clay	Very low	Well drained
Zone7386	Tropical forest	Clay to very clay	Medium to high	Well drained
Zone78NE	Tropical forest	Clay	Low to medium	Well drained

Farmer schooling (SCHOOL)

Schooling levels of farmers (measured in average years completed) are effectively determined by the time farmers enter the occupation. While migration patterns can change schooling levels, such changes are relatively minor over the period of our study. Schooling levels do influence varietal and seed choices as well as having effects on the general efficiency of farm managers.

Extension services (EXTSTK)

Brazil provides public extension services to farmers, and these services differ by state over time. These services are designed to inform farmers of technological opportunities and to advise farmers on management practices. This variable is defined as the cumulated number of extension staff per farmer over the past 3 years. This variable is not specific to particular crops and is thus the same for each crop.[7]

Crop/fertilizer price ratios (PRPRF)

Since we are treating our yield equation as our productivity equation and including the other two endogenous variables (VTC and RUS) in that equation, it is essential that we have a variable measuring the use of inputs, particularly fertilizer, that may affect yields. We do not have crop-specific input variables (e.g. fertilizer use on cotton).

[7] See Avila and Evenson (1995), for the procedure for developing this variable.

However, cost minimization theory tells us that variable input use per hectare responds to prices. We use the producer price of the crop to the price of nitrogen fertilizer to control for these effects. Note that we are also including the agro-ecological zone dummies for this purpose as well.

Crop breeding research, public sector (RES)

We have a measure of the crop-specific research in public sector programmes (EMBRAPA and state programmes). This variable is defined as the cumulated stock of EMBRAPA and state expenditures by year devoted to the crop. The cumulation begins in 1975 and is designed to measure the research service flow to the state (see Avila and Evenson, 1995). EMBRAPA scientist years are treated as having effects in more than one state. This variable is used as the key identifying variable in the VTC equation, where it is reflects the provisions of public sector adoptable variation to the state. It is also included in the RUS equation.

Varietal characteristics supplied (X in Table 20.2)

This set of variables is utilized in both the VTC and RUS equations. For each variety released for a state (most varieties are released only in selected states), breeders were asked to rate the variety as having low, medium, or high host plant resistance to insect pests and as having low, medium or high host plant resistance to major plant diseases and abiotic stress. Varieties are also rated for yield characteristics. These variables measure the proportion of seed with these ratings in the previous year.

Area share (ARSHARE)

This variable measures the share of a state's cropped area planted to the crop (lagged one period). This variable is used to identify the RUS equation.

IARC content (CG)

This variable measures the percentage of the seed sold that has IARC content. IARC content is measured as cases where the cross underlying the variety was made in an IARC programme and where an ancestor in a Brazilian-made cross was an IARC line. This variable is treated as being exogenous because the IARC programme supplies genetic resources to Brazil and to other countries. Brazilian plant breeders use such material if it is valuable.

Table 20.4 reports a summary of variable means.[8]

Parameter Estimates

We report a summary of estimates for key variables for each endogenous equation. The complete set of parameter estimates can be obtained from the authors.

Varietal turnover (VTC)

Table 20.5 reports parameter estimates and levels of statistical significance for the four major exogenous variables in the model. The full set of estimates includes the trait supply variables and the agro-ecological zone variables. Our expectations are that agricultural research, extension and schooling should have positive effects on varietal turnover.

Table 20.4. Variable means.

	Rice	Maize	Wheat	Beans	Potatoes	Soybean	Cotton
YIELD	2419	2,187	1,743	706	14,088	2,116	1,275
YINDEX	127.7	127.6	111.8	143.7	111.9	122.2	115.3
VTC	0.82	0.69	1.00	0.57	0.64	0.82	0.82
RUS	48.2	59.8	90.3	13.7	47.2	72.3	75.7
RES	15,520	14,472	14,356	2,665	6,320	22,509	18,788
EXTSTN	1,438	1,396	1,913	1,491	1,873	1,621	1,476
SCHOOL	3.39	3.11	3.59	3.42	3.59	3.46	2.92
CG	0.69	0.40	0.49	0.48	–	–	–
PRPRF	0.0007	0.0005	0.0006	0.0016	0.0009	0.0006	0.0013
ARSHARE	0.126	0.393	0.073	0.215	0.008	0.358	0.046

Table 20.5. Estimates: varietal turnover (VTC).

VARIABLE	Rice	Maize	Wheat	Beans	Potatoes	Soybean	Cotton
RES	0.000052***	–0.00003**	–0.00004***	–0.000022*	0.000116***	0.000074***	0.000041**
EXTSTK	0.00012	0.0004**	–0.00041*	–0.00023	0.00062***	–0.00144***	0.00064
SCHOOL	1.718***	2.298***	1.146***	1.596***	1.093***	0.756**	3.059***
CG	0.0432	–0.1183	0.2001***	0.1942***	–	–	–

* *t* between 1.0 and 1.5, ** *t* between 1.5 and 2.0, *** *t* greater than 2.0.

[8] We acknowledge that the identification of our structural model has elements of abritrariness. Our central concern here is to estimate VTC effects on other endogenous variables.

Consider first the variable measuring the agricultural research stock. The estimated coefficients for this variable are positive except for maize, wheat and beans. Since the public agricultural research system is the key supplier of adoptable varieties to different regions, we expect that stronger research programmes will stimulate more varietal turnover. In the case of maize, however, the private sector is the key supplier of varieties (hybrids). The public research programme concentrates on management and related activities and to some extent compensates for private sector limitations. For wheat over the period concerned, yield levels increased very slowly. In addition, wheat is grown in a limited area covering only a few states and dominated by production in Rio Grande do Sul.

The extension service in Brazil is not organized on a commodity basis. We generally expect regions with stronger extension services to adopt varieties more rapidly, although the effect on varietal turnover may be limited by the supply of new varieties. In this case, stronger extension services would not necessarily lead to more turnover. We find positive extension effects on varietal turnover in all crops except wheat, soybeans and beans.

The level of farmers' schooling is expected to stimulate more rapid adoption of varieties and more experimenting with varieties. Interestingly, we find positive and statistically significant effects of schooling in all crops. This is the major variable driving farmer adoption of new varieties.

Finally, the provision of IARC parental material is expected to provide a stimulus to the supply of adoptable varieties and should stimulate turnover. We find positive effects of IARC content on rice, beans and wheat, but not on maize, where private suppliers dominate.

The trait incorporation variables were significant explanatory variables in all crops.

For rice, high levels of host plant resistance to disease stimulated varietal turnover. For maize, high yield potential combined with disease, pest and stress resistance stimulated turnover. For beans, high yield potential was most important. For wheat, yield potential and disease resistance increased turnover. For potatoes, disease resistance was important. For soybean, traits were generally not important. For cotton, stress tolerance (to drought) increased turnover.

Seed utilization (RUS)

Table 20.6 reports estimates of key variables in the seed utilization equation.

The (endogenous) varietal turnover variable (VTC) is expected to have a positive impact on the proportion of seed purchased by farmers.

Table 20.6. Estimates: seed utilization rate (RUS).

VARIABLE	Rice	Maize	Wheat	Beans	Potatoes	Soybean	Cotton
VTC	0.122	−0.2766***	0.630	3.977***	−1.892***	0.00669	−1.164
PRPRF	−2537	2912	2384*	140	26	−2184	−3410*
ARSHARE	−16.747	−52.96***	74.52***	24.40***	−1505***	−2.21	−50*

* *t* between 1.0 and 1.5, ** *t* between 1.5 and 2.0, *** *t* greater than 2.0.

As more new varieties of value are made available, farmers will be purchasing more seed to take advantage of opportunities presented by more rapid flows of new varieties. We find little evidence for this – significant effects only in beans (rice, beans and potatoes have the lowest mean RUS – see Table 20.1) although we do find positive effects in rice, wheat and soybeans. Similarly, higher crop price to fertilizer price ratios might be expected to lead to higher rates of seed purchase, but we find little evidence for this. Higher area shares can lead to scale economies in local seed industries causing higher rates of seed purchase. We find this effect only for wheat and beans. Seed purchases are generally lower the larger the share of area planted to the crop.

Yield index (YINDEX)

Table 20.7 reports parameter estimates for the yield or productivity equation. These are the 'bottom-line' estimates for varietal turnover effects. Note that because we are using a yield index, we have taken out all cross-section differences in initial yields.

Our key variable is the varietal turnover variable (VTC). The VTC variable coefficient is positive in all crop regressions and statistically significant in all crops except wheat. The coefficients indicate the change in the yield index from a full turnover of varieties. (See below for further discussion.) This is solid evidence of varietal impacts on crop productivity. The estimates for rice are especially important because most rice in Brazil is upland rice. That is, most states produce upland rice and the estimates show that the EMBRAPA varieties have had important production impacts.

Table 20.7. Estimates: yield index (YINDEX).

VARIABLE	Rice	Maize	Wheat	Beans	Potatoes	Soybean	Cotton
VTC	18.245***	7.634**	0.0735	32.86***	10.04***	14.21***	18.91***
RUS	0.9498***	−1.763***	3.230***	−2.675**	0.211	0.897***	3.74***
PRPRF	3,014	31,301**	−57,474***	4,955**	3,966***	14,326***	26,811***

* *t* between 1.0 and 1.5, ** *t* between 1.5 and 2.0, *** *t* greater than 2.0.

The seed utilization rate is expected to have a positive impact on yield and this is borne out for all crops except maize and beans.

The ratio of crop prices to fertilizer prices should have a positive effect on yields because it should be reflecting fertilizer and related input use. This is borne out in all crops except for wheat. We consider this to be evidence that we are controlling for changes in input use per hectare.

Economic Implications

The coefficient estimates reported in Tables 20.5–20.7 can be used to compute the contribution of varietal turnover to yield improvement. Table 20.8 reports the yield index contributions of varietal turnover for the mean of the sample. This calculation takes into account the secondary effect of VTC on YINDEX via its effect on RUS.

These data show some variability in the contributions of varietal turnover to yield increases. For all crops in Brazil over the period studied, varietal improvement accounted for roughly 50% of the increased crop yields actually realized. The share of actual yield increases accounted for by varietal improvement varies across crops from 18 to 78%. These estimates are consistent with findings from other studies in this volume.

It is also possible to calculate the indirect effects of research, extension and schooling through their impacts on varietal turnover. It is important to note that these impacts are not the only impacts of these programmes on productivity. For all crops combined, the weighted impact of research programmes on yield via varietal turnover is positive but small. For extension this impact is negative, but small.

Table 20.8. Varietal turnover calculations.

	Coefficients						
Crop	VTC in YINDEX	VTC in RUS	RUS in YINDEX	Mean VTC	Mean change YINDEX	VTC contribution	VTC share
Rice	18.245	0.122	0.9488	0.82	2.1	15.06	0.72
Maize	7.634	–2.755	–1.763	0.69	18.5	8.62	0.46
Wheat	0.0735	0.630	3.230	1.00	11.8	2.11	0.18
Beans	32.86	3.977	–2.675	0.57	20.1	12.67	0.63
Potato	10.04	–1.892	0.211	0.64	11.9	6.17	0.52
Soybean	14.21	0.0069	0.869	0.82	22.2	11.66	0.52
Cotton	18.99	–1.164	3.74	0.82	15.3	12.00	0.78
All crops				0.74	18.8	10.61	0.51

Schooling, however, has a major impact on productivity through its effect on varietal turnover. Had schooling levels been 10% higher than they were, varietal turnover would have been 54% greater and this would have increased the varietal yield contribution by approximately 50%. This is a very large effect that could not be maintained unless the research system was producing new varieties at a high rate.

Concluding Remarks

Studies of Brazil's agricultural sector and of the research programmes' contributions to productivity growth in the sector have concluded that research programme contributions have been large (Evenson, 1984). The building of the federal research system (EMBRAPA) has contributed to a generally impressive productivity performance in Brazilian agriculture.

This study focuses on a central part of the research programme, crop genetic improvement (CGI). It shows that CGI programmes in Brazil have been important, contributing roughly 50% of the realized yield gains over the period of the study. Estimates of CGI impacts by crop show some variations in CGI contributions across crops. However, all crops benefited from CGI gains and these gains constituted a major part of productivity gains.

Brazil is one of the most advanced of the countries categorized as developing. In contrast to many countries with limited technological experience, it has an advanced research system. Most of the improved varieties in Brazil are crossed in Brazil, with limited IARC content. The estimates in this chapter show that crop genetic improvement has been a major contributor to productivity change in Brazil.

References and Further Reading

ABRASEM. Annuários Estatisticos de ABRASEM, 1987–1997.

Almeida, F.A. (1997) O melhoramento vegetal e a produção de sementes na EMBRAOB o desafio do futuro, Brasilia, EMBRAPA-SPI.

Almeida, F.A. *et al.* (1999) Impacto das Cultivares de Soja da EMBRAPA e Rentabilidade dos Investimentos em Melhoramento Brasilia, EMBRAOA.SEA (Temas para Discussão,2).

Avila, A.F.D. and Evenson, R.E. (1995) Total Factor Productivity Growth in Brazilian Agriculture and the Role of Agricultural Research. Anais do XXXIII Congresso Brasileiro de Economia e Sociologia Rural, (Vol. I), Curitiba, 31 July–3 August 1995, pp. 631–657

Cruz, E.R. and Evenson, R.E. (1997) Technological spillovers in Southern Cone agriculture *Economia Aplicada* I(4), 709–30. FIPE/USP, São Paulo.

Empresa Brasileira de Pesquisa Agropecuária. Serviço de Produção de Sementes Básicas (1999) *Banco De Informação de Sementes (BIS)*, Convenio EMBRAPA/ABRASEM/MA, Brasilia

Evenson, R.E. (1968) The Contribution of Agricultural Research and Extension to Agricultural Production. PhD Dissertation, University of Chicago.

Evenson, R.E. (1982) Observations in Brazilian agricultural research and productivity. *Revista de Economia Rural* 20, 368–401.

Evenson, R.E. (1984) Benefits and obstacles to appropriate agricultural technology. In: Heston, A.W. and Pack, H. (eds) *Technology Transfer: New Issues, New Analysis. The Annals of the American Academy of Political and Social Science* 458, November 1981. [Also Chapter 24 in *Agricultural Development in Third World*, Eicher, C.K. and Staatz, J.M. (eds) (1984). Johns Hopkins University Press, Baltimore.]

Evenson, R.E. and Avila, A.F.D. (1995) Productivity change and technology transfer in the Brazilian grain sector. *Revista de Economia Rural* 34, no. 2.

Evenson, R.E. and Kislev, Y. (1975) *Agricultural Research and Productivity.* Yale University Press, New Haven, Connecticut.

Griliches, Z. (1957) Hybrid corn: an exploration in the economics of technological change. *Econometrica* 25(4), 501–522.

Appendix 20.1

Brazil estimates: specifications test

1. Do 3SLS and obtain residuals for each equation (using predict, resid command).
2. Regress residuals against the excluded independent variables.
3. The test statistic is χ^2 with degrees of freedom equal to the number of excluded variables (reported at 5% significance).

Results

Most estimates pass the test. Out of the 21 equations, only five fail the test and one of these fails it narrowly. All four equations for which we reject the exclusion restrictions are for RUS (these results are shown in the table below: rice, wheat, potato and soybean).

		d.f. (no. of excluded vars)	χ^2 (critical)	nR^2
Beans	YIELD	12	21.026	7.7792
	VTC	3	7.815	1.2408
	RUS	11	19.675	7.0488
Maize	YIELD	10	18.307	1.4872
	VTC	3	7.815	4.6816
	RUS	9	16.919	8.1664
Rice	YIELD	11	19.675	8.8264
	VTC	3	7.815	1.5752
	RUS	10	18.307	33.7656
Wheat	YIELD	11	19.675	10.038
	VTC	3	7.815	1.2348
	RUS	10	18.307	27.3588
Cotton	YIELD	8	15.507	3.7512
	VTC	3	7.815	3.7656
	RUS	7	14.067	3.2904
Potato	YIELD	7	14.067	6.9132
	VTC	3	7.815	1.428
	RUS	6	12.592	16.0776
Soybean	YIELD	10	18.307	19.7424
	VTC	3	7.815	3.024
	RUS	9	16.919	21.276

21 Modern Variety Production: a Synthesis

R.E. Evenson

In this, the first of three synthesis chapters in this volume, the focus is on the production of modern varieties (MVs). First, a summary of annual rates of production of MVs is developed. This summary includes data on the source of the initial cross leading to the successful release of a variety and on the source of the initial cross for parents and other varietal ancestors. The second part of the chapter is directed toward an investigation of the relationship between international agricultural research centre (IARC) crop genetic improvement (CGI) and national agricultural research system (NARS) crop genetic improvement programmes. This relationship is one in which IARC CGI programmes provide germplasm to NARS CGI programmes in the form of breeding lines and finished varieties that are in turn utilized as parental materials in NARS CGI programmes.

The crop studies reported in Part II of this volume provide indicator data on the sources of crosses and on the sources of parental and other ancestral crosses in crop varietal releases. Several chapters also report data on the flow of genetic resources from IARC gene banks to NARS breeding programmes and on IARC training and related programmes. These activities are part of the design of the IARC system and reflect a view that IARCs should support NARS programmes by providing germplasm services that will complement NARS CGI programmes and make them more productive. In the third part of this chapter, estimates of IARC germplasm effects on NARS breeding programme productivity are presented.

Because IARCs do develop finished (or near finished) varieties suitable for release, they inevitably compete with NARS breeding programmes as well as complementing them through germplasm services. It is possible that the competition effect could outweigh the complementarity effect, leading to reduced NARS investments in varietal improvement (at least in some countries). It is also possible that the reverse is the case, i.e. that complementary germplasm effects outweigh the competition effects, so that IARC programmes stimulate additional NARS investment. This question has obvious policy importance. Because of the location-specificity of improved varieties, it is important that NARS CGI programmes be located in important producing regions. Reduced NARS CGI investments, in a setting where under-investment in CGI is the norm, are not in the public interest. The analysis is undertaken utilizing a statistical model recognizing both production and investment effects of IARC CGI impacts on NARS CGI productivity and on NARS CGI investment.

A Summary of Varietal Production and Varietal Content Indicators

Before turning to the estimation of germplasm effects on investment and NARS varietal production, it will be instructive to summarize the production of varieties and of IARC content by crop and period. Table 21.1 provides the relevant data by crop and by region for all crops for each 5-year period since 1965.

First, consider the production of varieties for all crops combined. Annual varietal production has risen steadily over time. Varietal production in the late 1970s was almost double the rate of the late 1960s. By the late 1980s it had increased further and by the late 1990s, varietal production was double that of the late 1970s and quadruple that of the 1960s. This is an impressive expansion of the production of CGI products.

Varietal production can be divided into varieties where the cross was made in an IARC programme (usually with selection in the IARC programme as well, although some varieties were selected in NARS programmes) and varieties where the cross was made in a NARS programme.[1] The data indicate that 36% of the approximately 8000 released varieties were crossed in an IARC programme. Since a number of IARC crosses produce multiple releases (i.e. a variety may be released

[1] In this chapter reference is made to IARC content indicators. These indicators are measures of the source of the cross leading to a successful varietal release and measures of the sources of parental and other ancestral crosses. They are indicators and do not by themselves measure germplasmic contributions.

Table 21.1. Average annual varietal releases by crop and region 1965–1998.

	Average annual releases							1965–1998 IARC content**			
Crop	1965–70	1971–75	1976–80	1981–85	1986–90	1991–95	1996–98*	IX	IP	IA	IN
Wheat	40.8	54.2	58.0	75.6	81.2	79.3	(79.3)	0.49	0.29	0.08	0.14
Rice	19.2	35.2	43.8	50.8	57.8	54.8	58.5	0.20	0.25	0.07	0.48
Maize	13.4	16.6	21.6	43.4	52.7	108.3	71.3	0.28	0.15	0.04	0.53
Sorghum	6.9	7.2	9.6	10.6	12.2	17.6	14.3	0.16	0.07	0.06	0.71
Millet	0.8	0.4	1.8	5.0	4.8	6.0	9.7	0.15	0.41	0.09	0.35
Barley	0.0	0.0	0.0	2.8	8.2	5.6	7.3	0.49	0.20	0.01	0.30
Lentils	0.0	0.0	0.0	1.8	1.8	3.9	(3.9)	0.54	0.05	0.01	0.40
Beans	4.0	7.0	12.0	18.5	18.0	43.0	(43.0)	0.72	0.05	0.01	0.19
Cassava	0.0	1.0	2.0	15.8	9.8	13.6	(13.6)	0.53	0.15	0.01	0.31
Potatoes	2.0	10.4	13.0	15.9	18.9	19.6	(19.6)	0.17	0.06	0.02	0.75
All crops											
Latin America	37.8	55.9	65.9	92.5	116.2	177.3	139.2	0.39	0.14	0.04	0.43
Asia	27.2	59.6	66.8	86.3	76.7	81.2	79.9	0.18	0.29	0.10	0.43
Middle East and North Africa	4.4	8.0	10.2	12.2	28.4	30.5	82.2	0.62	0.22	0.04	0.12
Sub-Saharan Africa	17.7	18.0	23.0	43.2	46.2	50.1	55.2	0.45	0.21	0.07	0.27
All regions	87.1	132.0	161.8	240.2	265.8	351.7	320.5	0.36	0.20	0.06	0.42

* Numbers in parentheses are simple repetitions of 1991–95 rates because of insufficient data.

** IX, Variety based on IARC cross; IA, variety based on NARS cross with at least one non-parent IARC ancestor; IP, variety based on NARS cross with at least one IARC parent; IN, variety based on NARS cross with no IARC ancestors.

in several countries), the importance of IARC crosses is in one sense overstated, but in terms of real effects in the field (see Chapter 22, where it is estimated that IARC crosses are planted on roughly 36% of the area planted to modern varieties in the late 1990s), multiple releases from an IARC may be just as important as NARS releases.

These data also show a high degree of location specificity for crop varieties crossed in both IARC and NARS programmes. Relatively few NARS-crossed varieties were released in another country. For rice (see Chapter 5), multiple release estimates have been made. The ratio of total releases to original unique varieties is 1.64 for IRRI and 1.04 for 12 Asian NARS programmes. Applying these ratios to all crops in the study indicates that IARC crosses accounted for roughly 24% of unique releases (see Chapter 22 for diffusion comparisons). There is little evidence that these multiple release ratios have changed over time.

NARS-crossed releases can be further classified according to whether one or both parents in the cross was based on an IARC cross. For all NARS-crossed varieties, roughly 17% had at least one IARC-crossed parent (these varieties accounted for 25% of MV hectarage in 1998). This attests to a strong germplasmic effect, since NARS breeders found success in using IARC parental material. When grandparents and other ancestors of NARS-crossed varieties are considered, IARC-crossed germplasm appears in 23% of all NARS-crossed varieties. This IARC germplasm proportion is also rising over time.

To put these figures in perspective, note that the IARCs account for only small proportions of the scientists working in crop improvement programmes in developing countries – roughly 3% of the developing world's maize researchers, about 4% of the developing world's wheat researchers and no more than 15% of the rice scientists in South and Southeast Asia, excluding China. The fractions of expenditures on crop improvement in developing countries are somewhat higher, since IARCs spend more money per researcher. Even by this measure, however, the IARC shares of research input are not large. In wheat, for example, international centres probably account for little more than 10% of the amount spent by developing countries' NARS, excluding China (Heisey *et al.*, 2002). Thus, the IARC cross shares reported in Table 21.1 suggest that the IARCs are contributing to modern varieties far out of proportion with those institutions' shares of scientific manpower or spending.

An additional point to note is that there were very few contributions from research organizations in developed countries. The agricultural research system in the USA, for example, contributed almost no crosses and very few ancestors to the Green Revolution. At an aggregate level, the same can be said for research programmes in Japan, France and other European countries. Developed countries did not create the Green Revolution!

There are significant differences in these patterns among crops. There are also differences in what might be termed the 'maturity' of the breeding programmes by crop. In wheat, the total number of releases has stayed relatively constant since 1985 but with a high proportion of varieties based on IARC crosses or parents. The varieties/breeder ratio has been constant since 1985. Wheat is produced over a narrower range of climate and biotic diversity than rice is and has a relatively high level of multiple releases. This might be termed a mature pattern.

Rice also exhibits a mature pattern, but of a different type. Total releases have also been roughly constant since the mid-1980s, but the IARC-cross proportion has declined from the 1970s Green Revolution levels. This appears to be an example of maturing and strengthened NARS programmes. Previous work by Gollin and Evenson (1998), and work reported in Chapter 5 supports this interpretation. Again, the ratio of varieties to breeders in NARS programmes does not appear to be falling and may be rising over time.

Maize, the third most important cereal grain in developing countries, offers another pattern. Public sector releases appear to be rising with a relatively low IARC germplasm component. Private sector varietal production, primarily of hybrids, is clearly increasing. It is also clear that IARC germplasm has been useful to private breeders, along with NARS germplasm. This is a case of public sector research creating a 'platform' on which the private sector can be productive. (Note that this also provides a platform for modern biotechnology products.)[2]

The pattern for sorghum is roughly similar to the pattern for maize, again with a growing proportion of hybrid sorghum varieties being produced by private sector breeders. The pearl millet pattern indicates relatively weak NARS production in the early years. This is a case where the IARC programme not only provided germplasm to NARS but initiated expanded CGI work generally. Until ICRISAT began its CGI work, there was little useful raw material for NARS programmes to work with.

The pattern where the IARC programmes effectively initiated breeding work on a crop holds for barley, lentils, groundnuts and cassava as well as for millet. In each of these crops the IARC-cross proportion is high and total varietal production is generally rising. The NARS CGI programmes for these crops are not very mature at this point.[3]

2 This platform creation element is important to the general breeding process. Traditional farmer-selected varieties are often not responsive to fertilizer and not suited to hybridization.

3 A number of NARS programmes on those commodities were in the early stages of development. The IARC programmes facilitated these efforts by building germplasm collections and making them readily available to NARS programmes.

For beans, IARC programmes have also stimulated increased varietal production with IARC crosses accounting for high proportions of released varieties. In sub-Saharan Africa, the IARC programmes are dominant in beans, cassava and potatoes.

The conditions for potato CGI differ from those of other crops because of different management factors (see Chapter 15). IARC-cross shares are low but have been rising.

Table 21.1 also reports release data for all crops by region. These data show that the highest rate of increase in varietal production in the 1980s and 1990s occurred in the Middle East and North Africa and sub-Saharan Africa regions. These regions were also the most dependent on IARC crosses and germplasm, reflecting different levels of NARS development. By contrast, the more mature NARS programmes in Asia were very productive in the 1960s and 1970s, but they grew more slowly. They are less dependent on IARC crosses but have high dependence on IARC germplasm.

The Statistical Specification for Estimation of IARC Germplasm Impacts on NARS Productivity and Investment

The conceptual foundation for an empirical study of the impact of IARC germplasm on NARS CGI productivity and investment is in two parts. The first is a model of germplasm impacts on NARS CGI productivity based on the 'search' model of research. The second is a model of NARS CGI investment. We are constrained by data limitations to measure NARS CGI investment in the form of scientist man-years. This variable is also an important component of the NARS CGI productivity specification (Rosebloom and Pardey, 1989). This has implications for the estimation procedure.

The first part of the model calls for a specification of the following form:

$$V_N = a_0 + a_1 \ln(B_N) + a_2 \ln(B_N) \times G_I + a_3 \ln(B_N) \times G_N + \text{climate dummies} + \text{period dummies} \tag{1}$$

where V_N is the number of varieties produced by a NARS CGI programme in a given period, B_N is the NARS plant breeding resources employed during the period, and G_I is a measure of IARC germplasm available to the programme, G_N is a measure of NARS germplasm available to the programme.

The logarithmic form for the B_N variable is based on the concept of search within a distribution (or distributions) of potential varieties (see Evenson and Kislev, 1975, and Appendix 21.1 for a technical development). Diminishing returns to search dictates the logarithmic specification.[4]

The G_I and G_N variables are not in logarithmic form because they are not part of the current NARS search. That is, IARC germplasm affects NARS productivity, but NARS do not produce IARC germplasm (except indirectly). G_I is thus a linear shifter of the search distribution for NARS and has the specific form noted in equation (1). Similarly, G_N NARS germplasm produced in prior periods is also a linear shifter interacted with ln (B_N).

An alternative general specification that does not rely on search theory might use a general production function of the form:

$$V_n = \ln(A) + a_1\ln(B_N) + a_2\ln(G_I) + a_3\ln(G_N) + \text{climate variables} \qquad (2)$$

Equation (1) could be estimated with data by country, period and crop. Data for the relevant variables can be assembled for five crops – wheat, rice, maize, beans and potatoes – for a varied number of countries by crop for three periods: 1965–1974, 1975–1984, 1985–1994. Table 21.2 provides variable definitions and means for the relevant variables in equation (1).

The measurement of B_N in terms of scientist man-years (SMY), while subject to some variability in the definition of an SMY, is in principle feasible (see below for some issues). The measurement of G_I and G_N is not as straightforward. Consider G_I, the IARC germplasm stock. Each IARC maintains gene bank collections as well as breeding programmes. Typically, there is free exchange between IARC and NARS programmes of gene bank accessions (landraces) as well as 'advanced' breeding materials, in the form of breeding lines (many of which are released by national programmes). IARCs often maintain international nurseries both to test advanced lines and to make them readily available to NARS breeders.

However, these genetic resources are not of equal value in different countries and different periods. One 'test' of the value of advanced breeding lines is whether the national programme sees fit to release an IARC-crossed line as a national variety. Using this as a guideline, we define G_I as the cumulated number of IARC crossed releases in the country including releases in the current period (NARS breeders use advanced lines before they are released as varieties).

[4] The functional form implications of the search model applied to plant breeding (see Appendix 21.1) provide a test of the search model against more general specifications.

Table 21.2. Variable definitions: means by crop.

Variable	Definitions	Wheat (66)	Rice (54)	Maize (32)	Beans (45)	Potatoes (51)
I. Endogenous variables						
B_N	Number of scientist man years in NARS CVI programme (see Appendix 21.1)	298	206	126	25.6	61.1
V_N	Number of NARS-crossed varietal releases (source: Chapters 4–16)	30.6	19.5	10.5	4.36	11.6
II. Exogenous variables						
G_I	International germplasm stocks: cumulated number of IARC-crossed varieties released in the countries	19.0	6.31	1.88	4.09	9.98
G_N	National germplasm stocks: cumulated NARS-crossed varietal releases (V_N) in previous periods	21.1	15.4	6.15	1.06	6.61
HA	Hectares ('000) planted to the crop at the beginning of the period	2847	4613	244	377	1470
RPOPDEN	Population density at the beginning of the period, Rural population/area in crops and pasture (FAO)	256	411	523	268	230
GDP/c	GDP per capita in US dollars beginning of period, World Bank Atlas Method (World Bank tables)	2784	2820	1954	2577	3410
Tech 2	Technology capital indicators from Evenson (2000);	0.19	0.11	0.29	0.27	0.08
Tech 3	Measures of technological capacity (see Appendix 21.2)	0.48	0.38	0.54	0.37	0.51
Tech 4		0.31	0.28	0.21	0.35	0.37
Climate 2	Climate class indicators from Evenson (2000)	0.26	0.39	0.64	0.62	0.58
3	(see Appendix 21.2)	0.40	0.22	0.27	0.25	0.24
4		0.31	0.16	0.18	0.19	0.24
5		0.31	0.06	0	0	0.06
6		0.18	0.11	0.18	0.13	0.12
7		0.09	0.06	0	0.06	0.06

The national germplasm stock, G_N, is defined as the cumulated stock of releases in the country based on NARS crosses from the countries' own programmes. This does not include the current period, because the dependent variable is NARS-crossed releases in the current period. The nature of this specification is that the G_N variable is at least partially a lagged dependent variable. However, it is in some genuine sense a measure of NARS-produced breeding germplasm.

The investment equation specification is given by equation (3):

$$\ln(B_N) = b_0\ln(G_I) + b_2\ln(G_N) + b_3\ln(HA) + b_4\ln(\text{Popden}) + b_5\ln(GDP/c) + b_6\ln(G_I)\text{x}\ln(HA) + b_7\ln(G_I)\text{x}\ln(Popden) + b_8\ln(G_I)\text{x}\ln(GDP/c) + \text{technology capital dummy} + \text{climate class dummies} + \text{period dummies} \quad (3)$$

where: B_N is the measure of SMYs in the NARS CGI programme,
G_I is the stock of IARC germplasm,
G_N is the stock of NARS germplasm,
HA is the total hectares planted to the crop at the beginning of the period,
GDP/c is GDP per capita at the beginning of the period,
$Popden$ is rural population density at the beginning of the period.

This specification is a general specification and is not guided by the restrictions imposed on equation (1) by the search model. The dependent variable is actually the same independent variable specified in equation (1) (see below for estimation implications). The independent variables include G_I and G_N and three 'economic' or policy variables. The variable ln(*HA*) is the primary demand variable, reflecting the units over which a country benefits from CGI research. The ln(*GDP/c*) variable reflects constraints on public spending. Most NARS CGI programmes are publicly supported, and this variable is measuring the capacity to raise government revenues for this purpose. The ln(*Popden*) variable is a measure of 'policy awareness'. Countries with policy awareness recognize the implication of population growth and associated food security issues and respond to this through public investment in food production. Many countries with perceived reservoirs of land available for expansion simply ignore investments in agricultural technology (Boserup, 1981; Bloom and Williamson, 1998).

The interaction terms are designed to capture the association of IARC germplasm with policy variables. Time period dummy variables and climate dummy variables were included in both equations (1) and (3). Technology capital dummy variables (see Appendix 21.2) were included in equation (3). Table 21.2 provides further definitions and means.

The fact that the dependent variable in (3) is an independent variable in (1) raises the question of simultaneity bias. One could agree that (1) is effectively a technical relationship and that NARS plant breeders are technically efficient. If so, (1) could be estimated independently of (3). But given the nature of public investment in CGI programmes and the fact that NARS CGI programmes differ greatly in terms of technical efficiency and capability, it is prudent to utilize econometric techniques to incorporate the implied simultaneity between equations (1) and (3).

Thus, two procedures are employed. In the first (described as two-stage least squares, 2SLS), equation (3) is first estimated. Then predicted values from equation (3) are used in estimating (1). In the second, where pooled observations over all five crops are included (with crop dummy variables), three-stage least squares (3SLS) are directly applied to the joint estimation of a four-equation system where equations for the $\ln(B_N) \times G_I$ and $\ln(B_N) \times G_N$ are included.

Estimates

Table 21.3 reports relevant estimates of equation (1), the NARS CGI investment specification (3), and elasticities of relevant variables, calculated at mean values

It is convenient to discuss the estimates of specification (1) first, noting that in the 2SLS estimate the predicted value of $\ln(B_N)$ is from the first stage estimates of equation (1). The 3SLS estimates of equation (2) are from the system of four equations. One reason for discussing results in this order is that we can first determine whether germplasm (G_I and G_N) actually affects NARS varietal productivity. Then we can turn to the question of germplasm effects on NARS investment.

Table 21.3 reports 2SLS and 3SLS coefficient estimates for the three key variables in equation (1). It also reports elasticity calculations evaluated at sample means.

Before turning to the estimates, two statistical tests can be noted.

The first is a test of specification (2) against specification (1). This test shows that in all cases the mean square error of (1) is lower than (2) (though this is not significant for beans and potatoes) lending support to the functional form dimensions of equation (1) and to the search concept as a foundation for specifications.

The second set of tests is a test of exclusion restrictions in equation (1). This test entails obtaining residuals from equation (1) and regressing these residuals on the exogenous variables excluded from (1) (i.e. included in (3) but not (1)). The statistic nR^2, where n is the number of observations, is distributed as χ^2 with one degree of freedom (d is the number of excluded variables in c). This test shows that the excluded variables are reliably excluded.

Now consider the coefficient estimates for $\ln(B_n)$, $\ln(B_n) \times G_I$ and $\ln(B_n) \times G_N$. All are positive except for $\ln(B_n) \times G_N$ for beans, where the

Table 21.3. Estimates: NARS varietal production specification dependent variable: NARS varietal releases.

Independent variables	2SLS (second stage)						3SLS
	Wheat	Rice	Maize	Beans	Potatoes	Pooled	Pooled
ln(B_N)	4.813	11.926	6.742	1.287	2.490	4.918	7.702
	(1.54)	(2.32)	(2.79)	(0.71)	(0.69)	(2.94)	3.26
ln(B_N) x G_I	0.0966	0.1443	0.8919	0.3496	0.3314	0.1985	0.2395
	(1.85)	(1.53)	(4.21)	(4.55)	(6.24)	(6.61)	(6.29)
ln(B_N) x G_N	0.0141	0.0835	0.0236	–0.00015	0.0427	0.0600	0.0146
	(0.58)	(3.45)	(0.72)	(0.05)	(1.47)	(4.50)	0.73
D beans						–2.32	1.69
D rice						–3.69	–9.48*
D potatoes						–6.78	–8.42*
D maize						–0.35	2.09
No. observations	66	54	32	45	51	248	248
R^2	0.711	0.593	0.849	0.742	0.741	0.533	0.596
P-values							
Equation R^2 (2)	0.0000	0.0000	0.0000	0.0000	0.0000	0.0000	0.0000
$-R^2$ (3)	0.0000	0.0214	0.0032	0.369	0.429	0.0000	0.0000*
Excluded variable χ^2	5.04	16.6	1.17	0.83	16.4		
0.05 value	11.07	12.59	11.07	12.59	16.84		
MV production elasticities							
B_N	0.23	0.72	0.81	0.62	0.52	0.45	0.60
G_I	0.23	0.20	0.50	0.73	0.92	0.38	0.46
G_N	0.04	0.28	0.04	0	0.08	0.14	0.04
Sum	.50	1.20	1.35	1.35	1.52	0.97	1.10

coefficient is effectively zero. All are statistically significant except for the $\ln(B_n) \times G_N$ coefficients for wheat, maize, beans and potatoes, and for the 3SLS pooled results. The statistical significance of the $\ln(B_n)$ coefficient entails the sum of three coefficients and this test is significant except for beans and potatoes.

The elasticity calculations reported in Table 21.3 vary by crop but are generally consistent across crops. The varietal production elasticities for B_N, breeding resources, are low for wheat, but are generally in the 0.45 to 0.8 range and for the pooled estimates which are probably the best estimates, they range from 0.45 to 0.6.

The G_I elasticities are high. They are quite variable by crop and are in the 0.38 to 0.45 range for the pooled estimates. This provides considerable evidence that IARC germplasm does enhance NARS CGI productivity. (Interestingly, the elasticities are similar to the IARC ancestor (IP + IA) shares shown in Table 21.1 for these crops.)

The G_N elasticities are lower and not as significant, but the 2SLS pooled estimates show significant productivity from cumulated NARS-created germplasm.

These elasticities indicate that germplasm stocks form a kind of capital in plant breeding. They also show a large indirect contribution of IARC germplasm in addition to a large direct contribution through IARC-crossed national releases. The sum of the elasticities suggest that with germplasm capital, plant breeding has roughly constant returns to programme scale.

Table 21.4 reports the key coefficient estimates and elasticities combinations for the investment equation. The elasticities in Table 21.3 show large IARC germplasm effects on NARS MV production indicating a complementary relationship between IARC and NARS programmes. Table 21.1 shows a high proportion of IARC-crossed national MV releases, indicating a significant competitive effect between IARC and NARS programmes. The estimates in Table 21.4 should enable us to determine which effect dominates.

We first note that the statistical fits for the investment equations are generally good, with all specifications being highly significant. Since the computed elasticities entail more than one coefficient, *P* values for the relevant coefficient sums are reported with the calculated elasticities.

The HA elasticities are consistently between 0.5 and 0.8 with the pooled estimate being 0.5 to 0.52 at mean levels and 0.63 to 0.67 when evaluated at high G_I levels (one standard deviation above the mean). The strong positive interactions between G_I and HA are important (see below). All HA elasticities are based on coefficient sums that are statistically significant. These HA elasticities indicate perceived scale economy in support of NARS CGI programmes, with larger countries investing less than proportionally as area planted to the crops increases.

Table 21.4. Estimates – Elasticities NARS CGI investment. Dependent variable: $\ln(B_N)$ by period.

Independent variables	2SLS (second stage)						3SLS
	Wheat	Rice	Maize	Beans	Potatoes	Pooled	Pooled
$\ln(G_I)$	–1.82 (1.35)	1.164 (0.56)	0.1884 (2.40)	–2.921 (1.13)	–1.074 (0.37)	–2.061 (2.36)	–2.041 (2.56)
$\ln(G_N)$	–0.87 (85)	0.150 (1.33)	0.251 (1.04)	–0.018 (0.09)	0.013 0.09	–0.03 (0.43)	–0.055 (0.97)
ln(HA)	0.495 (3.45)	0.745 (4.06)	0.581 (1.83)	0.533 (2.08)	0.390 (2.19)	0.347 (6.03)	0.332 (6.24)
ln(Popden)	1.341 (5.88)	0.708 (2.44)	0.313 (0.52)	0.281 (0.63)	–0.022 (0.05)	0.367 (2.39)	0.534 (2.39)
ln(GDP/c)	–0.339 (0.74)	0.863 (2.01)	–2.76 (0.61)	–0.275 (0.86)	1.065 (1.73)	117 (0.76)	0.125 (0.89)
$\ln(G_I) \times \ln(HA)$	0.108 (1.87)	0.029 (0.46)	0.170 (1.02)	0.066 (0.67)	0.069 (0.84)	0.103 (4.39)	0.122 (5.63)
$\ln(G_I) \times \ln(Popden)$	–0.095 (0.75)	0.237 (1.59)	0.971 (1.57)	0.122 (0.63)	0.149 (0.69)	0.112 (1.65)	0.094 (1.52)
$\ln(G_I) \times \ln(GDP/c)$	0.112 (0.78)	–0.015 0.10	1.409 (2.16)	(0.150) (0.17)	–0.092 (0.34)	0.007 0.09	–0.022 (0.32)
No. observations	66	54	32	45	51	248	248
R^2	0.711	0.593	0.849	0.742	0.741	0.533	0.496
F	22.9	13.16	6.36	5.70	6.82	29.5	
NARS CGI investments elasticity calculations							
HA***	0.75 (0.95)	0.78 (0.81)	0.67 (0.86)	0.59 (0.66)	0.51 (0.59)	0.50 (0.63)	0.52 (0.67)
	(0.000)	(0.000)	(0.026)	(0.026)	(0.007)	(0.000)	(0.000)
Popden***	1.12 (0.93)	0.99 (1.26)	0.84 (1.56)	0.39 (0.52)	0.24 (0.46)	0.49 (0.63)	0.47 (0.58)
	(0.000)	0.044	(0.005)	(0.279)	(0.704)	(0.000)	(0.000)
GDP/c***	–0.08 (0.14)	0.84 (0.83)	0.49 (1.50)	–0.14 (0.02)	0.80 (0.79)	0.13 (0.14)	0.09 (0.07)
	(5.26)	(0.641)	(0.032)	(0.653)	(0.022)	(0.319)	(0.330)
G_I****	–0.05 (0.18) (0.10)	2.80(2.85)(3.08)	–0.68 (–0.39) (0.41)	–0.28 (–0.02) (0.10)	–0.16 (–0.02) (0.09)	–0.11 (0.11) (0.21)	–0.14 (0.12)(0.40)
	(0.143)	(0.612)	(0.028)	(0.255)	(0.704)	(0.015)	(0.009)
G_N*****	–0.087	(0.150)	(0.251)	–0.018	0.013	–0.03	–0.066
	(0.40)	(0.192)	0.315	(0.930)	(0.920)	(0.67)	(0.334)

*Evaluated at mean G_I and mean + 1 SD G_I; ***P* values sum of coefficient; ***evaluated at mean HA, mean + 1 SD HA, mean + 1 SD Popden; ****evaluated at mean.

The population density elasticities, interestingly, are also generally significant, both economically and statistically (except for beans and potatoes). The elasticities for pooled estimates are roughly 0.5 and rise to 0.6 with high levels of G_I because of the strong interaction terms. This indicates that governments (and aid agencies) are responding to population pressure in supporting NARS CGI programmes. This is quite strong evidence for a 'Boserupian' effect where population pressure induces investments to offset the 'Malthusian' effects of population growth.

The GDP/c effects on investment, by contrast, are low both in elasticity size and statistical significance. In general, the income effects are not significant. Nor are they strongly related to IARC germplasm. This may be due in part to a correlation with the prices of scientists, where higher GDP/c is proxying for higher real prices of scientists. This negative price of scientist effect may then effectively produce a positive income effect (in any case, it is not obvious that a strong income effect should be observed for investments in production enhancement).

The G_I elasticities are of particular interest here. The individual crop elasticities are generally not statistically significant and are quite low (except for rice). The pooled estimates are statistically significant, however. G_I elasticities are evaluated at the means for interactive variables, at the mean plus 1 standard deviation for the HA variables, and at the mean + 1 SD for both HA and Popden. When evaluated at mean values for the HA and Popden varables, G_I impacts are low and negative, indicating that the competition effects (or dependency effect) outweigh the complementary effects for the mean country. However, for the large countries, the G_I elasticities are positive (but low, 0.11–0.12), indicating that the complementarities outweigh the competition effects. When large countries also have high population densities, the G_I impacts are higher (0.21 to 0.4).

Conclusions

This statistical study yielded the following conclusions:

1. That IARC germplasm services provide a very important input to NARS CGI programmes. The elasticities of IARC germplasm capital were significant (approximately 0.4) indicating a germplasm contribution even higher than IARC ancestry indicates.
2. That IARC germplasm impacts on NARS CGI investments consist of two opposing components. One is the complementarity effect on NARS CGI productivity as reflected in the ancestry indicators and the statistical estimates. The second is the competition effect reflected in the high proportion of national MV releases used in IARC crosses.

For the mean country in the study, the competition effect was slightly higher than the complementary effect. For countries with small hectarage planted to the crop, the competition effect is dominant, while for the largest countries, the complementarity effects dominated. For countries with lower population densities in rural areas, the competition effect dominated. For countries with high population densities in rural areas, the complementarity effects dominated.

These estimates thus show both dependency on IARC programmes, where the competitive effects dominate, and enhancement where complementing germplasm effects dominate. When weighted by population, enhancement effects dominate. For all countries weighted by hectares planted, the net complementarity effect produced roughly 15% more NARS CGI investment.

References

Bloom, D.E. and Williamson, J.G. (1998) Demographic transitions and economic miracles in merging Asia. *World Book Economic Reviews* 12, 419–55.

Boserup, E. (1981) *Population and Technological Change: a Study of Long Term Trends.* University of Chicago Press, Chicago.

Evenson, R.E. (2000) *Productivity and Agricultural Growth in the Second Half of the 20th Century: State of Food and Agriculture.* FAO, Rome.

Evenson, R.E. and Gollin, D. (1997) Genetic resources, international organizations, and rice varietal improvement. *Economic Development and Cultural Change* 45, 471–500.

Evenson, R.E. and Kislev, Y. (1975) *Agricultural Research and Productivity.* Yale University Press, New Haven, Connecticut.

Gollin, D. and Evenson, R.E. (1998) Breeding values of rice genetic resources. In: Evenson, R.E., Gollin, D. and Santaniello, V. (eds) *Agricultural Values of Plant Genetic Resources.* CAB International, Wallingford, UK, pp. 179–194.

Heisey, P.W., Lantican, M.A. and Dubin, H.J. (2002) Wheat. In: Evenson, R.E. and Gollin, D. (eds) *Crop Variety Improvement and Its Effect on Productivity: the Impact of International Agricultural Research.* CAB International, Wallingford, UK.

Kortum, S. (1997) Research, patenting and technological change. *Econometrica* 65, 1389–1419.

Rosebloom, J. and Pardey, P.G. (1989) *A Global Data Base on National Agricultural Research Systems.* Cambridge University Press, Cambridge.

Appendix 21.1

Search models applied to plant breeding

Plant breeders at any point in time are constrained by their knowledge of techniques, their experience and germplasm stocks available to them. These constraints define a probability distribution for the discovery of a 'trait' (the trait may be a 'quantitative' trait expressed in the crop yield, or as a 'qualitative' trait such as best plant resistance to a plant disease). Suppose that probability distribution $f(x)$ has the exponential form (after Evenson and Kislev, 1975)

$$f(x) = \lambda e^{-(\lambda - \theta)} \quad (1)$$

The cumulative distribution $F(x)$ is:

$$F(x) = 1 - e^{-\lambda(x - \theta)} \quad (2)$$

with mean and variance

$$E(x) = \theta + {}^{1}/_{\lambda} \quad (3)$$

$$\text{VAR}(x) = {}^{1}/_{\lambda}{}^{2} \quad (4)$$

The cumulative distribution of the largest value of x (defined as z) in a sample of size n is the central concept of the search model because plant breeders are searching for the largest x. This distribution is the order statistic.

$$H_n\,(z) = [1 - e^{-\lambda(z - \theta)}]^n \quad (5)$$

and the probability density function for (z) is:

$$H_n\,(z) = \lambda_n[1 - e^{-\lambda(z - \theta)}]^{n-1} \quad (6)$$

The expected value of z is then a type of trait production function. It is:

$$E_n(z) = \theta + \frac{1}{\lambda} - \sum_{\lambda=1}^{n} \frac{1}{n} \quad (7)$$

This expression is well approximated by:

$$E_n(z) = \theta + B\ln(n) \quad (8)$$

Kortum (1997) has shown that the approximation (8) holds not only for the exponential distribution $f(x)$ but for the uniform distribution and virtually all classes of distributions except the 'fat-tailed' Cavehy type distribution. Viewing inventions as proportional to $E_n(z)$ provides a functional form for the breeding invention function:

$$V_n = a + b\ln(B_n) \quad (9)$$

Germplasm enters this model by shifting $f(x)$ to the right. Suppose that this shifts the mean but not the variance, then the expression for $E_n(z)$ will be:

$$E_n(z) = \theta_1 + \theta_2 + (B_1 + B_2) \ln(n) \tag{10}$$

This yields the germplasm enhancement term:

$$V_N = a + b\ln(B_N) + cG_I \ln(B_N) \tag{11}$$

Appendix 21.2

Climate zones and technology classes

Climate zones are the Trewartha types. *Goode's World Atlas*, Rand-McNally and Company, 1960.

These climate types are:

1. Af, Sm: Tropical Rainforest
2. Aw: Tropical Savannah
3. Bs: Steppe
4. Bw: Tropical and Subtropical Desert
5. Cs: Mediterranean
6. Ca: Humid Subtropical
7. Cb, Cc: Marine West Coast
8. Da: Humid Continental, Warm Summer
9. Db: Humid Continental, Cool Summer

Dummy variables for the two leading climate types were created.

Technology capital classes

The table provides the criteria for classifying countries into Technology Capital Classes, TCI to TCIV (Evenson, FAO State of Food and Agriculture, 2000).

Dummy variables for Technology Classes I–IV were created.

Technology capital indexes

Developing countries were classified according to technology capital (TC) classes in three different periods (1961–1976, 1971–1986, 1981–1996). Four TC classes were defined, based on eight indicators as shown below. The criteria ensure that countries are included in only one TC class in each period. Most countries achieved TC improvements in recent decades.

Indicators[a]	TC-1	TC-II	TC-III	TC-IV
Adult male literacy	= 50%	<50%	<50%	<35%
Proportion of labour force in industry	<10%	<15%	>15%	>15%
Foreign direct investment/GDP	Little or none	<0.5%	= 0.5%	0.25% or more
R&D in manufacturing firms/value added	None	None	<0.25%	= 0.25%
Royalties and licence fees paid	None	None	None	Substantial
Royalties and licence fees received	None	None	None	Minor
Agricultural research investment intensity	Low, <0.25% of agricultural production	Moderate, 0.25–0.5% of agricultural production	High, 0.25% of agricultural production	High, 0.5% of agricultural production
Intellectual property rights	None	None	Weak protection	Moderate protection

[a]Indicator data are drawn from the World Bank Development Indicators Database.

22 Production Impacts of Crop Genetic Improvement

R.E. Evenson

This chapter, the second of three synthesis chapters, provides estimates of the magnitude of crop genetic improvement (CGI) impacts on production in developing countries. Two estimates are provided. Both are expressed in terms of annual contributions to productivity growth by decade and region. A range of estimates (high, low) is provided to reflect the uncertainty in the estimates.

The first estimate provided is for all CGI improvements since 1965 in developing countries. The second estimate provided is for the IARC CGI contributions. These estimates are utilized in the final chapter of this volume, where the economic consequences of CGI on prices, production, trade and welfare are analysed.

The productivity measure used in this chapter is a 'Total Factor Productivity' (TFP) measure. This measure has two interpretations. The first is as a measure of additional output or product for a given set of inputs. Thus, when farmers adopt improved crop varieties, they realize increased production from given land, water, fertilizer, labour and other inputs. The TFP measure adjusts for changes in input use that may have been associated with the adoption of improved varieties. Thus, a CGI contribution of 1% per year for the decade of the 1970s in Asia means that after taking into account changes in input use, farmers increased production by 1% per year over the decade.

The second interpretation of the TFP measure is that it is also a measure of real cost reduction (RCR). For given input prices, i.e. the rental prices of land and water, fertilizer prices and labour wages, a 1% per year change in TFP over a decade essentially means that if input prices did not change, the real costs of producing 1 t of rice or other crops was declining by 1% per year. The TFP-RCR measure adjusts for actual changes in input prices, but the RCR interpretation is important because it has implications for equilibrium prices of commodities. We have observed significant declines in the real world equilibrium prices of most agricultural products over recent decades. This is in part due to high RCR gains in economies where high rates of population growth have kept real wages from rising.

It is important to note, however, that there are, in principle, many sources of TFP-RCR growth. CGI contributions are only one source. Other sources include improved skills of farmers and farm managers, improved functioning of markets, improved transport and communication infrastructure. These other sources also interact with CGI to produce TFP-RCR gains (see Chapter 19).

In order to measure CGI contributions to TFP-RCR gains, we first require data on actual adoption of modern varieties (MVs; varieties produced after 1965). Part 2 of this chapter reviews MV adoption data and draws comparisons with MV production data. This comparison shows a considerable disparity in rates of MV production and MV adoption by region. A statistical analysis of this relationship is also reported. This analysis, combined with the analysis in the previous chapter, attests to a long 'pipeline' in the delivery of CGI to farmers.

The second step in computing the CGI contribution is to estimate the productivity gains associated with the conversion of land area from traditional varieties (TVs; i.e. pre-1965 varieties) to MVs. Estimates of these gains from crop studies (Chapters 4–16) and country studies (Chapters 18–20) are reviewed in part 3 of the chapter.

Part 4 reports TFP calculations for the three major crops in developing countries – rice, wheat and maize – and relates these measures to MV adoption. Part 5 summarizes CGI gains by decade and region.

Part 6 develops two estimates of the IARC contribution to CGI gains. The two estimates differ according to the assumptions regarding NARS varietal production in the absence of IARC competition in the production of varieties.

Part 7 compares estimated CGI contribution with actual yield changes and 'tests' the validity of the CGI contribution estimate.

Estimates of Modern Variety Adoption (MV/TV conversion)

Chapters 4–16 present data on farmers' adoption of 'modern' or improved crop varieties (MVs) for different periods of time. These chapters also review studies of the production impacts associated with the replacement of traditional varieties (TVs) by MVs. In some cases, estimates of the production impact of modern variety 'turnover', i.e. the replacement of each generation of modern or improved varieties with successive generations of modern varieties, were reported.

Table 22.1 summarizes estimates of MV diffusion by region and crop for 1970, 1980, 1990 and 1998. These estimates are not of equal reliability, being most reliable for wheat and rice, but on the whole they offer a reasonably accurate picture of modern variety diffusion. That picture is one of unevenness by region and crop. This is particularly apparent for the Middle East and North Africa (MENA) and sub-Saharan African regions where MV adoption rates were low for all crops in 1970 and were still low for most crops in 1980. By contrast, Latin America and Asia have significant MV adoption by 1980. As of 1998, MV adoption was still low for cassava, beans and lentils in all regions and for sorghum, millets and maize in sub-Saharan Africa.

Table 22.1 also clearly shows that MV diffusion for aggregate crops differs greatly by region. Sub-Saharan Africa had less than one-third the level of MV adoption attained in Asian economies in 1998. In the 1960s and 1970s sub-Saharan Africa had a little over 10% of the MV adoption levels of Asia.

A statistical analysis of the relationship between MV production and MV adoption (as diffusion) is reported in Table 22.2. The observations on which these estimates are based are for countries where both MV production and MV adoption data could be obtained. For rice and wheat, these data were available for three decadal periods. For maize, data on MV adoption were available only for the last two decades. For other crops (sorghum, millet, beans, cassava, potatoes) only current (1998) MV adoption levels were available.

The variables utilized in the specifications are defined as follows:

ln (CMVA): logarithm of the cumulated (i.e. end of decade) area of the crop planted to modern varieties.

ln (CMVR): logarithm of the cumuluated (i.e. end of decade) number of varieties released in the country. (This includes both IARC-crossed and NARS-crossed varieties.)

ln (CMVR) × ln (HA): ln (CMVR) times total hectares planted to the crop at the end of the decade.

ln (CMVR) × Age: ln (CMVR) times the average age of MV varieties (calculated as the share of MVs by decade times the decadal lag).

Climate dummies (see Appendix 21.2 for a definition); Climate A_F is the reference climate.

Period dummies: P_{75-85}, P_{86-95} (the first period is the reference period).

Crop dummies: used in the pooled specification, sorghum and millets, cereals and all crops.

Table 22.1. Modern variety (MV) diffusion 1970, 1980, 1990 and 1998. Percentage of area planted to modern varieties.

	Latin America				Asia (including China)				Middle East–North Africa				Sub-Saharan Africa			
	1970	1980	1990	1998	1970	1980	1990	1998	1970	1980	1990	1998	1970	1980	1990	1998
Wheat	11	46	82	90	19	49	74	86	5	18	38	66	5	22	32	52
Rice	2	22	52	65	10	35	55	65					0	2	15	40
Maize	10	20	30	46	10	25	45	70					1	4	15	17
Sorghum					4	20	54	70					0	8	15	26
Millets					5	30	50	78					0	0	5	14
Barley									2	7	17	49				
Lentils									0	0	5	23				
Beans	1	2	15	20									0	0	2	15
Groundnut					0	15	20	50					0	0	20	40
Cassava	0	1	2	7	0	0	2	12					0	0	2	18
Potatoes	25	54	69	84	30	50	70	90					0	25	50	78
All crops	8	23	39	52	13	43	63	82	4	13	29	58	1	4	13	27

Table 22.2. Estimated relationship: MV production and MV adoption; international data: 1965–75, 1976–85, 1986–95 OLS estimate: dependent variable ln (CMVA) weighted by crop area.

Independent variable	Rice	Wheat	Maize	Sorghum/ millets	Cereals	All crops
ln(CMVR)	–1.296	–3.144	–3.608	–0.778	–2.739	–2.572
	(2.17)	(7.93)	(4.77)	(3.49)	(8.86)	(9.31)
ln(CMVR) × ln(HA)	0.113	0.237	0.266	0.246	0.193	0.189
	(3.92)	(12.13)	(6.27)	(11.52)	(12.50)	(13.46)
Climate A_N	–0.33	–0.11	–0.74	–0.42	–0.33	–0.45
	(1.03)	(0.45)	(1.91)	(1.46)	(2.11)	(2.95)
Climate B–	–3.02	0.12	–0.62	0.12	–0.01	–0.06
	(2.84)	(0.50)	(1.12)	(0.19)	(0.02)	(0.36)
Climate B–	–1.12	–0.02	0.11	–1.11	0.19	0.16
	(0.73)	(0.07)	(0.16)	(1.80)	(0.93)	(0.79)
Climate C–	3.56	0.56	–0.87	0.41	0.65	0.67
	(2.36)	(2.37)	(0.77)	(0.90)	(2.92)	(3.78)
Climate C_a	–0.04	–0.50	1.04		0.16	0.19
	(0.18)	(1.51)	(1.67)	NR	(1.05)	(1.30)
Climate C_b	0.37	0.35	0.01		0.70	0.74
	(0.83)	(0.72)	(0.02)		(3.37)	(3.78)
Period 76–85	–0.176	1.16			0.511	0.618
	(0.86)	(4.38)	NR	NR	(3.74)	(4.31)
Period 86–95	0.084	1.59	–0.70		0.685	0.969
	(0.32)	(4.67)	(2.82)	NR	(4.20)	(4.73)
Ln(CMVR) × Age						–0.009
						(2.69)
Constant	13.18	11.02	13.15	7.15	12.80	
	(23.54)	(23.92)	(19.60)	(13.74)	(39.69)	
No. observations	51	66	43	27	187	205
R^2	0.904	0.904	0.912	0.984	0.834	0.853
Elasticity of CMVA wrt. CMVR at mean HA	(0.438)	(0.378)	(0.153)	(0.959)	(0.105)	(0.215)

The estimates reported in Table 22.2 show that the ln (CMVR) × ln (HA) variable is highly significant and positive in all specifications, showing that MV production is related to MV adoption rates. This may seem to be obvious, but note that this means that as more MVs are produced, MV/TV conversion is taking place. If it were the case that the first MVs were responsible for conversion of TV areas to MV areas and that subsequent MVs were responsible for MV/MV turnover, the elasticity of MV adoption with respect to MV production would be approximately zero. When these elasticities are computed at the mean HA levels (and considering the negative ln (CMVR) coefficients), they are low but positive, indicating that as more MVs are produced they are creating MV/TV conversion as well as MV/MV conversion.

The elasticities of CMVA with respect to cropped area are less than 1 in all cases for periods 1 and 2 and approximately 1 in period 3.

For crop specifications, the ln (CMVR) × Age variable was not statistically significant. For the pooled crop specification this variable was negative, indicating a decline in the MV/TV conversion elasticity as MV experience increases.

Table 22.2 thus supports the conclusions of the country chapters, where the strategy of qualitative trait breeding for host plant resistance to plant diseases and insect pects and for host plant tolerance to abiotic stresses enabled the high yielding quantitative trait plant types to be planted on more hectarage; i.e. to create MV/TV conversion. These estimates also attest to the long 'pipeline' associated with plant breeding. In Chapter 21, the pipeline for MV production was shown to include germplasm effects. In this chapter, further time lags between MV adoption and MV production are identified.

Estimates of Productivity Impacts of MV/TV Conversion from Crop Studies, Country Studies and Crop TFP Estimates

In this section three sets of evidence are used to evaluate the productivity impacts of MV/TV conversion (and in some cases of MV/MV conversion as well). The first set of evidence is reported in the crop chapters (Chapters 4–16) in this volume. The second set of evidence is reported in the three country study chapters (Chapters 18–20). The third set of evidence is based on crude crop TFP calculations based on FAO country data. These calculated TFP growth rates are statistically related to MV/TV conversion data for rice, wheat and maize, where data are available.

Each set of evidence is subject to limitations and each taken separately may not be regarded to be 'consensus' estimates of MV/TV or

MV/MV turnover impacts on crop productivity. But taken together, all three sets of evidence are in substantial agreement and this agreement supports the consensus interpretation.

Estimates from the crop studies and country studies

Crop study evidence is of two types. The first type is experimental evidence, where MV/TV yield comparisons (and MV/MV comparisons as well) are made under conditions where experimental controls are utilized. These experiments may be in field-station locations or they may be on farm sites with some degree of farm management. In the absence of a statistical design to farm site experiments, however, this evidence is subject to the criticism that real farm experience is not being replicated.

The second type of crop study evidence is based on secondary data (e.g. at the province or district level) on production, area and yield. In some cases data on other inputs, fertilizer, labour, machines are available (as in the China study, Chapter 18) to enable crop TFP calculations.

Productivity impacts, whether based on MV/TV conversions or MV/MV turnover, are not necessarily constant as MV/TV ratios change. For rice, and to some degree for other crops as well, MV 'generations' have been defined. The first-generation MVs are based on quantitative high yielding plant type traits. This generation, once established may have high MV/TV impacts but these are often transitory because of susceptibility to plant diseases and insect pests. The second-generation MVs are based on direct responses to these susceptibilities. Host plant resistance to diseases and pests is sought through qualitative trait breeding. As these varieties are adopted, they replace first-generation MVs and expand MV areas to new regions where first-generation susceptibility limited MV adoption.

Third-generation MVs in rice have incorporated host plant tolerance to abiotic stresses (drought, salinity, submergence, etc.). These traits have also enabled expansion of MV area as well as MV/MV turnover.

Byerlee and Traxler (1995) have argued that first-generation impacts are larger than second and third generation impacts in wheat. For rice, however, the evidence is less clear.

A study of the productivity impact of rice by Evenson (1997) estimated that improved rice varieties had contributed 13.4% to production by 1984 when 41% of rice area was planted to modern varieties. A second study for rice (Evenson, 1998) utilizing district data for the 1956–1987 period, estimated modern variety impacts in a multi-equation model where the adoption of MVs was treated as an endogenous variable. Determinants of MV adoption included the availability (in MVs suited to the district) of host plant resistance (HPR) traits for disease and insects and host plant tolerance (HPT) traits for drought and salinity. The study concluded that the incorporation of these traits into MVs increased the MV coverage from under 40% to over 60% by 1987. The yield effect was unrelated to the MV coverage variable, indicating that the new area covered achieved yield gains that were roughly of the same order of magnitude as those achieved in the earliest adopting regions. The estimated yield effect was 1 t ha^{-1} (i.e. rice yields would have risen from 1.5 t ha^{-1} to 2.5 t ha^{-1} with 100% MV adoption).

Table 22.3 reports a summary of estimates of yield impacts of MV adoption and of MV turnover on productivity from both crop studies and country studies. Most of the estimates are of MV adoption effects, i.e. the replacement of traditional varieties by MVs. The 'percent' estimates are estimates of full (i.e. 100%) replacement of TVs by MVs. Some studies are based on statistical studies of micro farm-level data and some are based on aggregate panel data of the type utilized in the India chapter.

Several of the statistical studies treated the area planted to modern varieties as an endogenous variable to be predicted as a function of variables such as extension service, farmer schooling and of agricultural research services suited to the area. In the Evenson (1998) study, variables measuring the availability of traits for drought and submergence tolerance and the number of landraces in the suitable released varieties were also included in the MV adoption specifications.

These studies did not fully resolve the comparison between MV/TV versus MV/MV effects, because the HPR and AST traits were incorporated into the second- and third-generation MVs that were replacing first-generation MVs as well as in the MVs replacing TVs. However, the country studies summarized in Table 22.3 do provide some evidence on the matter of MV/TV versus MV/MV conversion because most of the turnover in Brazil and China was MV/MV conversion, i.e. of new MVs replacing older MVs. These turnover estimates (for 100% replacement) are roughly one-third of the gains associated with replacement of TVs.

Table 22.3 reports mean 'consensus' estimates of full MV/TV replacement by crop. These are relatively conservative estimates based on the available evidence. The strategy in the CGI contribution to productivity reported later in this chapter is to apply two-thirds of the consensus estimate to the increments in MV hectarage by decade (see Table 22.1). The remaining one-third is applied to cumulated MV hectarage from past and current decades, so that the total effect at the end of each decade is the present MV at that time multiplied by the consensus factor.

Table 22.3. Synthesis: estimates of MV/TV and MV/MV impacts on yield crop studies and country studies.

	MV/TV estimates (full replacement)				MV/MV (turnover) estimates (full turnover)				
Crop	Country	t ha^{-1}	Per cent increase	Source	Country	t ha^{-1}	Per cent increase	Source	Consensus MV/TV estimated
Wheat	India	0.98	46	Evenson, 98	China	0.74	24	Chapter 18	45%
	India	0.98	45	Chapter 19	Latin America	0.2	10	Chapter 4	
Rice	India	0.50	33	Gollin & Evenson, 99	China	1.6	29	Chapter 12	47%
	India	0.98	65	Evenson, 98	Brazil	0.5	20	Chapter 20	
	India	0.67	43	Chapter 19					
	Sub-Saharan Africa (upland)		24	Chapter 6					
Maize	India	0.98	65	Chapter 19	Brazil	0.41	20	Chapter 20	50%
	Sub-Saharan Africa	0.60	45	Chapter 8	Latin America		5–15	Chapter 7	
Sorghum	India	1.38	80	Chapter 9					45%
	India		37–40	Chapter 19					
	Sub-Saharan Africa		7–63	Chapter 9					
Pearl millet	India	0.48	45	Chapter 19	India		40–45	Chapter 10	45%
	India		45	Chapter 10					
	Sub-Saharan Africa		38	Chapter 10					
Barley	Middle East–North Africa		25	Chapter 11					41%
Lentils	Middle East–North Africa		41	Chapter 13					41%
Beans	Latin America	0.21	35	Chapter 12					25%
	Sub-Saharan Africa	0.4	55	Chapter 12					
Cassava	Sub-Saharan Africa	3.74	48	Chapter 16					48%
	Latin America	3.29	29	Chapter 16					
Potatoes	Global	2.5	35	Chapter 15					35%

TFP and MV/TV Conversion Estimates for Rice, Wheat and Maize

In this section, evidence that MV/TV conversion impacts directly on TFP growth is presented for the three major crop commodities in developing countries. This evidence is an important addition to the crop and country study estimates in two respects. First, it is based on TFP calculations rather than yield. Second, it is based on international comparisons as well as comparisons over time, thus adding an international dimension to the micro-crop studies and the regional country studies.

Calculations of TFP growth

The TFP growth relationship can be expressed as:

$$G_{TFP} = G_P - S_A G_A - S_W G_W - S_F G_F - S_{AP} G_{AP} - S_M G_M$$

where

G_P is the growth rate in production of the crop
G_A is the growth rate in land (and water)
G_W is the growth rate in work (human power) use
G_F is the growth rate in fertilizer use
G_{AP} is the growth rate in animal power use
G_M is the growth rate in mechanical power use.

The shares S_A, S_W, S_F, S_{AP} and S_M are cost shares and reflect the marginal products of each factor of production. Under conditions of scale neutrality, cost shares, i.e. the share of the factor in total cost, are the correct shares for this calculation. These shares can be changed from one period to the next if appropriate data are available.

FAO maintains a database for countries from 1961 to date, enabling the following calculations:

G_P and G_A for rice, wheat and maize
G_W, G_F, G_{AP} and G_M for all crops
S_F, S_{AP}, and S_M for all crops.

There are two issues, then, associated with calculating G_{TFP}.

First, is it reasonable to use G_W, G_F, G_{AP} and G_M measured for all crops as proxies for crop-specific measures?

Second, can one obtain measures of the missing shares, S_A and S_W?

There is no question that errors of approximation are made when G_W, G_F, G_{AP} and G_M are treated as crop-specific. But this error is lower for major crops than for minor crops. Rice, wheat and maize are the three major crops in most developing countries. In aggregate these three crops are planted on roughly two-thirds of cropped land in developing countries.

The second question is also important because land rent data are not available to compute S_A, and wage data are also not effectively available to compute S_W.

In view of the importance of the crops and the potential value of corroborating evidence from MV/TV impacts, a decision was made to calculate G_{TFP} measures for rice, wheat and maize in countries producing more than 1 million hectares of the crop. These calculations were made for three periods – 1965–1975, 1976–1985 and 1986–1995. Three-year averages were used for the growth measures. Shares were calculated by period for S_F, S_{AP} and S_M using international (dollar) prices for fertilizer, animal services, tractors and harvester–threshers, and the estimates of the crop agricultural value (in dollars). The shares of land and labour were arbitrarily set to equal half the residual ($1 - S_F - S_{AP} - S_M$). (This allocation is generally consistent with farm management cost studies.)

The reader should, of course, be aware that there are errors of attribution in these measures (note, however, that G_P and G_A are crop-specific measures).

Tables 22.4 and 22.5 report simple analyses of the G_{TFP} measures computed for 54 countries for rice, 32 countries for wheat, and 64 countries for maize. Table 22.6 reports estimates of MV/TV impact on G_{TFP} for the subset of countries for which MV/TV data are available.

Table 22.4 reports estimates of G_{TFP} measures by decade. These estimates are based on area weighted OLS regressions of G_{TFP} measure on time period (specification 1) and geographic region dummy variables (specification 2). The explanatory power of this regression estimate is low (although all meet the basic *F*-test requirement). This reflects the fundamental nature of international agricultural production data.

These data show that rice TFP growth was modest in the first two periods, then declined in the third period. For wheat the picture is one of very high TFP growth in the first period, high growth in the second period and modest growth in the third period. For maize, TFP growth has been high in all three periods.

These calculations then show high TFP growth rates for both wheat and maize of over 2% per year for 30 years and more modest TFP gains for rice (approximately 1.2% per year over the 30-year period). Growth in the first (original Green Revolution) period was highest and has slowed in the past two decades.

Table 22.5 is simply intended to show the relationship between these TFP growth measures and yield growth measures. For the pooled regressions these estimates show that TFP growth was 80% of yield growth and that TFP growth and yield growth are highly correlated (this is partly related to the computational procedures, however).

Table 22.4. TFP growth estimates, rice, wheat and maize. Dependent variable TFP growth by decade.

	Rice		Wheat		Maize		Pooled	
	(1)	(2)	(1)	(2)	(1)	(2)	(1)	(2)
1965–76	0.147**	0.083**	0.582**	0.431**	0.282**	0.254**	0.231**	0.253**
1975–86	0.159	0.094	0.384**	0.236**	0.186**	0.150**	0.148**	0.160**
1986–95	0.059	0.015**	0.096**	0.048**	0.308	0.279**	0.096**	0.122**
Wheat								0.143**
Rice								–0.131**
East–SE Asia		0.089**		0.309**		0.143**		0.134**
South Asia		0.043		0.193**		–0.217**		0.045*
Middle East–North Africa		0.104		–0.046		–0.128		–0.113*
Sub-Saharan Africa		0.102		0.192		0.018		0.170**
No. observations	162	162	96	96	192	192	450	450
R^2	0.084	0.119	0.304	0.436	0.034	0.157	0.034	0.288

*Significant at 10% level; **Significant at 5% level.

Table 22.5. TFP–yield relationships. Dependent variable: TFP growth by decade.

	Rice	Wheat	Maize	Pooled	
Yield growth	0.663	1.075	0.927	0.807	
	(10.44)	(11.05)	(13.18)	(25.08)	
Constant	−0.707	−1.155	−0.951	−0.793	
	(8.86)	(8.44)	(10.28)	(17.08)	
				d wheat	0.009
					(0.44)
				d rice	−0.093
					(4.79)
No. observations	162	96	192	450	
R^2	0.405	0.565	0.477	0.499	

Table 22.6 reflects the major objective of this exercise. It relates cumulated TFP growth to cumulated MV percent measures for 17 rice producing countries, 20 wheat producing countries and 19 maize producing countries where MV adoption data are available (see Table 22.1 for regional measures).

These OLS estimates (weighted by area harvested) should be interpreted in the context of a dependent variable with attribution errors as well as weather errors and other measurement errors.

The estimates do show that MV/TV conversion produces TFP growth. Note that time dummy variables also show that other factors are producing cumulated TFP growth over time as well. Some variation in the coefficients is apparent.

Consider the pooled regression,[1] however. For these three crops in the countries in the sample, MV adoption had reached roughly 65% of planted area even for the countries concerned. The MV/TV coefficient of 0.534 then indicates a CGI contribution to TFP growth of $0.534 \times 0.65 = 0.35$. This CGI contribution is approximately 55–65% of realized TFP growth and 44–52% of realized yield growth for these crops.

These estimates, while subject to error (note that the statistical procedure recognizes these errors in dependent variables), do corroborate the consensus estimate reported in Table 22.3 from the crop and country studies.

[1] The pooled regression (pooled (2) in Table 22.6) uses the square of CMVA to reflect MV/MV conversion (which should increase with CMVA squared). The evidence for MV/MV conversion is positive but weak.

Table 22.6. TFP–MVA relationships. Dependent variable: cumulated TFP growth.

Independent variables	Rice	Wheat	Maize	Pooled (1)	Pooled (2)	
Cumulated MV adoption (CMVA)	0.720	0.470	0.644	0.534	Sq (CMVA)	0.253 (0.73)
	(3.82)	(1.18)	(3.24)	(2.83)	SQRT (CMVA)	0.156 (0.82)
					HA × CMVA	4.88 (1.00)
d 1976–85	0.059	0.302		0.148	0.179	
	(0.66)	(1.23)	NR	(1.31)	(1.45)	
d 1985–95	−0.048	0.355	2.476	0.201	0.188	
	(0.42)	(1.17)	(5.43)	(1.57)	(1.32)	
Constant	−0.079	0.467	−0.019	1.203	0.231	
	(0.89)	(2.43)	(0.15)	(0.73)	(0.95)	
D Wheat				0.421	−0.302	
				(2.93)	(2.04)	
D Rice				−0.218	−0.306	
				(1.13)	(2.05)	
No. observations	51	60	38	149	149	
R^2	0.339	0.152	0.497	0.390	0.337	

CGI Yield Growth Contribution Estimates

Table 22.7 reports a summary of annual CGI contributions to yield growth by crop by decade. The estimates are produced from the MV adoption data in Table 22.1 and the consensus MV/TV conversion estimates reported in Table 22.3 (and supported by Tables 22.4–22.6). (Note that the 1960–2000 estimates include projections for 1999 and 2000.) These growth components are reported by crop and region in Table 22.9. Since these estimates are based on MV adoption levels and on the consensus productivity estimates, it is not surprising that they are largely determined by MV adoption patterns. The highest growth contributions over the 40-year period are realized in the 'Green Revolution' crops, wheat and rice. Interestingly, contributions in potatoes are also high. Maize contributions have been important as well. Growth contributions in lentils, beans and cassava have been low, although they are rising rapidly for beans and lentils.

Table 22.7 also enables a comparison of the IARC content of adopted varieties with the IARC content of released varieties over the entire period. For all crops, IARC crosses accounted for 36% of releases and 35% of area under MVs. It should be noted that IARC crosses have higher levels of multiple releases than NARS crosses (see Chapter 21), and when this is considered, IARC crosses have a higher proportion in adoption than in releases. This is particularly pronounced in crops other than wheat, maize and potatoes. It can also be noted that both proportions are very high in barley, lentils, beans and cassava, where IARC programmes effectively initiated CGI work on the crop in most regions.

In any case, it is quite extraordinary that IARC crosses account for 35% of MV hectarage in 1998. This proportion is considerably greater than the IARC shares of scientists or of research expenditures on these crops.

It is also impressive that another 34% of MV hectarage is planted to NARS-crossed varieties with IARC parents and other ancestors (25% parents and 9% grandparents and other ancestors). This proportion is higher than the comparable proportion in varieties released and attests to the success of the IARC strategy of producing first-generation MVs that are then utilized as parents in subsequent generations of MVs (it also suggests that the production impact of subsequent generations of MVs may be below that of the first-generation MVs). The 35% figure for IARC ancestors for all crops is approximately the same as the IARC germplasmic contribution estimated in the previous chapter.

Table 22.8 reports the CGI growth estimates for aggregated crops by region and period. The growth picture that emerges here is quite impressive in terms of regional differences and their timing. Many observers have noted that the agricultural productivity performance of sub-Saharan Africa, and to some extent of the Middle East and North

Table 22.7. CGI contributions to yield growth by crop.

						Contribution shares			
						Adoption (1998)		Varieties (1965–2000)	
Crop	1960s	1970s	1980s	1990s	1960–98	IX	IA	IX	IA
Wheat	0.514	0.981	1.125	0.975	0.960	0.32	0.32	0.49	0.37
Rice	0.342	0.940	0.959	0.747	0.794	0.29	0.29	0.20	0.32
Maize	0.311	0.481	0.733	0.906	0.665	0.23	0.32	0.28	0.19
Sorghum	0.055	0.391	0.716	0.676	0.504	0.22	0.16	0.16	0.11
Millet	0.228	0.428	0.537	0.854	0.565	0.27	0.38	0.15	0.50
Barley	0.073	0.199	0.424	1.01	0.490	0.50	0.30	0.49	0.20
Lentils	0.0	0.0	0.193	0.750	0.283	0.70	0.20	0.54	0.65
Beans	0.022	0.027	0.367	0.331	0.208	0.80	0.20	0.72	0.05
Cassava	0.0	0.006	0.087	0.636	0.222	0.74	0.19	0.53	0.16
Potatoes	0.708	0.711	0.749	0.846	0.739	0.08	0.09	0.17	0.08
All crops	0.321	0.676	0.832	0.823	0.718	0.35	0.30	0.36	0.22

IX, varietal cross made in IARC programme; IA, varietal cross in NARS programme with IARC ancestor.

Table 22.8. CGI contributions to yield growth by region.

						Contribution shares			
						Adoption (1998)		Varieties (1965–2000)	
Region	1960s	1970s	1980s	1990s	1960–98	IX	IA	IX	IA
Latin America	0.312	0.600	0.781	0.751	0.658	0.28	0.30	0.39	0.18
Asia (including China)	0.452	0.932	1.030	0.890	0.884	0.30	0.37	0.18	0.39
Middle East–North Africa	0.141	0.270	0.681	1.228	0.688	0.51	0.31	0.62	0.28
Sub-Saharan Africa	0.017	0.142	0.358	0.497	0.280	0.44	0.27	0.45	0.28
All regions	0.321	0.676	0.832	0.823	0.718	0.35	0.34	0.36	0.19

IX, Varietal cross made in IARC programme; IA, varietal cross in NARS programme with IARC ancestor.

Africa region, has been disappointing when compared with expectations and when compared with Asian and Latin American performance. While the CGI component is not the only component contributing to productivity growth, it is the major component in most developing countries. One need look no further than Table 22.4 for an explanation of regional differences in growth performance. Research systems were simply not delivering MVs that merited adoption to sub-Saharan and MENA farmers in the 1960s and 1970s. (Note that they were producing MVs but their MVs did not merit adoption.) It was not until the 1980s that MENA farmers realized high growth from CGI programmes and not until the 1990s that sub-Saharan African farmers realized modest growth from CGI programmes. Over the 40-year period, sub-Saharan African farmers received only 30% of the CGI growth delivered to Asian farmers. They received only 10% of the CGI growth delivered to Asian farmers in the 1960s and 1970s.

Table 22.8 also provides IARC content indicators for adopted and released varieties. IARC crosses make up higher proportions of both releases and adoption in the MENA and sub-Saharan Africa regions than in Asia and Latin America. This attests to the relative strengths of NARS programmes. The delivery of CGI growth to Asia and Latin America reflects stronger, i.e. better organized and managed, NARS. It also reflects differences in institutional settings, as well as in basic biological factors underlying the production of CGI growth itself. There is little question that CGI growth has been more difficult to obtain in cassava, lentils and beans than in rice and wheat. Much of this is related to the fact that temperate zone-developed country CGI systems had achieved gains before 1950 in rice and wheat that were brought to the tropical and sub-tropical regions by IARC programmes. (It should also be noted that there are differences in CGI growth achievement between countries in regions and within countries in each region.)

IARC CGI Contribution Estimates

The estimation of IARC CGI contributions is complex, but it can reasonably be related to the data on both IARC crosses and NARS crosses, and IARC ancestors. Estimations made in Chapter 21 reported that IARC programmes have a germplasmic contribution to NARS CGI programmes that in the aggregate was roughly equivalent to the NARS cross–IARC ancestor proportion in varietal releases. IARC programmes were estimated to make NARS programmes 30% more productive over the period studied. In Chapter 21 it was also estimated that NARS CGI investment responded positively to the availability of CGI germplasm. This effect was approximately 13% and would have led to 7–8% more NARS varieties.

The complexity in calculating the IARC effect is that in the absence of IARC programmes, stronger regional and other coordinating programmes would have provided some IARC services. In addition, there is a competition effect (noted in Chapter 21) between IARC crosses and NARS crosses. In the absence of IARC crosses, more NARS crosses would have been released and adopted. These crosses, however, would have been affected by the loss of IARC germplasm.

Table 22.9 presents calculations of two alternative IARC CGI growth contributions by crop and region. The IARC CGI calculations are made as follows.

> '1/4 Substitution' = (0.75 IX + IA (1 – 0.75 IX)) × 1960–99 Total CGI Contribution when IX is the proportion of IARC crosses in adopted varieties and IA is the proportion of NARS crosses with IARC ancestry in adopted varieties

and

> '1/2 Substitution' = (0.5 IX + IA (1 – 0.5 IX)) × 1960–99 Total CGI Contribution

The 1/4 substitution computation postulates that in the absence of IARC programmes, NARS programmes would have produced 25% more varieties that would have been adopted by farmers with the same yield impact as the IARC crosses would have had. It also presumes that the germplasm loss (proxied by IA) applies to the 25% expansion.

The 1/2 substitution computation postulates a 50% substitution of NARS varietal production for the IARC-crossed varieties. Again, it is presumed that the loss of the IARC germplasmic effect (IA) applies to this substitution proportion. As a result, the differences between the two substitution cases are muted (for all crops in all regions, the 1/2 substitution calculation is 89% of the 1/4 substitution case).

The calculations reported in Table 22.9 are intended to reflect CGI contributions in economic settings where other factors contribute to productivity growth. CGI growth contributions are not independent of other sources of productivity growth. The three country studies (Chapters 18–20) each report estimates of CGI contributions and generally show that the contribution of non-CGI agricultural research, particularly agronomy research and of extension and related programmes, are enhanced by CGI contributions. The last of these enhancement factors is not considered in the IARC growth contribution calculations.

Consistency: CGI Contributions and Actual Yield Growth Tested at the Regional Level

The CGI calculations reported in Table 22.9 are based on MV adoption data and the synthesis MV/TV impact estimates. These CGI estimates should be generally consistent with actual yield changes in these crops.

Table 22.9. CGI and IARC contributions to yield growth.

	Annual yield growth contribution for CGI					Adoption shares		IARC growth contribution	
Crop/region	1960s	1970s	1980s	1990s	1960–98	IX	IA	$\frac{1}{4}$ Substitution	$\frac{1}{2}$ Substitution
Wheat									
Latin America	0.394	1.320	1.563	0.768	1.059	0.54	0.30	0.620	0.518
Asia	0.678	1.118	1.168	0.846	1.006	0.23	0.35	0.465	0.427
MENA	0.189	0.531	0.861	1.388	0.829	0.50	0.32	0.477	0.406
SS Africa	0.183	0.838	1.093	0.855	0.531	0.37	0.26	0.285	0.254
All regions	0.514	0.981	1.125	0.975	0.960	0.32	0.32	0.464	0.412
Rice									
Latin America	0.077	0.787	1.315	0.876	0.818	0.30	0.30	0.374	0.331
Asia	0.375	0.998	0.966	0.713	0.868	0.30	0.30	0.370	0.327
SS Africa	0.000	0.085	0.572	1.219	0.545	0.20	0.20	0.174	0.153
All regions	0.342	0.940	0.959	0.747	0.794	0.29	0.29	0.352	0.312
Maize									
Latin America	0.402	0.474	0.547	0.862	0.625	0.10	0.27	0.203	0.192
Asia	0.407	0.694	1.016	1.377	0.959	0.30	0.32	0.454	0.405
SS Africa	0.041	0.131	0.481	0.197	0.224	0.20	0.50	0.129	0.123
All regions	0.311	0.481	0.733	0.906	0.665	0.23	0.32	0.291	0.265
Sorghum									
Asia	0.148	0.622	1.403	0.976	0.847	0.05	0.20	0.195	0.186
SS Africa	0.000	0.257	0.316	0.514	0.304	0.50	0.10	0.133	0.122
All regions	0.055	0.091	0.716	0.683	0.504	0.22	0.16	0.151	0.127

Millets								
Asia	0.515	0.963	0.954	1.392	1.043	0.27	0.41	0.552	0.510
SS Africa	0.000	0.000	0.205	0.425	0.184	0.26	0.26	0.075	0.066
All regions	0.228	0.428	0.537	0.854	0.565	0.27	0.58	0.286	0.262
Barley									
MENA	0.073	0.199	0.424	1.010	0.490	0.50	0.30	0.278	0.235
Lentils									
MENA	0.000	0.000	0.193	0.750	0.283	0.70	0.20	0.144	0.112
Beans									
Latin America	0.034	0.041	0.463	0.281	0.222	0.70	0.10	0.127	0.092
SS Africa	0.000	0.000	0.188	0.426	0.180	0.80	0.20	0.122	0.094
All regions	0.022	0.027	0.367	0.331	0.208	0.75	0.15	0.131	0.098
Cassava									
Latin America	0.000	0.043	0.055	0.238	0.100	0.05	0.01	0.005	0.003
Asia	0.000	0.000	0.091	0.485	0.174	0.80	0.20	0.118	0.091
SS Africa	0.000	0.000	0.093	0.771	0.249	0.80	0.20	0.169	0.129
All regions	0.000	0.006	0.087	0.636	0.222	0.74	0.19	0.142	0.109
Potatoes									
Latin America	0.672	0.885	0.631	0.694	0.752	0.07	0.09	0.104	0.092
Asia	0.811	0.672	0.759	0.846	0.825	0.05	0.07	0.086	0.077
SS Africa	0.000	0.716	0.864	1.099	0.739	0.55	0.17	0.379	0.294
All regions	0.708	0.711	0.749	0.846	0.807	0.08	0.09	0.117	0.102
All crops									
Latin America	0.312	0.600	0.781	0.751	0.658	0.28	0.27	0.279 (0.42)	0.245 (0.37)
Asia	0.452	0.932	1.030	0.890	0.884	0.26	0.31	0.393 (0.44)	0.353 (0.40)
MENA	0.141	0.270	0.681	1.228	0.688	0.50	0.31	0.391 (0.57)	0.332 (0.48)
SS Africa	0.017	0.142	0.358	0.497	0.280	0.38	0.24	0.128 (0.46)	0.108 (0.33)
All regions	0.321	0.676	0.832	0.823	0.718	0.30	0.30	0.328 (0.46)	0.291 (0.41)

In this section the consistency between CGI growth contribution calculations and actual yield growth by region and decade is evaluated.

Table 22.10 reports two regression specifications relating actual yield growth, Y_g, to the Y_{CGI} growth rates in Tables 22.9. Actual yield growth rates for each crop and decade were estimated from FAO data. For each decade the Y_g was estimated as

$$\ln(Y_t) = a + Y_g \text{ YEAR}$$

The first specification in Table 22.10 simply relates Y_g to Y_{cgi} in a regression with crop and time period fixed effects. This regression indicates that the CGI component changes in concordance with actual yield growth. The coefficient of 0.01 indicates that a change in Y_{cgi} of 1% is associated with actual yield increases of 1%. This is what these estimates should show if the 'left-out' contributions are not correlated with CGI contributions.

The second specification adds the yield level (YL) at the beginning of the period and its interaction with the CGI growth component. This is a relatively weak test of a 'yield ceiling' effect. If, as yield levels rise, the CGI component as calculated actually overstates the real ceiling-

Table 22.10. Actual yield growth and CGI component. Dependent variable, yield growth, by regions, 1960s, 1970s, 1980s and 1990s.

	(1)	(2)
(1) CGI component	0.0108***	0.0120***
(2) Beginning yield level (YL)		−1.01E−08
(1) × (2)		−3.30E−08
D 60s	0.0096***	0.0091***
D 70s	0.0076***	0.0070**
D 80s	0.0051**	0.0047*
D Maize	0.0032	0.0023
D Sorghum	−0.0087**	−0.0097***
D Millet	−0.0111***	−0.0121
D Barley	0.0045	0.0050
D Rice	−0.0052*	−0.0062*
D Lentils	−0.0064	−0.0072
D Beans	−0.117**	−0.0123***
D Cassava	−0.0063*	−0.0075*
D Potatoes	−0.0083**	−0.0083**
Constant	0.0071*	0.0079*
R^2	0.33	0.34
F	3.12	2.74

constrained CGI component, the interaction term should be negative. The second regression suggests a yield ceiling effect, but it is not statistically significant.

A Summary of CGI Impacts on TFP-RCR Productivity Measures

This chapter sought to provide estimates of CGI contributions to TFP-RCR growth by crop, decade and region. The review of MV adoption rates by crop, decade and region showed wide variation in MV adoption rates. The review also showed important disparities between MV production rates and MV adoption rates. Two regions, sub-Saharan Africa and the MENA (Middle East and North African) region, had very low MV adoption rates in the 1960s, 1970s and 1980s, even though they were releasing significant numbers of MVs.

Evidence was reviewed regarding the production impact of conversion from traditional varieties to modern varieties. This evidence was of three types. First, the crop chapters (Chapters 4–16) reported evidence from experimental and other crop studies. Second, the country studies (Chapters 18–20) reported estimates based on aggregate data within the three largest developing countries, China, India and Brazil. Third, international calculations for rice, wheat and maize were made and these were related to MV/TV conversion data. All three bodies of evidence supported the consensus MV/TV estimates. When these estimates were combined with MV adoption estimates, CGI growth contributions were computed. These CGI growth contributions were further tested for consistency with actual crop yield growth estimates at the regional level. A high degree of consistency was found.

The CGI growth contribution calculations offer important policy insights. It appears that a relatively high proportion of actual TFP-RCR growth is accounted for by CGI contributions. The data suggest that at least half of all TFP-RCR gains are due to CGI gains. Perhaps more important, this implies that countries not realizing CGI gains are also unlikely to be realizing TFP-RCR gains from other sources. This is consistent with the observation of T.W. Schultz in his classic work, *Transforming Traditional Agriculture*, published in 1964. Farmers employing traditional technology are not inefficient. Programmes to reduce their 'inefficiency', therefore, will not 'transform' them. Transformation requires new technology, and CGI technology appears to be the dominant form of transformation of agriculture in developing countries.

References

Byerlee, D. and Traxler, G. (1995) National and international wheat improvement research in the post-Green Revolution period: evolution and impacts. *Journal of Agricultural Economics* 77, 268–278.

Evenson, R.E. (1998) Modern varieties, traits, commodity supplies and factor demands in Indian Agriculture. In: *Agricultural Values of Crop Genetic Resources.* CAB International, Wallingford, UK.

Evenson, R.E. (2001) Economic Impact Studies of Agricultural Research and Extension. In: *Handbook of Agricultural Economics.* North Holland Publishing Company, Amsterdam.

Evenson, R.E. and Gollin, D. (1997) Genetic resources, international organization and rice improvement. *Economic Development and Cultural Change* 45(3), 471–500.

Griliches, Z. (1957) Hybrid corn: an exploration in the economics of technological change. *Econometrica* 25(4), 501–522.

Appendix

A Note on Rates of Return

In the classic work of Griliches (1957) on hybrid maize, a benefit/cost analysis was performed. This requires a cost series (c_τ) over time and a benefit series (b_τ) over time. It is possible to construct a cost series for each region from ISNAR data on research expenditures and estimates of the CGI share of the expenditures. This cost series can be constructed for the 1950–2000 time period and denominated in units that are expressed as a percentage of crop production.

The data in Tables 22.8 (and 22.9) can be used to construct a benefits series, because they are also denominated as a percentage of crop production. This benefits series is effectively zero prior to the realization of MV gains.

The c_τ and b_τ series (note that the b_τ series is cumulative) can then be utilized to calculate the following:

PVB: The percentage of the benefits stream computed at a specific interest rate (we use 6%)

PVC: The present rate of the costs stream computed at the same specific interest rate

B/C = PVB/PVC. The benefit–cost ratio

IRR: The rate of interest where PVB = PVC

Table 22.11 reports the computed B/C ratios (using 6% as the external interest rate) and IRRs for both NARS CGI programmes and IARC CGI programmes by region. These are computed for median estimates and 'low' estimates of benefits. Note that these estimates include long periods of costs where few benefits are achieved. For example, for sub-Saharan Africa, costs incurred beginning in 1950 did not produce significant benefits for many years. Benefits exceeded costs in sub-Saharan Africa almost 15 years later than was the case for Latin America and Asia.

A recent review of the rate of return for agricultural research (Evenson, 2001) reported regional 'median' IRRs by region. These median IRRs were:

Latin America	47%
Asia	67%
Africa	37%

These IRRs are considerably higher than those computed in Table 22.11. This is primarily because individual studies tend to ignore the research costs required to reach the stage where benefits are produced. Some of this research is 'unproductive' but a considerable part of it is required to build the germplasm stocks and to enter the staging area where 'platform' MVs can be produced.

The IARC programme IRRs are very high, even when lower range benefits are used in the computation. These high IRRs appear to be very real and they reflect the 'leveraging' associated with the high production of IARC crosses and the high volume of IARC germplasm.

Table 22.11. Costs–benefits and internal rate of return.

	NARS CGI				IARC CGI	
	Estimated benefits		$\frac{3}{4}$ Estimated benefits		Estimated	Lower range
	IRR	B/C	IRR	B/C	IRR	BCC
Latin America	31	56	28.4	42	39	34
Asia	33	115	31	86	115	104
West Asia–North Africa	22	54	20	41	165	147
Sub-Saharan Africa	9	4.0	7	3.0	68	57

23 The Economic Consequences of Crop Genetic Improvement Programmes

R.E. EVENSON AND M. ROSEGRANT

In this final synthesis chapter, the economic consequences of CGI (crop genetic improvement) programmes are assessed. The methodology for this assessment requires a multi-market, multi-country model where crop supply and crop demand factors determine market-clearing prices, quantities produced and consumed, and international trade volumes. For this purpose, the IMPACT model of the International Food Policy Research Institute (IFPRI) is utilized to create the 'counterfactual' or 'what if' simulations. The two counterfactual simulations ask the following questions:

1. How would food prices, food production, food consumption and international food trade have differed in the year 2000 if the developing countries of the world were constrained to have had no CGI after 1965, while developed countries realized the CGI that they historically achieved? (This is the 1965 CGI counterfactual in this chapter.)
2. How would food prices, food production, food consumption and international food trade have differed in the year 2000 if the IARC system had not been built (and thus the IARC CGI contributions had not been realized), but NARS CGI gains in both developed and developing countries would have been realized? (NARS include both public and private research programmes.) (This is the no IARC CGI counterfactual.)

Agricultural development policy has been dominated by an emphasis on 'food security'. This emphasis has both global and local dimensions. The global dimension takes a 'feeding the world' form and global food prices are the key index of success or failure for this dimension. The local dimension takes the form of country-specific food security and the key indexes of success or failure for this dimension are 'food available per capita' or 'calories available per capita' and percentage of children with adequate diets. As this chapter will show, the global indicators of food security are impressive, if not spectacular. Global food prices (at the farm level) have declined relative to the prices of other goods. The decline has been of major proportions, with real food prices at the end of the 20th century being less than half their levels at mid-century. This was achieved in the face of historically unprecedented increases in population and food demand.

At the local level, the key indicator, calories (or food) available per capita, also shows broad-scale improvement. Only a few small countries have failed to realize an improvement in this index. Most countries have realized an improvement and many studies show that this improvement in food available per capita is linked to improvements in infant and child mortality and in measures of child malnutrition (weight for height).

The Green Revolution, as the IFPRI-IMPACT simulations will show, contributed importantly to both global and local indexes of food security. However, there is a further dimension of agricultural development that tends to be submerged in the broader context of food security. It is commonplace to hear that progress has been made on both global and local indicators of food security, *but* that at the 20th century's end more than 820 million people still had inadequate diets (as reflected in height and weight indicators and dietary standards). Of these, 780 million were in developing countries and most were in family circumstances where 'incomes per capita are from US$1 to US$2 per day'.

The 'submergence' question relates not to the 820 or 780 million people with inadequate diets, because malnutrition is linked to food availability (as in the IFPRI-IMPACT model). It relates instead to the 'incomes from $1 to $2 per day'. In today's world, when large numbers of people in a country earn only US$1–2 per day, the country is in a condition of 'mass poverty'. And the question of mass poverty requires direct confrontation, not the submerged confrontation associated with food security.

The second part of this chapter provides a discussion of multi-market models and of the important effects on farmers of access to cost-reducing technology. The following sections provide details regarding the IFPRI-IMPACT model, discuss the counterfactual specifications and simulations, and finally address the question of CGI and mass poverty reductions.

TFP and CGI Contributions

The CGI contributions calculated in the previous chapter are utilized in the IFPRI-IMPACT simulations. Recall that the CGI contributions are measured in terms of total factor productivity (TFP) growth rates. They measure the tonnes of added crop production made possible by CGI programmes, for a given level of production inputs (land, water, labour, fertilizer, machinery, etc.). That is, they measure increased product per unit of input.

The TFP measures also have an additional interpretation, as noted in Chapter 22. They measure the rate of change of real cost reduction (RCR) per unit of product, holding input prices constant. RCR is an important objective for farmers because they benefit directly from RCR whether crop prices are high or low. Their income depends, of course, on their assets and on both input and product prices, but RCR always benefits them. RCR realizations are also important determinants of supply and have important effects on prices.

The economic consequences of CGI are realized through markets and changes in market equilibria. CGI effects are both direct and indirect. The direct effects are the RCR effects, where farmers realize cost reductions from yield improvements. These direct effects, as noted in the previous chapter, vary by crop, region and period. The indirect effects are CGI-induced price effects. These effects tend to be crop-specific to some degree (although with substitutability in demand, CGI-induced price effects for one crop are transferred to other crops) but they are global in today's globalized economy.

The evaluation of the economic consequences of crop genetic improvement requires a market equilibrium model relating supply and demand in determining prices, ideally with a factor market specification enabling the determination of incomes. Such a model would compute an initial equilibrium with equilibrium prices, incomes (from land, labour and other resources) and consumption levels. These equilibria in the typical developing country may be judged by all to be less than ideal because they almost certainly would have large numbers of families in poverty and with inadequate diets. Then CGI gains can be introduced into this model. They would come in the form of shifts in cost and production functions, hence in supply functions of crops and in demand functions for labour, land, etc. The economy would then move to a new equilibrium with different levels of prices, quantities, incomes and consumption. This new equilibrium can then be compared to the original equilibrium to infer economic consequences of the CGI gains. Alternatively, CGI gains can be subtracted from an equilibrium in a counterfactual simulation.

This comparison of equilibria is a meaningful way to evaluate economic consequences. By understanding market forces and the way markets impinge on prices and incomes, one can better understand the mechanisms by which CGI gains affect consumption. It is important to distinguish between people as demanders of food and people as suppliers of food. CGI effects lower costs of production and increase the incentives for producers to supply more food. For given demand conditions this will mean a lower price in the new equilibrium. In a dynamic version of a market model a 'base case' rate of growth in demand and in supply is posited. Then a decrease in the CGI contribution will result in less supply and higher prices than in the base case scenario. The extent of the price change will depend on the localization or globalization of the market.

If the market is a local autarkic market with little trade between regions and countries, the price response associated with CGI improvements can be quite severe. This is because, in a local market, food demand elasticities can be quite inelastic. Suppose, for example, that an RCR of 0.5% is produced by CGI programmes. This would induce farmers to produce 0.5% more under 'neutral technical change' conditions. With a demand elasticity of minus one (–1), prices will fall by 0.5%. But if demand is inelastic, this will result in a price decline of more than 0.5%. If this happens, the production economy must make long-term structural adjustments, which in this case means that some producers will exit from production. Thus, in this local market situation, consumers will gain (including farmers who are also consumers), but producers will actually lose and may be forced into costly adjustment.

Now suppose that producers have differential access to CGI within this localized region. For example, suppose only half of the farmers in the region have the natural resource conditions to benefit from the CGI. Then the supply increase will be half as much as in the case where CGI is available to all. The price effect will be half as large, so consumers will gain half as much. But now the consequences for producers become very different for those with access to CGI and those without access. Those with access will realize RCR gains of 0.5% so their costs may fall by as much or more than prices fall. This may produce a net gain in income for them. The producers without access to CGI will unequivocally lose. Their costs will not fall, but prices will. Thus, their incomes will fall.

This phenomenon of differential delivery of CGI then has important welfare implications. A study of differential CGI delivery by David and Otsuka (1994) for rice farmers noted that agricultural workers can escape the burden of unfavourable delivery if they are mobile. But to the extent that they are mobile they shift more of the burden on the owners of non-mobile assets (family labour and land).

This localized economy is increasingly less relevant in a globalized economy. We observe that most countries today have integrated national markets in grains and agricultural products and, increasingly, international or global markets are emerging for most commodities.

When a local economy opens itself up to trade, there are two consequences. The first is that it can enjoy higher demand elasticities. This means that price effects (both for increases and decreases) will be smaller, easing the burden on producers. In fact, for a small open trading economy, CGI or RCR gains may have little or no price effects, enabling producer incomes to increase for those with access and for producer incomes to remain unchanged for those without access.

The second consequence of opening to trade is that the local economy is now 'exposed' to competition from abroad. If farmers in other countries have realized CGI gains that are not delivered to the local economy, the local economy will be in the same position as local producers without access are. Thus, if China is realizing CGI in rice, this will have a negative effect on the incomes of rice farmers in Indonesia and vice versa. However, consumers in both China and Indonesia will benefit from CGI in China.

In a globalized economy, the issue of delivery of CGI is not only an issue within countries, but between countries as well. There are gains from CGI, but the distribution of these gains depends on the nature of CGI delivery. In the previous chapter we noted that CGI delivery has been very uneven regionally, with farmers in sub-Saharan Africa realizing only 10% or so of the gains (per hectare) that farmers in Asia were realizing in the 1960s and 1970s. This had serious negative consequences for the region. Fortunately the situation is more balanced in the 1990s.

Another phenomenon is likely to exist in global markets where developing countries realize high rates of CGI gains. Most developing countries are experiencing high rates of population and labour-force growth. Only a few are realizing rapid industrial growth. Under these conditions, agricultural wage rates will tend to be low and to rise slowly. When these countries realize CGI gains, their supply response is large because wages will rise slowly and because wages are an important part of costs (in developed countries wages are likely to rise faster). Over the past four or five decades, CGI gains in developing countries have been rapid, as noted in previous chapters. The supply response to these gains has been large, contributing to extraordinary declines in the real prices of crops.

The IFPRI IMPACT Model

The International Model for Policy Analysis of Agricultural Commodities and Trade (IMPACT) developed at the International Food Policy Research Institute (IFPRI) is a partial equilibrium model covering 17 commodities and 35 country/regions. It computes global equilibria in real prices. It is synthetic in that it uses price elasticities and non-price parameters from other studies. The model incorporates non-agricultural sector linkages but does not compute equilibria for markets other than for the 17 agricultural commodities.

Each country/region sub-model has a set of equations for supply, demand and prices for each commodity and for intersectoral linkages with the non-agricultural sector. Crop production is determined by area and yield response functions. Area functions include price responses (own and cross-price terms) and a non-price trend reflecting remaining land availability and technology. Yield is a function of the prices of commodities and prices of inputs and a non-price total factor productivity (TFP or RCR) term. (This term is discussed further below.)

Livestock commodities are similarly modelled.

Domestic demand is the sum of food, feed and industrial use demand. Food demand is a function of prices (of all commodities), per capita income and population. Income growth is partially endogenous to the model and agriculture–non-agriculture links are specified. Feed and industrial use demands are derived from final demands.

Prices, production and trade volumes are endogenously determined in the model. Domestic prices are linked to global equilibrium prices via exchange rates, and producer–consumer subsidies and trade restrictions are allowed. Other policy instruments (hectarage restrictions) are considered. Trade is determined by net supply–demand equilibrium conditions and global market conditions.

National population growth is exogenously based on UN projections (*World Population Prospects UN*).

The non-price terms in the area and yield functions were developed for each commodity and country/region as follows.

First, an accounting structure based on experience in India and Brazil (Agcoali *et al.*, 1993; Avila and Evenson, 1999) was developed. The accounting components were:

1. Public (IARC-NARS) research contributions
   a. Management research (non-CGI) contributions
   b. Conventional plant breeding (CGI) contributions
   c. Wide-crossing-marker-aided breeding (CGI) contributions
2. Private sector agriculturally related R&D spill-in contributions
3. Agricultural extension contributions
4. Markets development contributions
5. Infrastructure contributions
6. Irrigation (interacting with technology) contributions.

The yield growth contribution of modern inputs such as fertilizers is accounted for in price effects in the yield response function.

The growth accounting contributions of both the public and private agricultural research components include both CGI and non-CGI contributions. CGI contributions affect the value of non-CGI contributions and vice versa. The CGI calculations reported in the previous chapter, however, do not include the complementarity between CGI and non-CGI components.

Table 23.1 summarizes TFP growth accounting studies for the USA, Brazil and India. These studies actually do consider CGI and non-CGI complementarity. The accounting framework is a 'partial' framework where the contribution of one component, holding constant other components, is computed.

Additional evidence on growth contributions is reported in Table 23.2, where a summary of rates of return to research and extension (IRR) is reported. The TFP or RCR component is basically synthesized from these two tables and from historical evidence on yield and area growth. The India and Brazil studies were used to partition growth into public sector (CGI and non-CGI) and private sector research contributions, extension and schooling contributions as well as market infrastructure contributions. The contributions were related to investments and growth in investments in each country and then scaled to approximate historical experiences. Thus, the contribution of the private sector depended on whether the private sector invested in relevant R&D in the countries. Commodity differences were scaled using the median rate of return by commodity reported in Table 23.2.

These computations were used to simulate a 'base case'. This base case was actually a forward projection. For our purposes we are using this forward projection to compute a 'backcast' or counterfactual simulation. To do this, we need first to check the base case for consistency with the CGI calculations. Then we can 'subtract' CGI contributions from the base case and compare the equilibrium calculations with the base case to create the 'counterfactual' simulation.

The consistency between the CGI reductions requires that the CGI components represent roughly the proportion of RCR growth that were indicated in Chapter 22. In addition, the population and related demand growth conditions should be similar between the backcast period and the projection period.

The CGI components account for different proportions of RCR growth in Chapter 22 and in the IFPRI-IMPACT model. Because the model is constrained to be consistent with historical experiences, there is consistency between the model and the CGI estimates. Also the population and demand growth elements in the base case are historically consistent.

However, the reader should note that the counterfactual simulations are not simulations of the base case where many complex variables are relevant. The counterfactual simulations compare equilibria where CGI growth components are removed from the base case equilibria. Thus, this exercise does not address the base case itself. What is required for these calculations is that the alternative equilibria can be meaningfully compared to the base case.

Table 23.1. TFP growth accounting.

	US agriculture (1950–1982)		Brazilian agriculture (1970–1985)			Indian agriculture		
	Crops	Livestock	Crops	Livestock	Aggregate	Crops 1956–65	Crops 1966–76	Crops 1977–89
Annual TFP growth	0.63	0.51	1.11	0.09	1.00	1.27	1.49	1.14
Proportion due to:								
a. Public sector research	0.36	0.09	0.23	0.55	0.30	0.22	0.38	0.45
(high yield varieties)	No	No	No	No	No	0	0.20	0.04
b. Industrial R&D	0.24	0.54	0.17	0	0.31	0.07	0.18	0.07
c. Agricultural extension	0.25	0.17	0.07	0.05	0.02	0.66	0.16	0.43
d. Farmer schooling	–	0.08	No	No	No	0.01	0.01	0.01
e. Government programmes	0.02	0.06	No	No	No	–	–	–
f. Markets	No	No	No	No	No	0.04	0.04	0.05
g. Other	0.30	0.13	0.45	0.40	0.37	0	0.19	0

Sources:
US agriculture: Huffman and Evenson (1993).
Brazil: Avila and Evenson (1998).
India: Evenson *et al.* (1999).

Table 23.2. Internal rate of return estimates summary.

	Number of IRRs	Percentage distribution						Approx. median IRR
		0–20	21–40	41–60	61–80	81–100	100	
Extension								
Farm observations	16	0.56	0	0.06	0.06	0.25	0.06	18
Aggregate observations	29	0.24	0.14	0.07	0	0.08	0.06	37
Combined research & extension programmes	36	0.14	0.42	0.28	0.03	0.08	0.06	37
By region								
OECD	19	0.11	0.31	0.16	0	0.11	0.16	50
Asia	21	0.24	0.19	0.9	0.14	0.09	0.14	47
Latin America	23	0.13	0.26	0.34	0.08	0.8	0.09	46
Africa	10	0.40	0.30	0.20	0.10	0	0	27
All extension	81	0.26	0.23	0.16	0.03	0.18	0.13	41
Applied research								
Project evaluation	121	0.25	0.31	0.14	0.18	0.6	0.07	40
Statistical	254	0.14	0.20	0.23	0.12	0.10	0.20	50
Aggregate programmes	126	0.16	0.27	0.29	0.10	0.09	0.09	50
Commodity programmes								
Wheat	30	0.30	0.13	0.17	0.10	0.13	0.17	51
Rice	48	0.08	0.23	0.19	0.27	0.08	0.14	60
Maize	25	0.12	0.28	0.12	0.16	0.08	0.24	56
Other cereals	27	0.26	0.15	0.30	0.11	0.07	0.11	47
Fruits and vegetables	34	0.18	0.18	0.09	0.15	0.09	0.32	67
All crops	207	0.19	0.19	0.14	0.16	0.10	0.21	58
Forest products	13	0.23	0.31	0.68	0.16	0	0.23	37
Livestock	31	0.21	0.31	0.25	0.09	0.03	0.09	36

Continued

Table 23.2. *(continued)*

	Number of IRRs	Percentage distribution						Approx. median IRR
		0–20	21–40	41–60	61–80	81–100	100	
By region								
OECD	146	0.15	0.35	0.21	0.10	0.07	0.11	40
Asia	120	0.08	0.18	0.21	0.15	0.11	0.26	67
Latin America	80	0.15	0.29	0.29	0.15	0.07	0.06	47
Africa	44	0.27	0.27	0.18	0.11	0.11	0.05	37
All applied research	375	0.18	0.23	0.20	0.14	0.08	0.16	49
Pre-technology science	12	0	0.17	0.33	0.17	0.17	0.17	60
Private sector R&D	11	0.18	0.09	0.45	0.09	0.18	0	50
Ex ante research	0.83	0.11	0.36	0.16	0.07	0.01	0.05	44

Specifying the Counterfactuals and Ranges

We require two counterfactuals with reduced CGI contributions. We would also like to provide a range of these contributions to enable the reader to assess the sensitivity of alternative counterfactuals.

The first counterfactual is the *1965 CGI counterfactual.* This counterfactual is intended to simulate conditions where developing countries are constrained to 1965 levels of genetic technology. For the lower end of the range of this counterfactual, we subtract the CGI components averaged for the 1965–2000 period reported by crop and regions in Chapter 22 (Table 22.8). These are our best estimates of the CGI components ignoring CGI–non-CGI complementarity. For the upper end of the 1965 CGI counterfactual we subtract 1.3 times the CGI components in the lower end of the range to reflect CGI–non-CGI complementarity. This estimate is consistent with the IARC-NARS germplasm complementarity estimates and roughly consistent with growth accounting studies evidence.

The second counterfactual is the *NO IARC CGI counterfactual.* For this counterfactual, we subtract the IARC CGI contributions calculated in Table 22.8 in Chapter 22. We use the $\frac{1}{2}$ substitution case as the lower end of this range and the $\frac{1}{4}$ substitution case as the upper end of this range. We also subtract one-quarter of the $\frac{1}{2}$ substitution case for wheat and rice in developed countries to reflect the IARC contribution to developed country production (see Alston and Pardey, 1999).

Note that, in the 1965 CGI counterfactual, developed countries realize their actual CGI gains. In the NO IARC CGI case, we subtract a small component for wheat and rice from developed country CGI gains.

Counterfactual Simulations

Table 23.3 reports global aggregate simulations for the two counterfactual scenarios. The simulation results are the percentage differences between the base case (i.e. the simulation representing actual changes) and the counterfactual case.

Thus for equilibrium prices (which are global equilibrium prices in US dollars per tonne with allowances for country price differentials because of tariffs) the 1965 CGI counterfactual indicates that equilibrium wheat prices would have been 29–61% higher than they actually were in 2000. For rice, 2000 prices would have been 80–124% higher (note the range). Price increases from CGI reductions in developing countries depend both on actual CGI gains, which varied by crops, and on the proportion of the crop produced in developing countries. Price increases for rice, which is produced mostly in developing countries, thus exceed those for wheat, half of which is produced in developed countries.

Table 23.3. Price production area and trade effects. Alternative counterfactual scenarios.

	Wheat	Rice	Maize	Other grains	Potatoes	Other root crops	All food crops
Price effects (positive)							
1965 CGI	29–61	80–124	23–45	21–50	13–31	28–52	35–66
No IARC CGI	19–22	30–35	13–15	14–16	2–3	15–32	18–21
Production effects (negative)							
1965 CGI	9–14	11–14	9–12	5–9	12–18	2–3	8–12
No IARC CGI	5–6	4–5	4–5	3–4	3–4	1–2	4–5
Area effects (positive)							
1965 CGI	3.2–5.6	7.5–9.4	1.1–1.9	0.4–2.2	0.0–0.0	2.2–3.2	2.8–4.6
No IARC CGI	2.1–2.1	2.9–3.3	0.5–0.6	0.5–0.6	0.0–0.0	1.4–3.2	1.5–2.7
Trade effects (positive)							
1965 CGI	31–19	0–2	45–46	25–19	190–192	21–65	27–30
No IARC CGI	7–6	0–2	16–18	1–2	16–33	11–12	6–9

For all crops (weighted by production) prices in the 1965 CGI counterfactual would have been from 35 to 66% higher. Since prices actually fell by 35% or so from 1965 to 2000, this would have more than offset the price fall. Some readers may be surprised that these price differentials were not larger. It should be noted, however, that the counterfactual does not posit lost CGI in developed countries and, with a supply response to price increases, production increases in developed countries partly offset production decreases in developing countries (see below).

For the more realistic NO IARC CGI counterfactual, the price effects are smaller, but they are significant. For all food crops, prices without IARC CGI contributions would have been 18–21% higher. This suggests that, even in the absence of IARC programmes, world prices of food crops would have fallen in real terms from 1965 to 2000. This, again, may appear unrealistic to many observers who credit the IARCs with creating the 'Green Revolution'. But, as noted in this volume, the Green Revolution is largely a joint product of NARS, IARCs and, in some countries, the private seed companies. However, much of the reason for the food price decline in the absence of developing country IARC contributions is that developed countries were realizing high rates of CGI gains.

Global production decreases under the 1965 CGI counterfactual are also more modest than many would expect. For all food crops, production would have been 8–12% lower, but this is misleading because it would have increased for developed countries (because of higher prices, see below).

Production decreases under the NO IARC CGI counterfactual would have been between 4 and 5% of production. This is roughly 45% of the decrease under the 1965 CGI counterfactual. The decline in production in the counterfactual is moderated by the strong rise in cereal prices. These price increases induce farmers in both developing and developed countries to expand area and increase the use of other inputs, partially compensating for the loss of crop yield growth.

Area effects under the counterfactuals would have been substantial. This is because, if yields are lower and prices higher, farmers would have planted more area to crops with attendant environmental consequences. These area effects are particularly large for rice. For all food crops, area under crops would have expanded by 2.8–4.6% in the 1965 CGI counterfactual. For the NO IARC CGI, counterfactual area under crops would have expanded by 1.5–2.7%.

As developing country regions lose competitiveness, they import more of their food crops from developed countries, which have gained competitiveness. For all food crops, developed country exports to developing countries would have risen by 27–30% under the 1965 CGI counterfactual. Note that this would have been in addition to the expansion in this trade that actually took place over the 1965–2000 period.

To provide further insight into the processes underlying the aggregate data; Area, Yield and Production effects are reported for developed countries (including the transition economies) and developing countries (including China) in Table 23.4.

Consider the yield effects. These include the direct losses of CGI and the indirect CGI-induced price effects. For developed countries, the 1965 CGI counterfactual is entirely the indirect price effect. This effect is substantial for wheat and maize, but not for other crops that are produced predominantly in developing countries. The NO IARC CGI case includes both direct and indirect effects. For developing countries, crop yields would have been significantly lower in 2000 in spite of the positive indirect price effects. The NO IARC CGI effects on yields are also substantial.

Area effects, interestingly, are approximately the same for developed and developing countries. This is because they depend on the indirect price effects and these occur globally. The NO IARC CGI area effects are a substantial part of the 1965 CGI effects in developing countries (especially in rice).

Production effects then show that, under the 1965 CGI counterfactual, developed countries would have produced approximately 5–7% more food crops and developing countries would have produced from 16 to 19% less. The NO IARC CGI case would also have resulted in 1–2% more production in developed countries and 7–8% less production in developing countries.

Tables 23.5 and 23.6 provide further detail for area and production effects for developing country regions. Table 23.5 shows that area effects differ by crop and region. The relatively small area effects in sub-Saharan Africa, for example, are due to the fact that this region had relatively low CGI gains, less than one-third those of other regions. Accordingly, the lost CGI counterfactuals are lower. Had sub-Saharan African CGI gains been as large as those in Asia, area increases under both counterfactuals would have been more than double those in Asia.

It is important to note, however, that the implications of area effects in the 3–4% range are significant from an environmental perspective. This increased cropland amounts to 9–12 million ha in developed countries and 15–20 million ha in developing countries under the 1965 CGI case (5–6 million ha in developed countries, and 11–13 million ha in developing countries for the NO IARC CGI case). This would constitute an expansion of croplands on marginal areas with higher environmental sensitivity (erodability, etc.) than cropland currently under production.

Table 23.6 shows production effects. Again we note that these are lower in sub-Saharan Africa because that region had the lower CGI gains over the period. Thus the counterfactuals based on taking these gains away have lowest effects in this region.

Table 23.4. Yield, area and production effects – developed and developing countries: counterfactual scenarios.

	Wheat	Rice	Maize	Other grains	Potatoes	Other root crops	All crops
A. Yield effects							
Developed countries (positive)							
1965 CGI	4.4–7.5	0.0–6.7	1.4–3.1	0.0–1.8	1.5–2.0	nc	2.32–4.77
No IARC CGI	2.7–5.1	0.0–1.0	0.5–2.5	0.0–1.8	0.5–1.0	nc	1.35–2.45
Developing countries (negative)							
1965 CGI	26.2–31.3	18.3–22.9	21.5–25.9	15.0–17.1	23.5–28.3	4.3–4.4	19.45–23.50
No IARC CGI	11.6–12.9	7.8–8.7	8.7–9.5	5.6–5.8	3.4–3.9	2.5–4.0	8.07–8.91
B. Area effects							
Developed countries (positive)							
1965 CGI	4.5–7.5	11.8–15.8	2.2–3.4	0.4–1.8	0.0–0.1	nc	2.82–4.92
No IARC CGI	2.7–3.1	4.8–5.5	0.9–1.1	0.3–0.4	0.0–0.1	nc	1.59–1.86
Developing countries (positive)							
1965 CGI	1.7–3.6	7.3–9.3	0.6–1.2	0.5–0.6	0.0–0.1	2.2–3.3	2.82–4.92
No IARC CGI	1.4–1.5	6.1–6.5	0.3–0.4	0.4–0.5	0.0–0.1	1.4–3.3	1.59–1.86
C. Production effects							
Developed countries (positive)							
1965 CGI	8.3–11.0	15.7–19.3	2.0–5.3	1.8–2.3	1.2–4.9	nc	4.43–6.93
No IARC CGI	1.6–2.1	3.1–5.5	1.6–1.3	1.3–1.4	1.2–1.6	nc	0.96–1.68
Developing countries (negative)							
1965 CGI	25.0–28.6	12.1–15.2	21.0–24.9	14.0–14.6	24.5–29.1	2.0–2.5	15.85–18.63
No IARC CGI	10.4–11.6	5.1–5.7	8.5–9.3	4.9–5.2	4.9–5.4	1.1–2.1	6.48–7.30

nc, not computed.

Table 23.5. Area effects (positive except where noted) by region and crop: counterfactual scenarios.

	Wheat	Rice	Maize	Other grains	Potatoes	Other root crops	All food crops
Latin America							
1965 CGI	5.1–9.6	9.1–11.7	2.1–3.6	0.4–0.6	−1.2–0.0	0.3–0.5	3.10–5.12
No IARC CGI	3.1–3.6	3.5–4.0	1.0–1.2	0.4–0.6	−0.1–0.0	0.3–0.3	1.54–3.08
Sub-Saharan Africa							
1965 CGI	2.5–4.4	6.7–7.4	0.8–1.5	2.1–4.8	−0.2–2.0	2.5–3.6	2.19–4.00
No IARC CGI	1.7–2.0	2.3–2.6	0.4–0.5	0.2–0.3	0.1–0	1.6–3.6	0.63–1.01
Middle East–North Africa							
1965 CGI	4.0–7.1	12.5–14.3	1.2–1.3	0.9–3.1	−0.0–0.0		3.20–5.78
No IARC CGI	2.5–2.9	4.3–4.8	0.2–0.3	0.6–0.7	−0.1–0.0		1.84–2.14
Asia (including China)							
1965 CGI	1.4–1.8	7.2–9.2	−0.5–0.0	−0.4–0.7	−0.1–0.0	2.6–3.5	3.52–4.74
No IARC CGI	0.6–0.7	2.8–3.2	−0.0–0.0	0.5–0.6	−0.1–0.0	1.0–1.5	1.47–1.71

Table 23.6. Production effects by region and crop. Alternative counterfactual scenarios.

	Wheat	Rice	Maize	Other grains	Potatoes	Other root crops	All food crops
Latin America (negative)							
1965 CGI	25.6–29.6	9.6–12.0	15.8–18.3	26.8–31.1	23.8–28.2	1.5–4.0	15.41–18.32
No IARC CGI	12.3–14.6	3.8–4.3	4.0–4.1	9.8–10.3	4.6–5.0	1.5–4.0	5.41–5.62
Sub-Saharan Africa (negative)							
1965 CGI	9.3–10.1	1.6–2.0	2.0–3.4	2.0–5.0	22.5–26.3	1.8–2.5	2.04–3.32
No IARC CGI	3.6–3.8	1.6–2.0	1.6–1.9	1.1–1.9	10.8–14.0	0.9–1.5	1.15–1.73
Middle East–North Africa (negative)							
1965 CGI	27.1–31.5	3.0–3.5	3.3–3.9	3.5–5.1	22.5–26.9		17.56–20.66
No IARC CGI	10.9–11.6	3.0–3.5	2.4–2.5	1.6–1.7	10.6–15.0		7.36–7.87
Asia (including China) (negative)							
1965 CGI	26.7–30.8	12.9–16.3	27.5–33.3	27.5–32.1	24.1–29.8	0.6–1.6	20.12–22.8
No IARC CGI	10.7–11.4	5.3–5.9	12.0–13.3	10.0–10.6	3.9–4.1	0.6–1.6	8.3–9.1
Developed countries (positive)							
1965 CGI	8.3–11.0	15.7–19.2	2.0–5.3	1.8–2.3	1.2–4.9		4.43–6.93
No IARC CGI	0.6–2.1	3.2–5.5	1.0–1.3	1.3–1.4	1.2–1.6		0.96–1.68
Developing countries (negative)							
1965 CGI	25.0–28.6	12.1–15.2	21.0–24.9	14.0–14.6	24.5–29.1	2.0–2.5	15.85–18.63
No IARC CGI	10.4–11.6	5.1–5.7	8.5–9.3	4.9–5.2	4.9–5.4	1.1–2.1	6.48–7.30

CGI and Poverty

Ultimately the concern with CGI programmes and other publicly funded programmes is with their impacts on human welfare. These impacts are complex because they include both consumption and income effects. Consumption effects depend on prices and on the distribution of income. The distribution of income depends on the ownership of assets, and on the returns to assets such as land and labour.

CGI income effects also depend on the ownership of assets such as land and water rights and on returns to labour. Because of differential delivery of technology, CGI income effects can be positive or negative. Through time, CGI impacts are producing real cost reduction (RCR) at differential rates to farmers in different regions. They are also producing price effects to farmers that are generally not differential, i.e. they are global. Thus farmers for whom RCR effects exceed (negative) price effects benefit from CGI while farmers for whom (negative) price effects exceed RCR effects are harmed by CGI.

The income consequences of differential rates of CGI delivery are very important but difficult to model. The IMPACT model does not model these income consequences directly. However, it does address the consumption consequences indirectly. It does this in two ways. First, one of the endogenous equilibrium outcomes of the model is food available per capita. Specifically, the model measures calories available per capita per day in each country. Thus we can gain some insight into consumer welfare by investigating counterfactual simulations of this measure. These are presented in Table 23.7. Note that this measure covers both rural and urban populations and reflects equilibrium trade.

Malnourished children projections for children (aged 0–6 years) are based on weight-for-age standards set by the US National Center for Health Statistics. Data for 61 developing countries for 1980, 1985 and 1990 (World Nutrition Database, ACC-SCN, 1992) were used to link proportions of malnourished children in the population to per capita calorie consumption (determined in the model). Utilizing this statistical linkage, the IMPACT model can predict changes in the percentage of malnourished children by country. This procedure allows the distribution of income to be brought indirectly into the counterfactual scenarios.

Table 23.7 reports counterfactual scenario ranges for these two welfare measures.

Consider the counterfactuals for the percentage of children malnourished. (Note this is the change in the percentage, not a percentage change.) For the 1965 CGI simulation, 6–8% more children in all developing countries would have been classified as malnourished in 2000 than actually were malnourished. This means that 32–42 million more

Table 23.7. Food consumption consequences. CGI counterfactual scenarios.

	Increase in % of children malnourished		Decrease in % of calorie availability	
Region	1965 CGI	No IARC CGI	1965 CGI	No IARC CGI
Latin America	1.79–2.31	0.58–0.67	4.01–5.25	1.70–1.93
Sub-Saharan Africa	2.53–3.25	0.86–0.98	7.68–9.76	3.35–3.84
Nigeria	3.59–4.63	1.19–1.35	10.05–12.78	4.31–4.93
Middle East–North Africa	1.78–2.29	0.62–0.72	2.43–3.13	1.13–1.35
South Asia	11.24–14.60	3.66–4.07	13.53–17.25	5.79–6.49
India	12.26–15.95	3.98–4.43	13.71–17.51	5.80–6.54
South-east Asia	6.29–8.14	2.08–2.29	11.23–14.26	4.89–5.38
China	7.75–9.50	2.44–2.69	14.10–17.79	5.69–6.78
All developing regions	6.08–7.86	2.00–2.23	10.41–13.32	4.51–5.00
Number of children affected (millions)	32.1–41.6	13.3–14.8		

children would have been malnourished. The highest impacts of this scenario are in South Asia, particularly in India, where 12–15% more children would have been malnourished.

The NO IARC CGI counterfactual indicates that without the IARC CGI some 13–15 million more children would have been malnourished. These children are predominantly located in South Asia, where malnutrition incidence is highest. Note, however, that had sub-Saharan Africa actually realized more CGI than it did, this counterfactual would have shown larger effects for countries in that region. These effects are smallest in Latin America, where the incidence of malnutrition is lowest.

The counterfactual simulations for per capita calorie availability (this is closely related to per capita consumption) are even more impressive because they affect entire populations. Food consumption is vital not only for health but also for work performance. The 1965 CGI counterfactual indicates an 11–13% reduction in caloric availability for all developing countries with 14–17% for South Asia. Had this decline in food consumption actually occurred, it would have had important consequences for health, fertility and income of millions of people.

The NO IARC CGI counterfactual effect on food availability itself has major welfare implications. For all developing countries, a reduction in food consumption per capita of 5% is a major reduction. For the poorest regions this effect is 7%.

Development agencies, multilateral and bilateral, have long stressed poverty reduction. The health and nutritional improvement that accompany poverty reduction are their central objectives. These agencies support many programmes and projects to contribute to this objective. The counterfactuals in Table 23.7 clearly indicate that CGI contributes to that objective.

In fact, when development agencies are seeking, as they should, high leverage projects and programmes where poverty reduction and nutritional improvement per dollar expended is high, they are unlikely to find better investments than CGI investments. The IARC CGI investments, in particular, have to be regarded as being truly extraordinary in this regard. The IARC programmes involved in CGI have cost roughly US$125 million per year over the past three decades. It is difficult to imagine a more effective poverty reduction investment than that implied by the counterfactuals reported here.

As noted in the introduction to this chapter, the fundamental issue of ‘mass poverty’ associated with incomes of US$1–2 per day is often ‘submerged’ by linkage with food security. The poverty analysis in Table 23.7 is a case in point. The calculations do address poverty issues, but they do not confront the US$1–2 per day issue directly. The direct confrontation requires asking how an economy generating income of US$1–2 per day for large numbers of workers can be made to generate US$2–4 per day, then $5–10 per day? Escape from mass poverty requires an escape from extremely low wages.

Have CGI programmes contributed to the escape from mass poverty? This question is only partially addressed in Table 23.7. A more direct assessment is required. This chapter cannot provide the kind of full assessment required for this issue, but some general assertions can be made.

For the poor, most income is labour income. Growth models show that incomes are ultimately determined by technical efficiency or productivity, i.e. the quantity of goods produced per unit of capital (human, natural and reproduced (buildings, etc.)) and by the quantity of capital in an economy. Many general economists writing in the 1950s concluded that improvements in technical efficiency in industry were the key to development. The agricultural sector was seen as having little potential for productivity improvement, at least relative to industry, where industrial technology was not location-specific, as was the case in agriculture, and where mastery of industrial technology could be achieved without the kind of R&D investment associated with CGI programmes. Industrial enthusiasts hoped that the agricultural sector productivity could be improved sufficiently that agriculture would not be a burden to industrial growth. By this view, it would be industrial development, not CGI programmes that would 'transform' traditional agriculture.

This line of analysis was translated into a broad-scale bias against agricultural investments in the early years of development programming. Bilateral and multilateral aid agencies in the 1950s and 1960s typically directed only 10–15% of their portfolios to agriculture. It was not until the food security crises of the late 1960s that these portfolios were changed to include more attention to agriculture. The IARC-NARS CGI programmes were created as part of this shift in emphasis.

After a full half-century of concentrated development programming, much has been accomplished. The success in newly industrialized country (NIC) economics in poverty reduction has been impressive. The 'road to NIC-dom' has produced results, but these results have not been realized in all economies; in particular, NIC-dom policies have failed the US$1 and $2 per day economies. More relevantly, NIC-dom policies have not transformed traditional agriculture. Even more relevantly, every successful case of NIC industrial growth is characterized by earlier agricultural growth (and ironically, the agricultural sector has outperformed the industrial sector in productivity growth in all developed economies over the period). The road to NIC-dom appears to be highly dependent on the state of institutional development and on income levels themselves. The road may end in NIC-dom, but policies suited to the NICs are not suited to economies inheriting pre-NIC circumstances. For most pre-NIC economies, the agricultural sector dominates economic activities, and moving along the road to NIC-dom (i.e. out of mass poverty) requires agricultural gains.

One of the reasons for limited impact of NIC-dom policies in pre-NIC economies is that low wages themselves constitute a barrier to TFP-RCR growth. Consider grain harvesting machinery. Literally thousands of inventions have been made to improve grain harvesting and related machines for two centuries (the first reaper inventions were made almost 200 years ago). Today the real cost of harvesting and handling 1 t of grain in developed countries (given wages in developed countries), is lower as a result of this invention. This has benefited the high-wage developing economies (Brazil, Argentina), but has provided little or no benefit to low-wage economies like Bangladesh where, even with these improvements, grain is most efficiently harvested by low-wage labourers.

CGI programmes are not subject to this wage-mediated constraint. In fact, the phenomenon of low-wage entrapment excludes the poorest economies from many sources of TFP-RCR growth. In a simple sense, these economies are dependent on agricultural productivity to move them down the road to NIC-dom, and most have many miles to go.

T.W. Schultz, in his classic work *Transforming Traditional Agriculture*, noted that traditional farmers have two options for improving productivity. They can improve efficiency (i.e. they can move closer to the technological frontier), or they can move the technological frontier (and then move closer to the new frontier). He argued that traditional farmers were already efficient and that programmes focusing on efficiency improvement would not transform traditional agriculture. That would require developing new technology suited to their economic conditions and, given the problems of low-wage entrapment, this effectively leaves CGI programmes as the major, perhaps the only, source of transformation.

Does this study support this Schultzian view? In the main, it does. It explains why many economies with few CGI gains are still in the US$1–2 income categories. It explains why many economies in Asia enjoying CGI gains in the 1960s and 1970s have been able to move along the road to NIC-dom (where the NIC-dom-based policies finally work). CGI gains do not account for all TFP gains, but it appears that they account for most.

It is difficult to press this interpretation further without more systematically studying sources of growth (including TFP-RCR growth and the capital growth associated with TFP-RCR growth), but it does appear that CGI growth contributions are vital to the escape from mass poverty.

Summary

This concluding chapter reported two counterfactual economic impact scenarios. The calculations were based on global market equilibrium outcomes. Because of this, some of the results may appear less extreme than anticipated by many. For example, in the counterfactual where no CGI improvements after 1965 are realized, the conclusion is that world grain prices would have risen instead of falling by 35% or more. This probably would not be considered a 'World Food Crisis'. However, a more careful examination of the calculations would show that much of the shortfall in production in developing countries would be made up by increases in production in developed countries.

It is important then to interpret these calculations in an appropriate historical context. In the first half of the 20th century, biological invention in the form of science-based CGI (as opposed to farmer-based CGI) had already initiated green revolutions in a number of crops (notably sugar, rubber, bananas and coffee in developing countries and rice, wheat, maize and most other crops in developed countries). By the 1930s, crop prices were beginning to fall relative to non-farm prices in many countries.

In a broad historical sense the CGI gains evaluated in this volume could be seen as bringing predominantly developed-country gains in the first half of the 20th century to developing countries in the second half of the century. However, the gains set in motion in the first half of the 20th century in developed countries continued to be realized at high rates in developed countries in the second half of the century. The rates at which these gains were brought to developing countries were very uneven by crop and region, as documented in this volume.

A related phenomenon in many ways paralleling the gains in food production were gains in public and private health. These have been impressive. Their initial impacts were to reduce infant mortality rates. This was a large factor in the population increases occurring in the second half of the century; because birth rates were high at the beginning of these demographic transitions and even though they fell rapidly in many countries, the disparity in birth and death rates created major expansions in populations. In retrospect, it is truly remarkable that most developing countries were able to increase food availability per capita in the face of major population expansion.

Much of the credit for the food production expansion has to be credited to CGI programmes. The calculations in Chapter 22 showed that CGI gains were a large part of the productivity gains that were critical to achieving increases in per capita food availability. The regional computations as well as statistical tests showed that CGI gains were the transforming events in food production over these decades. Those transforming CGI gains were not realized evenly by any means. Nor were they sufficient to bring about 'convergence' in per capita incomes between developing and developed countries. But they did constitute the beginning stages of the convergence process, which ultimately depends on industrial performance.

Development agencies have maintained poverty reduction as their core concern in programme choices. The CGI programmes in NARS and particularly in IARCs were originally established to address broad food security issues as well as poverty reduction goals. At the end of the 20th century, crop prices were low (in fact, at their lowest point in all recorded history), in large part because of CGI. As a result, food security was seen in a less urgent light than was the case when these CGI programmes were built and supported in the 1960s and 1970s. When assessed as poverty reduction programmes and as programmes improving human welfare, CGI programmes have been outstanding investments. Few investments can come close to achieving the poverty reduction per dollar expended that the CGI programmes evaluated in this volume have realized.

The 'food security' strategy for building support for the IARC system in the 1960s and 1970s was quite effective in terms of providing financial support and an environment conducive to achievement of CGI goals. The Consultative Group mechanism has been a very productive and effective institutional mechanism for CGI programmes (as well as other IARC programmes). There was a concern, at century's end, however, that the global success of CGI programmes was contributing to a decline in support. Many development agencies have reduced their agricultural portfolios in the past decade.

The 'escape from mass poverty' focus, however, indicates that any reduction in support to agricultural projects, in particular to projects designed to improve productivity, will seriously limit and hamper efforts to reduce mass poverty.

References

Agcaoli, M.C., Oga, K. and Rosegrant, M.W. (1993) *Structure and Operation of the International Food Policy and Trade Simulation (IFPTSIM) Model.* Paper presented in the Second Workshop of the Research Project on Projections and Policy Implications of Medium- and Long-Term Rice Supply and Demand; IRRI, Los Banos, Philippines.

Avila, A.F.D. and Evenson, R.E. (1995) Total Factor Productivity Growth in Brazilian Agriculture and the Role of Agricultural Research. *Anais do XXXIII Congresso Brasileiro de Economia e Sociologia Rural* (Vol. I), Curitiba, 31/07 a 3/08/95, pp. 631–657.

Avila, A.F.D. and Evenson, R.E. (1998) Total return productivity growth in Brazilian agriculture and the role of Brazilian agricultural research. *Economia Applicada* (Table 13).

Bloom, D.E. and Williamson, J.G. (1998) *The World Bank Economic Review* 12, 419–456.

David, C.C. and Otsuka, K. (1994) *Modern Rice Technology and Income Distribution in Asia.* Rienner Publishers, Boulder, Colorado.

Evenson, R.E. and Westphal, L. (1994) Technological Change and Technology Strategy. In: Srinivasan, T.N. and Behrman, J. (eds) *The Handbook of Development Economics*, Vol. 3. North Holland Publishing Company, Amsterdam.

Evenson, R.E., Pray, C.E. and Rosegrant, M.W. (1999) *Agricultural Productivity Growth in India.* International Food Policy Research Institute, Research Report 109.

Huffman, W.E. and Evenson, R.E. (1993) *Science for Agriculture.* Iowa State University Press, Ames, p. 212, Table 7.10.

Rosegrant, M.W. and Evenson, R.E. (1993) Determinants of Productivity Growth in Asian Agriculture: Past and Future. Paper presented at the 1993 AAEA International Pre-Conference on Post-Green Revolution Agricultural Development Strategies in the Third World: What Next, Orlando, Florida.

Schultz, T.W. (1964) *Transforming Traditional Agriculture.* Yale University Press, New Haven.

United Nations (1992) *World Population Prospects.* United Nations, New York.

Index

Page numbers in *italics* refer to figures and tables, those with suffix 'n' refer to footnotes

Browse Read and Buy
www.cabi.org/bookshop
ANIMAL & VETERINARY SCIENCES
BIODIVERSITY CROP PROTECTION
HUMAN HEALTH NATURAL RESOURCES
ENVIRONMENT PLANT SCIENCES
SOCIAL SCIENCES
CABI Publishing
A division of CAB International
Online BOOK SHOP
Subjects
Search
Reading Room
Bargains
New Titles
Forthcoming
Order & Pay Online!
MasterCard
VISA
AMERICAN EXPRESS
Crop Pollination by Bees
A DICTIONARY OF Entomology
G Gordh and D H Headrick
Principles of CATTLE PRODUCTION
C.J.C. Phillips
Seeds
THE ECOLOGY OF REGENERATION IN PLANT COMMUNITIES
2nd EDITION
Edited by Michael Fenner
FULL DESCRIPTION
BUY THIS BOOK
BOOK OF THE MONTH
Tel: +44 (0)1491 832111 Fax: +44 (0)1491 829292